Studies in History and Philosophy of Science

Volume 42

AF685702

More information about this series at http://www.springer.com/series/5671

Elise Crull · Guido Bacciagaluppi
Editors

Grete Hermann - Between Physics and Philosophy

Springer

Editors
Elise Crull
Department of Philosophy
The City University of New York
New York, NY
USA

Guido Bacciagaluppi
Descartes Centre for the History and Philosophy of the Sciences and the Humanities
Universiteit Utrecht
Utrecht
The Netherlands

ISSN 0929-6425 ISSN 2215-1958 (electronic)
Studies in History and Philosophy of Science
ISBN 978-94-024-0968-0 ISBN 978-94-024-0970-3 (eBook)
DOI 10.1007/978-94-024-0970-3

Library of Congress Control Number: 2016949571

Printed on acid-free paper

This Springer imprint is published by Springer Nature
The registered company is Springer Science+Business Media B.V.
The registered company address is: Van Godewijckstraat 30, 3311 GX Dordrecht, The Netherlands

Preface

Our project of a volume on Grete Hermann was born quite fortuitously out of a larger project on the historical debates surrounding the 1935 paper by Einstein, Podolsky, and Rosen (EPR), and our collaboration on that project was born even more fortuitously out of a breakfast conversation in Berlin between one of us [GB] and Don Howard, then Ph.D. advisor of the other [EC].[1] GB was mentioning to Don that he wanted to publish a book centred around Erwin Schrödinger's correspondence about EPR from the summer and autumn of 1935, specifically with Einstein, Bohr, Born, Pauli, and Teller. GB mentioned it might be fun to include also a translation of a little-known reply to EPR that Heisenberg had drafted but never published and was tucked away, in German, in a not-so-perfect transcription, in Volume 2 of Pauli's scientific correspondence (Pauli 1985). Don—always keen to launch new avenues of exploration and collaboration—said that as a matter of fact he had a graduate student [EC] who was just then working on a translation of the 'other' response to EPR, and why did we not pool our efforts together?

Despite some obvious geographical obstacles (EC then at the University of Notre Dame and GB at the University of Sydney), a fruitful collaboration started, resulting in a joint paper at HQ-2 (Bacciagaluppi and Crull 2009) and an online version of our translation of the Heisenberg manuscript (Bacciagaluppi and Crull 2011). GB's move to the University of Aberdeen, Scotland, and a generous grant from the Leverhulme Trust made things smoother: EC moved to Aberdeen on a two-year Leverhulme postdoc, and the trust funded three rich summers of our research at the Max Planck Institute in Berlin, where we enjoyed the wonderful hospitality of Christoph Lehner and his quantum group. To cut a long story short, the main end product of the project will soon be completed (Bacciagaluppi and Crull expected 2018), but our digging into EPR hit a rich vein (among several) that we had not quite expected, namely Grete Hermann. In his draft reply (and the letter to Pauli that accompanied it), Heisenberg

[1]This meeting occurred during HQ-1, the first of a great series of conferences on the History of Quantum Physics, part of a collaboration between the Max Planck Institute for the History of Science in Berlin and a number of other international researchers and groups.

mentioned the affinity of his position with that put forward by Hermann in her then very recent essay on 'Die naturphilosophischen Grundlagen der Quantenmechanik.' Looking into it and related material, we were struck by the importance of this material, and the Heisenberg side of our research soon included a substantial engagement with Grete Hermann.

Hermann's name continued to appear in a variety of contexts, and we decided to organise a workshop on her work and figure. The purpose of the workshop would be to bring together scholars—as well as colleagues—of Hermann in order to begin building a more coherent narrative about her life and work, and the relevance of these to an array of disciplines: philosophy, physics, ethics, pedagogy, and politics, to list but a few. The timing for such a workshop was also ideal in virtue of coinciding with the vibrant international research programme into the history of quantum mechanics and its interpretation. Additionally, we were seeing a marked resurgence in interest regarding Kant's treatment of the physical sciences and neo-Kantian approaches to philosophy of science, two key areas for understanding Hermann's wider philosophic concerns.

The present volume is in large part a product of this workshop, which was held at the University of Aberdeen in early May of 2012. As one can see from the list of contributors and the schedule of the workshop included with it, recent scholarly work on Hermann was presented by physicists, historians, philosophers of science, and philosophers and educators following in Hermann's steps. Not only were talks given shedding light on Hermann's academic and political work, but workshop attendees were privileged to hear about the personal side of Hermann from erstwhile colleagues and others who knew her, during the panel and general discussions. We invited Rene Saran and Dieter Krohn as people who had worked with Hermann herself and also invited Fernando Leal and Giulia Paparo as experts on Hermann's philosophical background. Léna Soler, Thomas Filk, Mélanie Frappier, and Michiel Seevinck were invited to speak on work they had done regarding Hermann's philosophy of physics.

Much good and lively discussion took place over the two days of the workshop, and we would be remiss not to extend warm thanks to non-presenting participants who brought extra richness to the discussion by offering different insights and knowledge of Hermann. In particular, and in no particular order, we wish to thank Patricia Shipley (Birkbeck College London) and Sally Redfern (King's College London) for their insights into the practical side of Hermann's teaching, Martin Jähnert from the Max Planck Institute's quantum group (MPIWG, Berlin), Gregor Schiemann (Bergische Universität Wuppertal), with his expertise on both Heisenberg and natural philosophy, Roberto Angeloni (SPHERE, Paris), and Tom Scott and Danny McShane (Aberdeen).

In addition to contributions from workshop participants, this book also includes translations of Hermann's main essay on quantum mechanics—that of 1935—as well as the translation of a hitherto 'lost' manuscript on indeterminism, from 1933. We have also included in this volume an English translation of a biographical essay on Hermann written by Inge Hansen-Schaberg.

We begin in Part I with this biographical sketch of Hermann, which aids in one's appreciation of Hermann's varied accomplishments and deep commitments to particular principles of thought, life, and action. Hers was a full life, indeed. After being introduced to the overall trajectory of Hermann's life, the volume continues with contributions by Fernando Leal and Giulia Paparo, who highlight (respectively) Hermann's training under Nelson as a member of the Friesian neo-Kantian school and the importance of the natural philosophical tradition in Hermann's work.

In Part II, we present contributions treating Hermann's philosophy of quantum mechanics. The first chapter in this part is by Léna Soler, one of the earliest scholars to recognise the importance of Hermann's work in this area and bring it to the eyes of other scholars—not only by editing (with introduction and postface) a French translation of Hermann's 1935 essay (Hermann 1935/1996), but also through her analyses of Hermann's philosophy (cf. Soler 2006, 2009). The two following chapters by Thomas Filk and Mélanie Frappier provide detailed analyses of Hermann's treatment of Heisenberg's famous γ-ray thought experiment, but in very different lights: whereas Filk compares Hermann's treatment of the microscope thought experiment to Weizsäcker's published account of the same, Frappier considers the γ-ray microscope as a dialectical tool used—to sometimes conflicting ends—by several authors writing at the time of Hermann.

Michiel Seevinck continues the careful investigation of Hermann's philosophy of physics in his contribution on Hermann's discussion of von Neumann's well-known proof against hidden variables. Seevinck compares Hermann's logical parsing of von Neumann's proof to that carried out by J.S. Bell in the 1960s. After Seevinck's chapter, we turn away from detailed analysis of Hermann's 1935 essay in order to introduce Hermann's newly discovered 1933 essay on determinism in quantum mechanics. In this chapter (by EC and GB), we provide an overview of this fascinating new work and compare it to her 1935 essay. We also situate this 1933 paper in historical context and consider the implications of her having sent the manuscript to Dirac, Heisenberg, Bohr, and Gustav Heckmann for their feedback.

After this brief foray into the 1933 paper, we return to the 1935 essay with a chapter by GB comparing Bohr's 'single-slit' thought experiment to Hermann's microscope and also considering the role of measurement in Bohr, Hermann, and alongside similar thinking about measurement from Pauli. In the final chapter of Part II, EC argues that if one considers as fundamental and central to the whole 1935 essay Hermann's specific thesis regarding the relative context of observation uniquely necessitated by quantum mechanics, novel aspects of the paper come to light—among these, a more nuanced, Kantian reading of Bohr's complementarity and correspondence principles, fascinating insights into the quantum–classical divide, and the thorough-going 'splitting of truth' stemming from observational contextuality.

Part III of the volume is comprised of the two transcripts of the panel discussion (with panellists Dieter Krohn, Rene Saran, and Fernando Leal) and the general discussion held during the Hermann Workshop.

Finally, Part IV presents, for the first time, English translations of the following: a letter from Gustav Heckmann to Grete Hermann discussing the latter's 1933 essay and its reception by Heisenberg and others in Copenhagen, Hermann's 1933 essay,

'Determinism and Quantum Mechanics,' and her 1935 essay, 'Natural-Philosophical Foundations of Quantum Mechanics.' All references and quotes from Hermann's 1933 and 1935 essays made elsewhere in this volume refer to our English translations, as given here.

The primary hope of this volume is to bring Hermann to the attention of a wider audience—that scholars from various disciplines relevant to her work will continue to explore her significance in their respective contexts. We are encouraged to see, even in the years since the workshop was held, increased interest in Hermann's work. In particular, a German volume put together by Kay Herrmann is forthcoming (Herrmann 2017), which will include further essays on Hermann as well as portions of her correspondence and all her main works in mathematics, foundations of physics, and philosophy of science.

Much interesting work remains to be done on the fascinating life and work of Grete Hermann, including the following:

- Exploring Hermann's as yet uncharted philosophical discussion of relativity (Hermann 1937) and bringing out the underlying unity between her philosophy of science and the rest of her work.
- The unifying framework within which Hermann situated her thinking is provided by Jakob Friedrich Fries' reading of Kant, as mediated by Leonard Nelson. Indeed, it appears that Hermann saw her analysis of modern physics as confirming Fries' own interpretation of Kant's transcendental idealism. In so doing, Hermann developed the Friesian approach well beyond Nelson himself, who died in 1927 and never addressed the challenges posed to Kantianism by modern physics. Future projects might fruitfully seek to substantiate these points in detail.
- Besides situating it within the Friesian lineage, the evaluation of Hermann's neo-Kantian position will further require comparing her with other prominent neo-Kantians of the day, such as Reichenbach, Cassirer, and Schlick (with the last of whom Hermann was engaging directly; cf. Hermann 1936). It also may be significant that Reichenbach, too, had studied with Noether and had substantial contacts with Nelson's circle (Milkov 2013).
- Hermann's position may be further applied to modern Kantian debates. For instance, Soler (Chap. 4) suggests that one of the lessons of Hermann's work lies in how to subject a system like Kant's to the test of history, i.e. in how transcendental idealism is compatible with the historical development of science. Hermann's work is also directly relevant to the modern debates about a Kantian-inspired unification of the physical sciences in their necessary and deterministic aspect with the life sciences in their teleological aspect (cf., e.g., Friedman 1992, 2013; Zuckert 2007; Massimi 2008; Massimi and Breitenbach 2016).
- Hermann's analysis of causation plays a crucial role both in her main papers on quantum mechanics and in her fundamental paper on ethics of 1953, 'Conquering Chance' (Henry-Hermann 1953). Regarding the latter, one-time Wittgenstein executor Peter Winch—who translated the paper (Henry-Hermann

1991)—reportedly stated it was better than anything Wittgenstein himself had written on the subject. This aspect of her work bears further investigation.
- Finally, Hermann's earliest years were spent doing research in pure mathematics under Emmy Noether, in the unique social and political environment of the Göttingen of the 1920s. Possible research questions on this aspect of Hermann include understanding her interactions with specific personalities during this time period (in particular Noether and indirectly Hilbert) and examining the direct connection between Nelson and Hilbert. Nelson claimed that his work was in part inspired by Hilbert, and Hilbert is known to have strongly supported Nelson within the faculty. The interplay between the mathematical and philosophical circles at the university is worth exploring and as yet has not been investigated in sufficient depth.

It goes without saying that a project of this scope is certain to generate a substantial list of thanks due. We extend heartfelt gratitude to the staff at the Churchill Archives Centre, Churchill College, Cambridge, where one of us [GB] located the 'lost' 1933 manuscript. Both editors wish to thank the Leverhulme Trust, as a large portion of this work was made possible through their generous support (Research Project Grant F/00 152/AN). The Hermann Workshop would not have been possible without financial assistance from the Centre for the History and Philosophy of Science, Technology and Medicine at the University of Aberdeen, the School of Divinity, History and Philosophy at the University of Aberdeen, the Scots Philosophical Association, the Mind Association, and the British Society for the Philosophy of Science; we are grateful for the support provided by each. We extend our thanks once again to the whole quantum group at the Max Planck Institute for the History of Science in Berlin and to Don Howard for suggesting that we begin this fruitful collaboration. Important work on Chaps. 8 and 10 was carried out while one of us [EC] held a postdoctoral research position from 2013 to 2014 at the Hebrew University's Edelstein Center. She wishes to thank Orly Shenker and members of the Edelstein Center, as well as folks at the Einstein Archives, for their support—both financial and academic—during that time. Lisa Frach and Tom Scott are owed thanks for their important behind-the-scenes contributions to the present volume: Lisa Frach produced a lovely translation into English of Hansen-Schaberg's biography of Hermann (Chap. 1), Fedde Benedictus prepared the index and helped with the proofs, and Tom Scott recorded and then tirelessly transcribed, not only the panel and general discussions from the workshop, but also the Q&A sessions after each talk. We are indebted to you for taking on this painstaking but important part of this project, Tom! Finally, we would like to thank Stephen Gaukroger as well as Lucy Fleet and her staff at Springer for their patience, encouragement, and tireless championing of the sort of crucial, interdisciplinary work we have aimed to provide in what follows.

New York City — Elise Crull
Utrecht — Guido Bacciagaluppi
July 2016

References

Bacciagaluppi G, Crull E (2009) Heisenberg (and Schrödinger, and Pauli) on hidden variables. Stud Hist Philos Mod Phys 40(4):374–382

Bacciagaluppi G, Crull E (2011) Translation, with introduction, of W. Heisenberg: 'Ist eine deterministische Ergänzung der Quantenmechanik möglich?' Preprint posted on the Phil Sci Archive, http://philsci-archive.pitt.edu/8509/

Bacciagaluppi G, Crull E (expected 2018) 'The Einstein Paradox': the debate on nonlocality and incompleteness in 1935. Cambridge University Press, Cambridge

Friedman M (1992) Kant and the exact sciences. Harvard University Press, Cambridge

Friedman M (2013) Kant's construction of nature: a reading of the metaphysical foundations of natural science. Cambridge University Press, Cambridge

Henry-Hermann G (1953) Die Überwindung des Zufalls: Kritische Betrachtungen zu Leonard Nelsons Begründung der Ethik als Wissenschaft. In: Specht M, Eichler W (eds) Leonard Nelson zum Gedächtnis. Verlag Öffentliches Leben, Frankfurt am Main/Göttingen, pp 25–111

Henry-Hermann G (1991) Conquering chance: critical reflections on Leonard Nelson's establishment of ethics as a science. Philos Invest (1):1–80. Translation of Henry-Hermann (1953) by Winch P

Hermann G (1935/1996) Les fondements philosophiques de la mécanique quantique. Vrin, Paris. Translation of Hermann (1935b) by Schnell A. Edited with introduction and postface by Soler L

Hermann G (1936) Zum Vortrag Schlicks. Erkenntnis 6:342–343

Hermann G (1937) Die Bedeutung der Modernen Physik für die Theorie der Erkenntnis. In: Drei mit dem Richard Avenarius-Preis ausgezeichnete Arbeiten von Dr G. Hermann, Dr E. May, Dr Th. Vogel, S. Hirzel, Leipzig, pp 3–44

Herrmann K (ed) (2017) Grete Henry-Hermann: Philosophie–Mathematik–Quantenmechanik. Springer, Berlin

Massimi M (ed) (2008) Kant and the philosophy of science today. Royal Institute of Philosophy Supplement 63. Cambridge University Press, Cambridge

Massimi M, Breitenbach A (eds) (2016) Kant and the laws of nature. Cambridge University Press, Cambridge

Milkov N (2013) The Berlin Group and the Vienna Circle: affinities and divergences. In: Milkov N, Peckhaus V (eds) The Berlin Group and the philosophy of logical empiricism. Boston Studies in the Philosophy and History of Science, vol 273, pp 3–32

Pauli W (1985) Scientific correspondence with Bohr, Einstein, Heisenberg a.o., vol II, 1930–1939. Springer, Berlin. Edited by Meyenn K von

Soler L (2006) Philosophie de la physique: dialogue à plusieurs voix autour de controverses contemporaines et classiques. Epistémologie et philosophie des sciences. L'Harmattan, Paris

Soler L (2009) The convergence of transcendental philosophy and quantum physics: Grete Henry-Hermann's 1935 pioneering proposal. In: Bitbol M, Kerszberg P, Petitot J (eds) Constituting objectivity: transcendental perspectives on modern physics, vol. 79. Western Ontario Series in Philosophy of Science, pp 329–344

Zuckert R (2007) Kant on beauty and biology. Cambridge University Press, Cambridge

CHPSTM Workshop

Grete Hermann: Between Physics & Philosophy

Divinity Library, King's College, University of Aberdeen

5-6 May 2012

Saturday, 5 May

Session 1 (morning): Intro & the Nelson School

9:30-10:00	*Tea/Coffee served in Divinity Library*	
10:00-10:30	Welcome & Introduction	Guido Bacciagaluppi (Aberdeen)
10:30-11:15	Hermann's Road to Leipzig & the 1935 Thesis	Elise Crull (Aberdeen)
11:15-12:00	Philosophical Background of Grete Hermann's Work	Fernando Leal Carretero (Guadalajara)
12:00-1:30	*Lunch served in Divinity Library*	

Session 2 (afternoon): Philosophy of Physics

1:30-2:10	Changing Perspectives on Heisenberg's Microscope Thought Experiment	Mélanie Frappier (Halifax)
2:10-2:50	C.F.v.Weizsäcker's Article on the Heisenberg Microscope and its Influence on Grete Hermann's Notion of "Relative Causation"	Thomas Filk (Freiburg)
2:50-3:30	Challenging the Gospel: Grete Hermann on von Neumann's no-hidden-variables proof	Michiel Seevinck (Nijmegen)
3:30-4:10	*Tea/Coffee break*	
4:10-4:50	Grete Hermann's Pioneering Contribution to the Philosophy of Quantum Physics: An Attempt to Reconcile Quantum Mechanics with Transcendental Philosophy	Léna Soler (Nancy)
4:50-5:30	Understanding Hermann's Philosophy of Nature	Giulia Paparo (Utrecht)
5:30-6:10	Grete Hermann and the 'Copenhagen Interpretation'	Guido Bacciagaluppi (Aberdeen)
7:30	*Workshop Dinner (Nargile Turkish Restaurant, 77 Skene St)*	

Sunday, 6 May

Session 3 (morning): Ethics & Politics

9:30-10:00	*Tea/Coffee served in Divinity Library*	
10:00-12:00	Panel Discussion	Dieter Krohn (PPA) Fernando Leal Carretero (Guadalajara) Rene Saran (SFCP)
12:00-1:30	*Lunch served in Divinity Library*	

Session 4 (afternoon): Between Physics and Philosophy

1:30-2:00	Complementarity in Ethics?	Fernando Leal Carretero (Guadalajara)
2:00-4:00	General Discussion	

Contents

Contributors

Guido Bacciagaluppi Universiteit Utrecht, Utrecht, The Netherlands; IHPST and SPHERE, Paris, France

Elise Crull The City University of New York, New York, USA

Thomas Filk Albert-Ludwigs-Universität Freiburg, Freiburg im Breisgau, Germany; Parmenides Center, Munich, Germany

Mélanie Frappier University of King's College, Halifax, Canada

Inge Hansen-Schaberg Technische Universität Berlin, Berlin, Germany

Dieter Krohn Philosophisch-Politische Akademie, Bonn, Germany

Fernando Leal Universidad de Guadalajara, Guadalajara, Mexico

Giulia Paparo Stiftung Deutsches Technikmuseum Berlin, Berlin, Germany

Rene Saran Society for the Furtherance of Critical Philosophy, London, UK

Michiel Seevinck Nijmegen, The Netherlands

Léna Soler Laboratoire d'Histoire des Sciences et de Philosophie—Archives Henri Poincaré, Nancy, France

Part I
Hermann's Background

Chapter 1
A Biographical Sketch of Prof. Dr Grete Henry-Hermann (1901–1984)

Inge Hansen-Schaberg

> The nature of discussion is philosophical. How could it be decided on what the development of human society depends without fundamental clarification of the aims and values against which political developments are to be judged?
> Hermann (1945)

Grete Hermann, ca. 1937. Photo courtesy of Prof. Dr Renate Tobies, Jena

---

Translated by Lisa Frach. This biographical sketch of Grete Hermann was originally published in German as Hansen-Schaberg (2011/2012).

I. Hansen-Schaberg (✉)
Technische Universität Berlin, Berlin, Germany
e-mail: hansen.schaberg@t-online.de

E. Crull and G. Bacciagaluppi (eds.), *Grete Hermann - Between Physics and Philosophy*, Studies in History and Philosophy of Science 42,
DOI 10.1007/978-94-024-0970-3_1

The fact that Grete Hermann received a doctoral degree in mathematics in 1925 is exceptional already. However, it is but a small piece in the mosaic of her life. Born on 2 March 1901 to Augusta Hermann (née Bigoldt) and the merchant Gerhard Hermann and baptised in the Evangelical Church, Margarete[1] Hermann was raised in Bremen in accordance with her father's motto: 'I train my children in freedom!' (Henry-Hermann 1985, p. 198). Grete attended the lyceum of A. Winter from autumn 1907 until Easter 1914, later entering the 'Untertertia' class at the 'Neues Gymnasium zu Bremen' (Hermann no date, Curriculum vitae). She was one of the few girls who were admitted by exception. Afterwards she became a student in the seminar class at the Oberlyzeum of August Kippenberg from Easter 1920 to Easter 1921, which was a typical interlude for young women. There she obtained the qualification to teach at lycea on 3 March 1921 with the classification 'very well qualified' (Hermann no date, Certificates and other personal documents). Nevertheless, Grete Hermann matriculated in mathematics, physics and philosophy at the University of Göttingen on 27 April 1921. Later she would become the first and only female doctoral student of mathematician Emmy Noether, who remained very vividly in her memories (cf. Tollmien 1990, 2011): 'Emmy Noether is standing at the blackboard, her head is tilted backwards, concentrated in thought; in front of her is only a small, though intensively participating, group of listeners' (Hermann 1982). However, it was not this academic teacher who became crucial to her life story, but philosophy professor Leonard Nelson (1882–1927).

1.1 Leonard Nelson as Teacher

> [...] the demand to never refrain from answering for fear of disgrace
> Henry-Hermann (1985, p. 180)

Already in her first semester Grete Hermann and her brother Carl Hermann, who was also studying at Göttingen, attended Nelson's course 'Typische Denkfehler in der Philosophie' ('Typical errors of reasoning in philosophy') in lecture theatre 16. She remembered them in her notes from 1928/1929: 'I was inspired by the logical sharpness of the lecture, but my overall impression was negative: "How full of himself he is! He believes he has a monopoly on truth"' (Henry-Hermann 1985, p. 179). Once she went to a lecture together with her father, who 'had made a break with civil life after a long religious-ideological quest and after several difficult personal experiences; he had transferred his business and all his money to my mother and wandered around with long hair and beard in loden coat, knee breeches and overshoes

[1]In the following I shall use the short form 'Grete' of the name, as preferred and used by her. She published mostly under the name Grete Hermann even after her marriage.

as an "itinerant preacher", as he called himself, or lived alone meditating' (ibid., p. 180). He correctly judged of Nelson: 'Finally a person who takes seriously what he has discovered'.

As can be seen from her 'registration form' for courses attended for credit (StAG no date, Registration form Margarete Hermann), she took part in 'Übung zur Religionsphilosophie' ('Tutorial in philosophy of religion') in the winter semester of 1921/22, even though she felt the preliminary discussion about attendance requirements was already 'embarrassing': 'Punctuality, speaking loudly, frequent participation in discussion surely were good things [...]. But why did we need to talk about that an entire evening! Only one demand that Nelson made got my total approval. It was the demand not to let oneself be prevented from answering for fear of disgrace' (Henry-Hermann 1985, p. 180).

Nelson's teaching methods were indeed unusual. They were based on the attempt to provoke independent acquisition of knowledge by the students. This caused Grete Hermann 'joy' but also 'fear of Nelson and his way of working': 'I internally reacted against Nelson even though I avidly absorbed his way of arguing, which the tutorials guided us towards—the use of self-evident examples, the sharpness of conceptualisation—and harnessed them for my own thinking' (ibid., p. 181). Her reaction against Nelson mainly referred to his insistence on drawing personal consequences out of philosophical insights, which she rightly held to be incompatible with her search for 'religious truths'. This led her to avoid Nelson for two years (ibid., p. 182).

After attending a lecture course by theologian Karl Barth and then staying in Freiburg for two semesters from autumn 1922 until 22 August 1923 (Hermann no date, Curriculum vitae) she returned to Göttingen, and in the summer semester of 1924 returned also to Nelson's course on 'Hauptprobleme der Ethik' ('Principal problems in ethics') (Hermann, Certificates and other personal documents). To her surprise he still remembered her name, but apart from that still behaved in a way she found disaffecting. In the usual discussion after the lecture, her question to Nelson whether '[i]t is possible to fulfill the precept of one's character with one single act of volition' remained unanswered. In the following lecture he again failed to answer her question; eventually she stopped asking and was 'angry—and silent' (Henry-Hermann 1985, p. 184). In the end, however, these experiences led to a lifelong engagement with Nelson's philosophical approaches: 'Through Nelson's challenges I gradually learnt to eke out, step by step, the courage for truth that is necessary if one is to utterly place one's trust, also within one's own thinking, in a method of thought recognised as cogent' (ibid., p. 182).

In the Winter semester of 1924/25 she attended Nelson's course on 'Philosophische Pädagogik' ('Philosophical pedagogy') (Hermann, Certificates and other personal documents) and was so fascinated by his way of dealing with the topic of 'authoritarian upbringing' (Henry-Hermann 1985, p. 186) that she requested Nelson as examiner when she registered for her doctoral examination 'in abstract mathematics as major with Miss Prof. Noether and in physics and philosophy as minors' (UAG 1923–1930, Graduation file of Margarete Hermann). She wrote her dissertation on the topic 'Die Frage der endlich vielen Schritte in der Theorie der Polynomideale. Unter Nutzung nachgelassener Sätze von Kurt Hentzelt' ('The question of finite steps in the theory of polynomial ideals. With the use of posthumous propositions by Kurt Hentzelt'). This was suggested by Emmy Noether, who also wrote the evaluation and assigned the mark 'very good' (cf. ibid.). As an exception, extraordinary[2] professor Emmy Noether was appointed main supervisor by the faculty after the evaluator Landau determined that 'colleague Miss Noether is *the* authority on these most abstract questions in mathematics. I can fully endorse her evaluation' (Landau 1925; emphasis in the original).

The oral examination took place on 25 February 1925 and was also marked 'very good'. But there were some obstacles to jump first: Grete Hermann's request to be examined by Leonard Nelson was denied with reference to the fact that as an extraordinary professor he was not authorised to adjudicate doctoral examinations; experimental psychologist Narziss Ach was to be examiner instead (Henry-Hermann 1985, p. 186). In her desperation, Grete sought out her doctoral advisor, 'who immediately took matters in her resolute hands and prevailed with the dean for Nelson to be the examiner, though only in the presence of Ach'. Nelson accepted, but on the day before the examination he realised that he would probably have to share examination time with Ach. He got angry and threatened to refuse to examine under these circumstances (cf. ibid., p. 187 f.). He demanded that Grete Hermann call on the dean about this matter to bring about clarification. The dialogue Hermann and Nelson had in this context highlights Nelson's personality: '"Are you confident you can see this through?"—"I don't think so".—"Learn to be. Then you will really have gained something from this examination"' (ibid., p. 188).

The upshot of her enquiry with the dean was that Nelson should be entitled to examine on his own. About the examination process she writes: 'He leaned back in his chair, crossed his legs and asked friendly and encouragingly: "Well?" Since I was not very responsive he said after a few preliminary questions: "Now tell me something about the theology of Karl Barth." Whereupon I explained that I would rather deal with what I thought was reasonable about Barth's inquiry. For this purpose I started from the doctrine of formal idealism, which I had taken from Kant and considered to be completely assured. The half hour passed with my recognising the collapse of the assumption of formal idealism as dogmatic under the short counter-questions from Nelson' (ibid., p. 189).

[2]Extraordinary professors are tenured professors whose positions correspond more or less to the English 'personal chairs', i.e. they do not fill established chairs (*eds.*).

The waiver declaration took place on 4 May 1926 (UAG 1923–1930, Graduation file of Margarete Hermann). However, prior to that Grete Hermann registered with the examination board for the state exam and was assigned Herman Nohl as examiner. Despite or because of the recent experiences she wanted to hold on to Nelson and asked him for help; Nohl as well as the examination board agreed to the requested change of examiner (Henry-Hermann 1985, p. 190). Her desired topic for the philosophy paper would have been the discussion of the 'Paradoxes of set theory', but Nelson rejected it: 'Your doctoral dissertation was already so abstract and formal-logical. Given that, you should choose a topic with a significant content. Otherwise you will be in danger of only being able to deduce and not to judge autonomously anymore. You should not neglect judgement, and that is something other than reason' (ibid., p. 190 f.). With reference to her interest in philosophy of religion he proposed the topic: 'On transcendental idealism'. 'There you can do whatever you want!' (ibid., p. 191). As she remembers, as a result of this she had a fruitful interaction, for 'through his occasional short questions and remarks, Nelson had managed to "force me into freedom"—at least in this field' (ibid., p. 192).

1.2 As Private Assistant to Leonard Nelson

> She studies mathematics for four years, and suddenly she discovers her philosophical heart!
> Emmy Noether[3]

Not only was Emmy Noether surprised, but also Grete Hermann herself when Nelson told her that he 'read her paper with delight, it was a good paper' (Henry-Hermann 1985, p. 194), because she had indeed intensely engaged with his proposed topic yet taken a contrary stance to Nelson's approach: 'I took the paper to be an attack on him and was astounded that he did not say anything about that. From that time on I noticed that Nelson was trying to get me to work together with him [...]. Soon we were agreed: right after my examination I should help Nelson with the edition of *Ethik und Pädagogik* (*Ethics and Pedagogy*) as his assistant' (ibid.). Before that her oral examination took place, in preparation for which Nelson asked her only to look at the marked passages in her paper, 'then we can talk about it tomorrow; otherwise it would be a shame about the half hour' (ibid.). On 10 December 1925 she 'passed with distinction' the academic exam for the teaching qualification for high schools, with main subjects mathematics and physics and additional subject philosophical propaedeutics (Hermann no date, Certificates and other personal documents).

After these experiences Grete Hermann decided in favour of philosophy, even though Emmy Noether had already tried to place her in an assistant job in mathematics at the University of Freiburg (cf. Henry-Hermann 1985, p. 195). Noether apparently reacted to this with the complaint cited above. Grete Hermann's decision immediately caused her new worries, especially about her independence: 'But on the evening after

[3]Grete Hermann provided this quotation from the year 1925 in Hermann (1982) when asked about her memories of Emmy Noether.

this agreement I had a bad anxiety attack and the feeling I was getting into something, in fact even I had brought about something, which was out of my depth [...], I realised that I was worried that I would not be strong enough in the face of Nelson's personality to preserve my own intellectual independence' (ibid., pp. 195 ff.).

What caused these fears? The influences she feared were with respect to Nelson's practical demands originating in his neo-Kantian insights, namely vegetarianism, exit from the Church, engagement in politics, just to name a few of the extensive 'minimum requirements' for members of the associations 'Internationaler Jugendbund (IJB)' ('International Youth League') and 'Internationaler Sozialistischer Kampfbund (ISK)(ISK)' ('International Socialist Combat League') (cf. Link 1964; Klär 1982), founded by Nelson and Minna Specht (1879–1961; cf. Hansen-Schaberg 1992).[4]

At that time Grete Hermann arrived at the 'decision to be watchful towards myself and to make it a rule not to follow any of Nelson's big and small lifestyle requirements known to me, unless I was myself convinced to do so independently of him advocating these views. I never talked with Nelson about this decision, but I soon got the impression that he knew and respected it' (Henry-Hermann 1985, p. 196). In January 1926 she became Nelson's personal assistant and worked with him on his volume on the *System der philosophischen Ethik und Pädagogik* (*System of philosophical ethics and pedagogy*). Nelson's interest in this collaboration was probably due to the extraordinary sharpness and logic of Grete Hermann's thinking, her insisting on argumentation and knowledge gain, her criticism of his writings, and her unfearful discussion of matters of dispute. Exemplary of this is Grete Hermann's 1929 write-up of a discussion (cf. ibid., pp. 203–209) on which she later had feedback from Minna Specht: 'Nelson came home happy the evening when we resolved our dispute, and said to her: "Grete Hermann has agreed with the doctrine of the ideal"' (ibid., p. 210).

In fact, she undertook the three steps mentioned above while Nelson was still alive; she kept her Sütterlin script despite his ironic remarks about her 'chauvinistic handwriting', and she became a member of the ISK only after his death (cf. ibid., pp. 196–199).

1.3 Continuation of Philosophical-Political Work with Minna Specht

> It was a deep break within my life when I became aware [...]
> Henry-Hermann (1985, p. 195)

Minna Specht was director of the 'Landerziehungsheim Walkemühle' near Melsungen, which combined a progressive educational section for children and a political training facility for adults, and was supported by the Philosophisch-Politische

[4] All statements about Minna Specht in this paper are based on results of research that was done for my dissertation.

Akademie e.V. (Philosophical-Political Academy, registered association) (PPA) instituted by Nelson. At the beginning of October 1927, Grete Hermann visited it for the first time in order to continue work with Nelson, who was teaching a class there. A few weeks later he died, and Grete Hermann moved into the Walkemühle, which had an extensive library, to continue with Minna Specht the work begun with Nelson. They published the planned volume *System der philosophischen Ethik und Pädagogik* (*System of Philosophical Ethics and Pedagogy*) posthumously in 1932. There started a life-long collaboration between Grete Hermann and Minna Specht, not only on academic work and administration of the Nachlass, but in all respects (cf. Hansen-Schaberg 2011/2012); most of the time they also lived together.

Grete Hermann was involved with the work of the Academy and also took over some classes at the Walkemühle. When the adult training centre was closed for the planned publishing of the daily paper *Der Funke*, she and Minna Specht, as well as other leading ISK members, moved to Berlin in November 1931. They moved into Inselstraße 8A, which became living quarters, headquarters of the ISK, and the editing, publishing and sales office. Minna Specht took charge of the foreign policy desk, and a great number of articles were authored by S.H. (=Specht/Hermann). *Der Funke* was published from 1 January 1932 until its prohibition on 7 February 1933, with the intention to build an alliance of all labour organisations, to confront National Socialism and to conduct resistance.

Due to this declared antagonism and scrutiny of the 'Nelson League' by the police already since 1925, a search was carried out by 30 to 40 Nazis who claimed to be special constables during the night of 14–15 March 1933, in Nikolausberger Weg 67 in Göttingen, which was owned by the 'Gesellschaft der Freunde der philosophisch-politischen Akademie e.V. Berlin' ('Association of friends of the philosophical-political academy, registered association Berlin') (cf. StAG 1925–1936, sheet 64). Further, Walkemühle was occupied, closed and cleared in March 1933. Minna Specht and Grete Hermann made arrangements to emigrate and took the most important documents to a safety deposit box in Kassel. However, the allegedly safe place was discovered, as emerges from files of the police administration: the courier Willi Warnke was intercepted, and the deeds and other documents he was carrying—materials about the purchase of the Walkemühle, capital and testaments as well as a key—were confiscated. Therefore it was possible to find the safety deposit box and empty it (cf. ibid., sheet 88). All attempts to prevent expropriation of property and capital failed and ended with the prohibition of the PPA on 27 February 1935 (cf. ibid., sheet 141).

1.4 Philosophy and Political Work Against the Nazi State

> There is no neutrality in the face of the legal and cultural demise of public life
> Hermann (1945, p. 46)

At first Grete Hermann remained in Germany and took the opportunity to visit the exile schools directed by Minna Specht in Denmark. Among the most important intellectual stimuli for her work were arguments about physical and philosophical questions in correspondence and encounters with Prof. Dr Werner Heisenberg and his assistant at that time, Carl Friedrich von Weizsäcker (Weizsäcker et al. no date). From a letter by Weizsäcker of 17 December 1933 from Copenhagen,[5] where he worked with Niels Bohr, it can be reconstructed that a first meeting between himself and Grete Hermann was arranged for the beginning of January 1934 in Berlin. In another letter of 30 January 1934 he refers to this meeting and mentions that Werner Heisenberg has told him that she will be coming to Leipzig for some time, and that he hopes it will be during the summer term because then he will be there as well. However, she was at the institute in Leipzig already from February, as is evident from a letter to Heisenberg of 9 February 1934 (with the autograph note: 'draft edited further'): 'Dear Mr Heisenberg! I was not up to following your physics example this morning. I ask you to let me repeat it again' (ibid.). Long considerations including mathematical formulae follow.

It is not possible to reconstruct how long exactly Grete Hermann stayed in Leipzig, but doubtlessly it was a stimulating and productive time for everyone involved, and resulted in some important publications by Grete Hermann. First 'Die naturphilosophischen Grundlagen der Quantenmechanik' ('The natural-philosophical foundations of quantum mechanics') with the preliminary remark:

> In the physics institute at Leipzig I had the opportunity to pursue the problems in natural philosophy raised by quantum mechanics by engaging with the physics circle there, and I thank therefore here above all Professor Heisenberg for his willingness to discuss the foundations of quantum mechanics, which was crucial in helping the present investigations. (this volume, Chap. 15, p. 239)

A few postcards from Werner Heisenberg to Grete Hermann have survived, e.g. one of 28 June 1935, in which he thanked Hermann for sending the publication mentioned above: 'Your book is a nice reminder of our animated discussions in the institute in Leipzig' (Weizsäcker et al. no date). In addition, he gave her the requested addresses of, among others, Courant, Institute for Mathematics of the New York University, told her about the death of Emmy Noether and mentioned Bohr, Einstein, von Laue and P. Jordan (in Rostock), who would certainly be interested in her contribution. Werner Heisenberg even dedicates a chapter to her in his memoirs (Heisenberg 1969) (English translation: Heisenberg 1971), in which he portrays the discussions between Grete Hermann, Carl Friedrich von Weizsäcker and himself about 'Quantum Mechanics and Kantian Philosophy' and comes to the conclusion: 'One of the requirements of Fries' school and hence of Nelson's circle was that all philosophical questions must be treated with the rigor normally reserved for modern mathematics. And it was by following this rigorous approach that Grete Hermann believed she could prove that the causal law—in the form Kant had given it—was unshakable. Now the new quantum mechanics seemed to be challenging the

[5]Several of the letters mentioned here will be published (in German) in a forthcoming book, Herrmann (2017) (*eds.*).

Kantian conception, and she had accordingly decided to fight the matter out with us' (Heisenberg 1971, pp. 117–118). The narration of the argumentation and questions conveys very clearly how persistently Grete Hermann must have voiced criticism and pointed out problems, allowing herself to be convinced but also convincing others, because Werner Heisenberg writes 'we had the feeling that we had all learned a good deal about the relationship between Kant's philosophy and modern science' (ibid., p. 124).

From 1934 until 1936, Grete Hermann lived on the Østrupgård estate on the Danish island of Fyn, where the work of the Walkemühle was continued, and where she was a member of the school board. Nevertheless she travelled to Germany time and again, gave lectures and worked further on problems of natural philosophy and physics. She took part in the 1934 prize competition of the Sächsische Akademie der Wissenschaften (Saxon Academy of Science) on the topic 'What are the consequences of the quantum theory and field theory of modern physics for epistemology?' Seventeen papers where submitted and three, among them Grete Hermann's, were awarded the Richard Avenarius Prize (Hermann 1937a, Preface). The prize money for each was 1000 Reichsmark and was paid out by the Avenarius Foundation in Leipzig on 22 June 1936, 'Signed Frings, Heisenberg, Krueger, le Blanc, Litt' (AdsD no date, Avenarius-Foundation). The papers were published by S. Hirzel in Leipzig in 1937.

Her work on philosophical problems was therefore publicly recognised, but her political opposition remained undetected: 'Of course I met up with illegally working friends on those trips when it was possible—not to participate in their activities, but to reason out with them the significance of their resistance in intensive discussions. Those were philosophical courses; they went deeper and were more vivid than any teaching I have ever done in my life' (letter Grete Henry-Hermann to Birgit S. Nielsen of 17 March 1981, quoted in Nielsen Nielsen 1985, p. 43). From 1935 until 1940 Grete Hermann also published numerous philosophical-political articles in the magazine *Sozialistische Warte* (*Socialist Watch*) edited by Willi Eichler in Paris, under the pseudonyms Leonore Bremer, Gerda Bremer and Peter Ramme,[6] to protect her identity and not to endanger herself and others during her stays in Germany.

From October 1936 until April 1937 she lived with seven of the older children from the school in Østrupgård and their teacher Gustav Heckmann in Copenhagen at Brogade 5, as can be concluded from her letters from Minna Specht (Hermann and Specht no date). There she took part in an international congress as well as in 1937 in Paris (cf. Hermann 1937b). At the end of 1937 or in the beginning of 1938 Grete Hermann moved to London when a wave of arrests put an end to a great part of the ISK resistance and flight was the only recourse. With her marriage with the technician Edward Henry on 1 February 1938 she acquired British citizenship (StAB no date, Personal file Prof. Grete Henry-Hermann). She was thereby freed from the usual reprisals and was not interned on the Isle of Man for being an 'enemy alien' as, for instance, Minna Specht had been.

[6] A list of the articles can be found in the volume of Hermann's writings, Henry-Hermann (1985, pp. 226 ff).

It was a marriage of convenience, even though they got divorced only on 1 March 1960, as becomes obvious for instance in an application letter for a job in teacher training in Bremen through the formulation: 'I am however in no way bound by my marriage, and am instead available at any time to work in Germany. I have the urgent wish to help to the best of my ability with the reconstruction work in Germany' (Hermann 1946). In this application letter, which she sent from 33 Green Lane, London, where she lived with Minna Specht, she indicates her diverse teaching experience, including the 'leading of Socratic work groups with adult students, for which I have had constant opportunity since 1933. In these we discussed philosophical questions of all kinds. I was able to continue this work also during the war years, in which I took up household management as my profession' (ibid.). This constitutes thus more evidence for her teaching activity within the resistance groups in Nazi Germany as well as in Great Britain.

In January 1945 she publishes through the ISK the 60-page paper 'Politik und Ethik' (Hermann 1945; English translation in Hermann 1947), and gives her view on the question of whether it is possible to make a place for oneself within National Socialism and to create 'islands in the political stream of events' for one's work and life:

> The pursuit of art and science or of relatively free human relationships, which is possible here, has contributed to misleading and has been abused to mislead the surrounding world about the circumstances of the entire society. Those who live in the Third Reich or otherwise come to terms with it but choose to close themselves off from the political process around them in order to concern themselves with things that are intrinsically beautiful and worthy, shield the system with their repute and that of their work. There is no neutrality in the face of the legal and cultural demise of public life. Those who do not confront it are part of it. Whatever else they may create that is beautiful and good is devalued by this participation in the social iniquity that burdens it. (Hermann 1945, pp. 46 ff.)[7]

More generally she criticises the one-sidedness of the scientific development in the previous century, which focuses on the scientist's direction of enquiry, into the causes and fundamental forces in the course of events and thereby pushes the ethical question into the background: 'This question is clearly not silenced by this. It arises wherever people seriously reckon with the meaning of their lives. This has happened and will happen, even though official representatives of science declare a priori such an involvement to be unscientific and a merely subjective position of the individual' (Hermann 1945, p. 58).

---

[7]In the English version, this passage reads: 'Devotion to art or science, to the creation of relatively free human relations which is possible in such a protected environment has helped to mislead the world about the real state of affairs in society. Indeed it has been assiduously exploited for this very purpose. Those who adapt themselves to a regime such as Hitler's and close their eyes to the political happenings around them for the sake of things which in themselves are valuable, support and strengthen the system in taking up such an attitude. There can be no neutrality when people stand face to face with the moral and cultural decline of a corrupt social order. Those who do not struggle against it grant it their support. However fine and noble the achievements otherwise obtained, they are rendered worthless by the share in the social injustices with which they are burdened' (Hermann 1947, p. 66) (*eds.*).

She further deals with the question of what guidelines a political education should be built upon: 'Such an education can and should tie in with the self-reflection stimulated in a person by significant social or personal experience. The task of education is to watch that this process of self-reflection is neither rashly aborted nor distorted' (ibid., p. 51). By connecting ethics, politics and pedagogy, Grete Hermann positions herself as a student of Leonard Nelson and develops her own critical approach in anticipation of historical developments.

1.5 Return to Germany and Professional Career

> The work that you did already 20 years ago on the philosophical significance of quantum mechanics was, in my opinion, the first fully valid answer that then-young theoretical atomic physics had on the side of philosophy, nor has much more followed since from others.
> von Weizsäcker (1956)

On 20 May 1946 Dr Grete Henry—as she called herself at that time—was employed by Bremen's secondary education and was assigned four hours of mathematics at the Oberschule for girls in Karlstraße, but was mostly engaged with the Pedagogical Seminar (cf. StAB no date, Personal file, sheets 6–8). On 1 October 1949 she became provisional head (kommissarische Leiterin) of the Pädagogische Hochschule (Teacher Training College) then being established. Since she had lost her German citizenship when she got married, she was refused a professorship. However, she had the support of the Senator for Schools and Education, who wanted to keep her in Bremen, and who mentions in his case to the personnel department of 18 February 1950 that she had rejected several calls from universities (Marburg and Tübingen, among others) and had proved herself well in her teaching activity; thus her professorship should not be held back any longer (cf. ibid., sheet 54).

After some legal disputes she managed to win the recognition of the dual citizenship she wished for, and after acquisition of German citizenship she was appointed full professor at the Pädagogische Hochschule on 1 July 1950 (cf. ibid., sheet 80). At the same time she applied to resign from her headship, 'because she is convinced that building up the college requires a different headship and she herself would like to be freer for her professional duties at the Pädagogische Hochschule than she has been so far' (ibid., sheet 69). She remained deputy head of the Pädagogische Hochschule when Hinrich Wulff took over the headship (cf. ibid., sheet 105).

Grete Hermann's commitment was to teacher training. In the series *Studium Generale* she published a text on 'Die Kausalität in der Physik' ('Causality in physics') (Hermann 1948) and took part in the discussions on school reform together with Minna Specht. She became head of the Pädagogische Hauptstelle der Gewerkschaft Erziehung und Wissenschaft (Central Office for Education of the Union for Education and Science), and out of this work she presented on 1 May 1952 a collection of materials with the title 'Die Schule in unserer Zeit' ('The school in our time'), in which she was concerned with the 'non-professorial teaching staff in differentiating unified schools' (Mittelbau der differenzierenden Einheitsschule).[8] In 1954 she was appointed to the Deutscher Ausschuss für Erziehungs- und Bildungswesen (German Committee for the Education System), which was a first collective step in the cultural policy of Federation and States. From 1957 Grete Hermann was also a member of the committee for cultural policy of the German Social Democratic Party (SPD), to which she had belonged since 1946.

What was most important for her, however, was the engagement with Nelson's philosophical insights, which she developed further in her essay 'Die Überwindung des Zufalls. Kritische Betrachtungen zu Leonard Nelsons Begründung der Ethik als Wissenschaft' (Henry-Hermann 1953; published as 'Conquering Chance: Critical Reflections on Leonard Nelson's Establishment of Ethics as a Science' in Henry-Hermann 1991) on the basis of the mathematical-scientific development in the twentieth century, seeing herself '(in agreement) with his own claim that each individual, appropriating and developing those teachings, enters the ranks of the researchers together with whom he is struggling with the same fundamental philosophical questions' (Henry-Hermann 1991, p. 1).

With the intention of deepening these lines of research through fresh work, Grete Hermann requested on 28 June 1956 a year of research leave from the Senator for Education (StAB no date, Personal file, sheet 91). She asked Weizsäcker at the Max Planck Institute for physics in Göttingen for a stay during the leave year 'to refresh my contact with modern physics and to have the opportunity to discuss questions in this field with you and other philosophically interested physicists. Would you agree with that and could you give me visiting status at your institute similar to the one I enjoyed at the Institute of Physics in Leipzig and in discussions with Mr Heisenberg and you?' (Henry-Hermann 1956).[9] He agreed with great pleasure and mentioned that he had 'nominated her several times [...] for chairs in philosophy' (StAB no date, Personal file, sheet 92). Her request for leave with the letter from Weizsäcker enclosed was presented on 8 August 1956 and resulted in her leave being granted. Because the Senator for Education, W. Dehnkamp, much later complained that she had not reported back and also had not handed in a report (cf. ibid., sheets 104–106), there exists a text by Grete Hermann from 22 November 1959 that casts light upon this period: she spent the 1957 summer semester in Göttingen, but noticed 'that the philosophical interpretation of modern physics must also draw on psychological research on perception and experience', and hence went to Marburg to Professor Düker at the Psychological Institute (ibid., sheet 107). With this an old contact was re-established, for Heinrich Düker had also belonged to the ISK, had been active in the PPA, had been active in the resistance during the Nazi period and survived prison terms and internment in a concentration camp.

At the beginning of 1954, Minna Specht had moved in with Grete Hermann at Am Barkhof 19 in Bremen. After Specht's death on 3 February 1961, Grete

[8]The concept of the 'Einheitsschule' refers to a comprehensive school for all pupils until school leaving, while 'differentiation' refers to the approach to pupils with different abilities and aptitudes (*eds.*).

[9]This correspondence is also included in Herrmann (2017) (*eds.*).

Henry-Hermann became president of the PPA and remained so until 1978. During this time, between 1970 and 1976, the *Collected Works of Leonard Nelson* (Nelson 1970–1977) were published under her editorship. Furthermore, she put together a study edition with his texts (Henry-Hermann 1975). On 1 April 1966 she went into retirement (StAB no date, Personal file, sheet 121) and died on 15 April 1984 in Bremen.

Grete Hermann had an extraordinary career as professor despite the adversities of the time and the forced exile for political reasons. She is now remembered in Bremen through the establishment of the Grete-Henry Research Programme at the University.

References

AdsD (no date) Archiv der sozialen Demokratie der Friedrich-Ebert-Stiftung, Bonn: Grete Henry-Hermann Collection

Hansen-Schaberg I (1992) Minna Specht: eine Sozialistin in der Landerziehungsheimbewegung (1918 bis 1951) Studien zur Bildungsreform (Book 22). Peter Lang International Academic Publishers, Frankfurt am Main

Hansen-Schaberg I (2011/2012) Prof. Dr Grete Henry-Hermann (1901–1984). Zeitschrift für Museum und Bildung Heft 74: Akten-Einsicht, pp 104–122

Heisenberg W (1969) Der Teil und das Ganze. Gespräche im Umkreis der Atomphysik. Piper Verlag, München

Heisenberg W (1971) Physics and beyond: encounters and conversations. Harper and Row, New York. Translation of Heisenberg (1969)

Henry-Hermann G (1953) Die Überwindung des Zufalls: Kritische Betrachtungen zu Leonard Nelsons Begründung der Ethik als Wissenschaft. In: Specht M, Eichler W (eds) Leonard Nelson zum Gedächtnis. Verlag Öffentliches Leben, Frankfurt am Main/Göttingen, pp 25–111

Henry-Hermann G (1956) Letter to C.F. von Weizsäcker, 30 June 1956. In: AdsD (no date), Box 1/GHAJ000009

Henry-Hermann G (ed) (1975) Leonard Nelson: vom Selbstvertrauen der Vernunft. Schriften zur kritischen Philosophie und ihrer Ethik. Meiner, Hamburg

Henry-Hermann G (1985) Die Überwindung des Zufalls. Meiner, Hamburg

Henry-Hermann G (1991) Conquering chance: critical reflections on Leonard Nelson's establishment of ethics as a science. Philos Inv (1):1–80. Translation of Henry-Hermann (1953) by Winch P

Hermann G (1937a) Die Bedeutung der Modernen Physik für die Theorie der Erkenntnis. In: Drei mit dem Richard Avenarius-Preis ausgezeichnete Arbeiten von Dr. G. Hermann, Dr. E. May, Dr. Th. Vogel. S. Hirzel, Leipzig, pp 3–44

Hermann G (1937b) Die naturphilosophischen Bedeutung des Übergangs von der klassischen zur modernen Physik. In: Travaux du IX. Congrès international de philosophie, Paris, 1–6 Août 1937, pp 99–101

Hermann G (1945) Politik und Ethik. Renaissance Publishing Co., London

Hermann G (1946) Letter from Grete Henry-Hermann, 15 January 1946. In: StAB (no date)

Hermann G (1947) Politics and ethics. International Publishing Company, Bombay. Translation of Hermann (1945)

Hermann G (1948) Die Kausalität in der Physik. Studium Generale 1(6):375–383

Hermann G (1982) Letter to B.L. van der Waerden, 24 January 1982. In: AdsD (no date), Box 1/GHAJ000012

Hermann G (no date) Certificates and other personal documents. In: AdsD (no date), Box 1/GHAJ000001
Hermann G (no date) Curriculum vitae Margarete Hermann. In: UAG (1923–1930)
Hermann G, Specht M (no date) Correspondence. In: AdsD (no date), Box 1/GHAJ000005
Herrmann K (ed) (2017) Grete Henry-Hermann: Philosophie–Mathematik–Quantenmechanik. Springer, Berlin
Klär KH (1982) Zwei Nelson-Bünde: Internationaler Jungendbund (IJB) und Internationaler Sozialistischer Kampfbund (ISK) im Licht neuer Quellen. IWK 18:310–361
Landau E (1925) Note to Grete Hermann, 30 January 1925. In: UAG (1923–1930)
Link W (1964) Die Geschichte des Internationalen Jungendbundes (IJB) und des Internationalen Sozialistischen Kampfbundes (ISK). Hain, Meisenheim/Glan
Nielsen BS (1985) Erziehung und Selbstvertrauen. Ein sozialistischer Schulversuch im dänischen Exil 1933–1938. Beltz, Wuppertal
Nelson L (1970–1977) Gesammelte Schriften. 9 volumes. Meiner, Hamburg
StAB (no date) Staatsarchiv Bremen: Personal file Prof. Grete Henry-Hermann, Collection personal files of the Senator for Education, signature 4,111 Pers. No. 2260
StAG (1925–1936) Stadtarchiv Göttingen: File Internationaler Sozialistischer Kampfbund. Police Administration Göttingen, section 155, no. 5, period 1925–1936
StAG (no date) Stadtarchiv Göttingen: Registration form Margarete Hermann
Tollmien C (1990). 'Sind wir doch der Meinung, dass ein weiblicher Kopf nur ganz ausnahmsweise in der Mathematik schöpferisch tätig sein kann...': Emmy Noether 1882–1935 zugleich ein Beitrag zur Geschichte der Habilitationen von Frauen an der Universität Göttingen. Göttingen Jahrbuch 38:153–219
Tollmien C (2011) Begegnung im Exil: Emmy Noether und Albert Einstein. In: Hansen-Schaberg I, Häntzschel H (eds) Alma Maters Töchter im Exil: zur Vertreibung von Wissenschaftlerinnen in der NS-Zeit. Edition text + kritik, München, pp 77–96
UAG (1923–1930) Universitätsarchiv Göttingen: Files of the Mathematisch-naturwissenschaftliche Fakultät, University of Göttingen, re. Graduations Specialia Litt. 'H'. Fol. I. 1923–1930. No. 33/869 Margarete Hermann
von Weizsäcker CF (1956) Letter to Grete Hermann, 2 July 1956. In: StAB (no date)
von Weizsäcker CF, Heisenberg W, Hermann G (no date) Correspondence. In: AdsD (no date), Box 1/GHAJ000009

Chapter 2
Grete Hermann as a Philosopher

Fernando Leal

2.1 Introduction

The first thing to be said about Grete Hermann[1] as a philosopher is that she was a Kantian. To say that may seem to locate her clearly as a scholar, given that today Kant is not as central to philosophy of science as he once was; but the label of a Kantian is not so distinctive if one considers that, historically, all German scholars after Kant have been Kantians. This is just a fact about the history of German thought; for Kant was and is the greatest success story in German intellectual life. He somehow managed to convert everyone to his particular way of seeing things and asking questions.[2] And so what we need to know is not whether Hermann was a Kantian but what *kind* of Kantian she was. She was a member of one very particular tradition of

[1] A.k.a. Grete Henry or Grete Henry-Hermann, the surname 'Henry' originating in a political marriage of the sort common during and after the Second World War to secure residence for German political refugees in Britain. Apart from occasional papers, Grete Hermann wrote some 100,000 words in the shape of four substantial contributions. Two of them belong to the philosophy of science: Hermann (1935, first English translation in this volume, Chapter 15) and Hermann (1937b). The other two belong to moral and political philosophy: Henry-Hermann (1953) (reprinted in Henry-Hermann (1985a, pp. 3–95); English translation as Henry-Hermann (1991)) and Henry-Hermann (1985b), a set of notes first published in Henry-Hermann (1985a, pp. 99–178)). To these we may add a short paper on the potential contribution that the new science of ethology might make towards a reformed Kantian philosophy (Henry (1973a); English translation Henry (1973b)). Finally, Hermann's unpublished correspondence is in my view also of great philosophical importance.

[2] Even the so-called logical positivists or logical empiricists, who vociferously opposed Kant, were Kantians in the sense that they wholly approved of the fundamental distinction between analytic and synthetic propositions upon which Kant's edifice is erected. Only Quine dared to question that assumption, but today—not least because of the logical work of Saul Kripke—it has come back with a vengeance.

F. Leal (✉)
University of Guadalajara, Guadalajara, Mexico
e-mail: ferlec@hotmail.com

E. Crull and G. Bacciagaluppi (eds.), *Grete Hermann - Between Physics and Philosophy*, Studies in History and Philosophy of Science 42,
DOI 10.1007/978-94-024-0970-3_2

Kantian thought. At the risk of oversimplification one could say that there are three great post-Kantian traditions. To avoid misunderstanding, let me hasten to add that none of these traditions was servile. All three raised objections to Kant's work, but all wanted to build upon it. They all wanted to *separate* the good parts in Kant from the bad parts. The differences among the three traditions have precisely to do with what each thought were the good and the bad parts. This is our first topic.

2.2 Kant and His Followers

In order to gain a little perspective, let me say briefly what Kant was all about. He was the heir of the two greatest revolutions in thought that transformed Europe and, in due time, the rest of the world. One was the Scientific Revolution, about which all I need to emphasise here is that the key to its success was the employment of mathematics (algebra and analysis) for scientific purposes. The other was what we may call the Liberal Revolution, whose purpose and result was to free people from the shackles of the *ancien régime*; most readers of this volume happen to live free in this sense because liberalism, after many struggles, triumphed completely in our societies, so that we (somewhat ungratefully) take the freedoms we enjoy for granted.[3] Although both revolutions were long in preparation, they came to fruition in the seventeenth and eighteenth centuries. And it was at the end of the eighteenth century that Kant began to develop his philosophy.

Because Kant was, and understood himself to be, the heir of these two tremendously successful revolutions, it is no surprise that his philosophy had two main goals.[4] One was to explain how human beings were capable of scientific (including mathematical) thinking.[5] The other was to develop a coherent (i.e. complete and

[3] A wonderful description of the Liberal Revolution is given by Robert L. Heilbroner in *The Worldly Philosophers* (Heilbroner 1999, Chap. II). There are still places on this earth where people are not free to think as they please, let alone to speak, write and publish what they have thought, to associate with others and to meet with them in public, to choose where to live, what places to visit or what to do for a living. For them, the Liberal Revolution has not yet arrived. For the rest of us it is almost impossibly difficult to put ourselves in their shoes, which are indeed Kant's own shoes.

[4] Kant's allegiance to the Scientific Revolution is clearly expressed in the preface to the two editions of his *Critique of Pure Reason* (cf. Kant 1781). His two essays against the cliché 'this may be true in theory but does not apply in practice' (Kant 1793) and for 'perpetual peace' (Kant 1795), bear witness to his allegiance to the Liberal Revolution, and so does his *Metaphysics of Morals* (Kant 1797), albeit more subtly.

[5] Although there are antecedents in Descartes, Pascal, Leibniz and Berkeley, it is no exaggeration to say that Kant invented what we now call the philosophy of science. Compared with him, famous philosophers such as Bacon, Hobbes, Locke, Spinoza, Vico and Hume were certainly aware that something quite big had happened in the seventeenth century at the hands of Galileo, Torricelli, Kepler, Huygens, Newton and others; but they did not attempt to build philosophical theories about it. After Kant, no philosopher could afford to content himself with just 'being aware' of the Scientific Revolution. An amazingly thorough historical proof of Kant's knowledge of, and dedication to, hard, detailed questions of the physics of his time can be found in Michael Friedman's long-awaited commentary on Kant's *Metaphysical Foundations of Natural Science* (Friedman 2013; compare Wilson 2010).

consistent) version of liberalism as a way of collective life on the basis of a new ethical, educational, legal and political order. The first goal was purely theoretical, the second goal mainly practical. However, Kant had one big problem: his theory of how science was possible seemed to make liberalism theoretically impossible. Science, in Kant's view, was and could only be causal and deterministic, yet causal determinism seems to undermine any idea of freedom and free will. This big problem still haunted Grete Hermann, as I shall try to explain later on, which is why her ideas on quantum mechanics cannot be separated from her ideas about ethics and politics. In fact, the problem still haunts contemporary philosophers.[6]

I said that there are basically three traditions of German post-Kantian thinking. To this I should add that two of them are scientifically minded while one is not. Given the likely interests of the readers of this volume, I am going to ignore the latter. For those who know something about the history of philosophy, the names Fichte, Schelling and Hegel may indicate the general direction of that particular tradition, which by the way is still alive and kicking.[7]

One of the two scientifically-minded post-Kantian traditions has been cultivated by, and belongs to, scientists and mathematicians. As I said before, Kant wanted to explain how science (including mathematics) was possible. In the course of his explanation he said many intriguing things—things that provoked and challenged German mathematicians and scientists. All through the nineteenth and twentieth centuries they responded to those challenges in different ways. Their responses gave birth to the German versions of Euclidean and non-Euclidean geometry, mathematical logic and the axiomatisation of arithmetic, algebra and analysis, cell biology, psychophysiology, and solutions to the foundational crisis in physics. Anybody who reads carefully the writings of German and German-influenced authors in these fields will see for themselves that Kant is a constant discussion partner, from Gauss to Einstein and beyond.[8]

---

[6]For a collection of contemporary papers on freedom and determinism see Campbell et al. (2004); for a pretty full panorama of views on free will see Kane (2002); and for very short introductions to such views see Pink (2004) and Kane (2005). The readers of this volume may have heard of musings about the connection between quantum theory and the free-will problem. A recent discussion on this by *Scientific American* senior editor George Musser has now started at http://www.scientificamerican.com/article.cfm?id=quantum-physics-free-will. It promises continuous updating and even a full special issue on the topic. A very promising project uniting philosophers and scientists is now underway at http://www.freewillandscience.com.

[7]There are two important exceptions to the above sweeping statement on the nonscientific character of this particular post-Kantian tradition. One is Schelling's conception of *Naturphilosophie* tracing back to Kant and Leibniz, which although remote from any practical scientific work seems to have fed the theorising of some scientists in their quest for the principle of the conservation of energy (Kuhn 1977, pp. 97–100). The other is Hegel's awareness of the new science of political economy as well as the new philological methods of modern critical history, both of which had been all but ignored by Kant.

[8]My favourite example is Hermann von Helmholtz, who referred to Kant during his whole life, even if the area of agreement he recognised seems to have become increasingly smaller (Schwertschläger 1883; Conrat 1903; Riehl 1904). However, more important than the direct agreement of views is the agreement on *questions* that are important for the understanding of human cognition. Thus Kant's second analogy of experience is at bottom an argument that the principle of causality depends on

The second scientifically-minded post-Kantian tradition belongs not to scientists or mathematicians but to philosophers who were not (or at least not to a great extent) practitioners of either mathematics or science, but who had studied mathematics *and* science (especially but not exclusively physics). This second tradition was started by a now almost forgotten author called Jakob Friedrich Fries (1773–1843), and there is a more or less continuous link between his work and that of Grete Hermann.[9] I say 'more or less continuous' because there is a curious historical gap of about 50 years between 1859 and 1904 in which only scientists and mathematicians, but no professional philosopher of note, declared their allegiance to the Friesian tradition.[10] I personally think that this was a great disgrace for the Friesian school, in that no proper engagement between Friesian philosophy, on the one hand, and science and mathematics, whether Friesian or not, could take place in that period. Anyway, from 1904 until his death in 1927, the German philosopher Leonard Nelson, Grete Hermann's teacher, resurrected the Friesian tradition in all its earlier aspects. Like a good Friesian, he studied philosophy, mathematics, physics and the law; he maintained relations with some of the best mathematicians and physicists of his time, notably with David Hilbert and his circle; and he developed a systematic revised critical version of Kant's philosophy in no less than nine solid volumes of work, recently completed by two additional, slimmer ones (Nelson 1970–1977, 2004, 2011, English translation of some of these in Nelson 1949a, 1971, 2016).

(Footnote 8 continued)
the human ability to distinguish between my own movements (the movements of my eyes, my head or my entire body) and any external movements. How *that* is possible from a physiological point of view was recognised as a fundamental question by the great scientist (see e.g. his 1855 address, 'Über das Sehen des Menschen', in von Helmholtz 1903, pp. 105–107).

[9]The considerable output of Fries has been made available to contemporary scholars in 33 volumes (four of them as yet unpublished) edited by Gert König and Lutz Geldsetzer at Scientia Verlag in Aalen, Germany (Fries 1967–2011). The main books relevant to philosophy of science are: *The Mathematical Philosophy of Nature: A Philosophical Exposition* (Fries 1822/1979), *A System of Metaphysics* (Fries 1824/1970), *A Handbook of Natural Science: Experimental Physics* (Fries 1826/1973), *A New Critique of Reason* (Fries 1828/1968, 3 vols.), *A Popular Course on Astronomy* (Fries 1833/1973), *A System of Logic* (Fries 1837/1967), *Handbook of Psychological Anthropology* (Fries 1837–1839, 2 vols.), *On the Optical Centre of the Eye* (Fries 1839/1975), *A Critical Essay on the Principles of the Calculus of Probabilities* (Fries 1842/1996).

[10]Fries had quite a few followers both in science and in philosophy, foremost among them Matthias Jakob Schleiden (1804–1881) and Ernst Friedrich Apelt (1812–1859). Although mainly known to historians as one of the discoverers of the theory of the cell as the basic unit of all organisms, Schleiden also wrote significant philosophical studies such as *The Theory of Visual Knowledge* (Schleiden 1861) and the important methodological introduction to his *Principles of Scientific Botany* (Schleiden 1842, pp. 1–112), which unfortunately is omitted in the English translation of that work (Schleiden 1849). Apelt was both a pioneer in the history of science—particularly of astronomy, conducting the first serious historical research on Kepler's discoveries (cf. Apelt 1849, 1851)—and a systematic philosopher of science, developing what is perhaps the first logical account of induction (Apelt 1854). Apelt also wrote an innovative exposition of critical philosophy in the Friesian tradition (Apelt 1857).

2.3 Grete Hermann's Rebellion

Nelson's most brilliant student, Grete Hermann—herself a trained mathematician—became his assistant and collaborator (and much later, one of the editors of his *Collected Works*, (Nelson, 1970–1977)). During the time of the collaboration between Nelson and Hermann, the latter came to question the two fundamental aspects of Nelson's philosophy corresponding quite precisely with Kant's two main goals.

On the *theoretical* side, Nelson was witness to the emergence of special and general relativity, although not so much of quantum mechanics, because that theory came to fruition during Nelson's last years of life. And although death thus prevented him from reacting to quantum mechanics, he—like other members of the Friesian school, including mathematician Paul Bernays (cf. Bernays 1914), and in fact many other authors—was quite sceptical of relativity in the beginning. In a course of lectures given shortly before his death, Nelson even regretfully suggests that 'natural philosophy and empirical physics will have to go their separate ways, perhaps for quite some time' (Nelson 1971, p. 253; German original in Nelson 1970–1977, vol. IX), until some form of reconciliation is found. In fact, his very last academic address confirmed his firm belief in Euclidean geometry against Minkowski's spacetime (Nelson 1949b; German original in Nelson 1970–1977, vol. III, pp. 187–220).

Now, Grete Hermann shared Nelson's and Bernays' doubts, which only became greater with quantum mechanics, with its even more radical challenge to the Kantian–Friesian view. And so she came to take part in Heisenberg's seminar and had protracted discussions with him and with Carl F. von Weizsäcker (see Heisenberg 1969, Chap. 10; compare Gilder 2008, Chap. 16). As a result of those discussions, she wrote two long papers on the foundations of modern physics in general and quantum mechanics in particular (Hermann 1935, 1937a). The more technical papers in this volume will allow readers to make their own judgement as to whether or to what extent what Grete Hermann proposes in these two papers is sufficient to effect some form of reconciliation between modern physics and the concepts and principles that underlie physics as a science in all Kantian philosophy, viz Euclidean space, causality and so on.

On the *practical* side, Nelson developed a causal account of how a special kind of interest—the so-called 'moral' interest—can gain the upper hand in the struggle against other kinds of interest, so that people as individuals and collectives of people (groups, communities, nations, humankind itself) can become moral. This causal account of moral decision-making was indeed posited by him as the ultimate basis on which individual ethics, the educational system, and the legal and political order of liberalism (or rather, a special combination of liberalism and socialism) had to be constructed. Nelson himself develops all those topics in three of the nine volumes of his *Collected Works* (see Nelson 1970–1977, vols. IV, V, VI). By the way, he enjoyed the assistance of Grete Hermann for part of this work.

So far, the whole thing looks very academic and dry. But it was not. Nelson's work in the aforementioned three volumes was certainly *theoretical* in the sense that it was a relentless sequence of philosophical argumentation that has few equals in contemporary philosophy; but it was not *only* theoretical. The whole point of the exercise was actually to realise those ideas—to make them work in the real world. Nelson's school was not just a philosophical school like all the others, and in order to understand Grete Hermann one has to understand the unique character of Nelson's school. Its members were supposed to try and create the society envisaged as the moral one.[11] So we have here to do not just with a school of thought—one more among many at the time, as well as before and after—it was an activist movement as well, with an educational branch and a political branch. The important thing is that the movement was grounded in a detailed philosophical argumentation, one of whose crucial elements was the said causal account of moral decision-making.

Now, Hermann had serious doubts about the practical part of Nelson's philosophy as well. Although a very young woman and probably in some awe of her teacher, she challenged him. Nelson's account of moral decision-making was crucial to what he, in the jargon of the Kantian Friesian tradition, called the 'deduction' of the moral law, and so also crucial to the whole philosophical project Nelson and Hermann shared.[12] This calls for some explanation. In the next paragraph I shall couch this explanation in second-person mode to convey the idea that the 'deduction' is about you, whoever you are.

According to Nelson, then, the moral law, or the categorical imperative, orders *you* in an unconditional manner—no excuses allowed—first, to take account of all the interests, both *yours* and those of others, that are involved in and might presumably be affected by an impending decision *you* have to make; second, to compare them on an equal footing—that is, to consider each interest as though it were *your own*; and thirdly, to make *your* decision and to carry out *your* action without in any way favouring *your own* interests over those of others. Now, to 'deduce' this moral law means, in the parlance of the Kantian Friesian tradition, that *you* manage to prove empirically—by some form of introspective method—that this moral law lives in the deepest recesses of *your* mind. This empirical proof was assumed by Nelson to transform *you* (who are carrying out the proof) into a moral being, and to do so in a lasting, indeed permanent, fashion.

The said empirical proof was conducted by Nelson as a *causal* investigation into what moves a human being to make a decision and to act. According to Hermann's own report (Henry-Hermann 1985a, p. 199), Nelson told her before the 1925 Christmas holidays—less than two years before his death—that nobody within or without

[11] At the end of the war, Hermann herself published a 30,000 word long pamphlet outlining the ethical and political philosophy which was at the basis of the practical endeavours of the group (Hermann 1945; English translation in Hermann 1947). A more detailed description of the peculiarities of Nelson's tradition is given in Leal (2013).

[12] Here and elsewhere I place scare quotes around the word 'deduction' to alert the reader not to take it in the contemporary logical sense of a formally valid derivation.

Nelson's circle had read his 'deduction', and he begged her to study it carefully. She did; in fact she never stopped doing so for the rest of her life. In a manuscript written in 1978, six years before her own death, she wrote that she was still working at it (see Henry-Hermann 1985a, p.119). But even when Nelson was still alive, she told him in no uncertain terms that she completely disagreed with his account of moral decision-making and of how the moral interest overcomes the other interests in the human heart. Nelson, always the true philosopher, told her immediately, 'Now you are on the right path' (ibid., p. 199). Before going on, please remember that the issue here is both theoretical and practical, because it was conceived as both, and Grete Hermann took both sides equally seriously.

2.4 What Is at Stake in the Deduction of the Moral Law?

'Now you are on the right path'. That is what Leonard Nelson told Grete Hermann shortly before his death when she expressed her complete disagreement with his account of moral decision-making. In my experience, people laugh when the anecdote is repeated. It is just too *cute*. But what does it mean?

You must remember that the whole issue is not, as with common or garden-variety philosophers, merely theoretical. Ever since Socrates, philosophers have elaborated a bewildering variety of ethical systems. Some systems tell us how one should live and what one should do either in the abstract (what is nowadays called 'normative' ethics) or in connection with concrete problems ('practical' or 'applied' ethics), whereas some systems concentrate on what is the *meaning* of ethical talk ('meta-ethics').[13] Most are a mixture of two or the three concerns. Rarely have philosophers found themselves unable to write up an ethical system of one or all of these kinds.[14] On the other hand, what is not only rare but extremely rare is a philosopher who actually lives the way he or she talks. Socrates was certainly one of them. Nelson and Hermann were Socratics in that particular sense—they walked the talk.

But how did it all start? How did they decide to make their lives agree with what their theories said? That is precisely what Nelson, in his 'deduction' of the moral law, was trying to give an account of: how does anyone choose to take seriously enough his or her ethical ideas to try and carry them through in real life? I said before that Nelson wrote three thick volumes on ethics. The first volume, dedicated to none other than his mentor and friend David Hilbert, was called, in conscious imitation

[13] 'Applied' ethics emerged, at least in recent times, as a reaction to the formalist excesses of 'meta-ethics' (cf. Warnock 1960), whereas the latter arose as a consequence of the many disagreements and the lack of clarity within 'normative' ethics (Copp 2006).

[14] Ludwig Wittgenstein is famous for having declared that ethical problems belonged to those things about which we cannot, and therefore should not, speak or write. However, he did not obey that injunction himself, and his works are chock full of moral sermons.

of Kant, *Critique of Practical Reason*.[15] It was completed and published in 1917, when Nelson was 36 years old. I'll come back to it in a moment. The second volume, although edited and published posthumously by Hermann in 1932, was dedicated to individual morality (the duties and ideals of a moral person) and to moral education. The third volume, published in 1924 because of the urgency felt in a politically chaotic Germany, developed Nelson's philosophy of law as well as his political philosophy. We could say that the second volume was about the individual and the third about society. But Nelson made clear that the ideas presented there were a consequence of the ideas meticulously developed in the first volume.

The first volume, the *Critique of Practical Reason*, has four parts. Here I only want to talk about the third part. What Nelson offers there is called 'deduction'. Any modern reader would understand this word in the sense of modern mathematical logic, i.e., as logical inference or logical derivation. That would be wrong. The word 'deduction' was borrowed by Kant from legal practice, and in fact from *liberal* legal practice. A *liberal* judge does not just mechanically apply whatever laws are given beforehand, but rather makes an effort to 'find the law' that applies to the case brought before him.[16] This process of 'finding the law' was apparently called 'deduction' in Kant's time, long before it began to be used in the general sense of reasoning. (The reader will know that it was Peirce who first brought some order in the terms European authors, from Francis Bacon to Sir Arthur Conan Doyle, used to differentiate between kinds of reasoning.)

The analogy Kant had in mind is that people were like judges who were trying to 'find the law' that could be applied to decision making. When people succeed, what they find is the law that has to be followed. Kant thought that every time we manage to have an experience, either in ordinary life or in science, we 'found the law', in this case the law of nature—physical law; and every time we manage to be moral, we also 'find the law'—the moral law—and we *act* upon it.[17] To illustrate my point, in the 1999 film *The Confession*, there is a scene in which an unctuous lawyer (played,

[15]The dedication to Hilbert was justified by describing the book as 'an attempt to extend the empire of strict science to a new province', viz the province of ethics or moral philosophy. What Nelson meant with such a dedication becomes clear when one comes to the last part of the book (Nelson 1970–1977, vol. IV, Part 4, pp. 619–661). There Nelson tries to use Hilbert's axiomatic thinking—itself a renovation of Euclid's method thanks to modern logic—in order to bring clarity and coherence to the different positions one could take in and on ethics (see Hilbert 1918; Peckhaus 1998). Details about Nelson's relationship to Hilbert and his circle are carefully set forth in Peckhaus (1991, 2001).

[16]An excellent discussion of the history of this process of 'finding the law', and the obstacles it had to struggle with, can be read in Leoni (1961).

[17]The analogy between an experience in the realm of theoretical reason and an action (particularly a moral action) in the realm of practical reason is extremely important for understanding both Kant and the Friesian tradition. We often make a loose use of the term 'experience', 'action', 'moral action', applying them to things that no Kantian or Friesian would consider worthy of the label. To take an example that may appeal to readers of this volume, many so-called 'experiments' are so poorly conceived or executed and the arguments from them so incoherent that they should not be considered experiments at all.

very appropriately, by Alec Baldwin) is talking to an accountant (wonderfully played by Ben Kingsley) who has murdered three people responsible for his son's death, has confessed to the murder, and is waiting for trial. Baldwin is trying to convince Kingsley to allege temporary insanity, whereas Kingsley says he acted deliberately and so has to be punished for his crime with the maximum penalty. At some point Kingsley says pointedly that all he wants is 'to do the right thing'. Baldwin sees his chance and airily says that 'it is very difficult to do the right thing'. Kingsley looks at his lawyer the way a father looks at a son who is too clever by half and retorts: 'Oh, it is not difficult to do the right thing. It is sometimes difficult to know what the right thing is. But once you know it, it is very easy to do it'. Well, this is precisely what Kant means. 'Deduction' or 'finding the law' means knowing what the right thing is in that Kingsleyan sense.

Now, Kant was somewhat fuzzy about what 'deduction' actually implies—how to go about it. In fact, the section called 'Transcendental Deduction' in his *Critique of Pure Reason* is probably the most discussed piece in all of Kant's work, and there is no end to the controversies. It is certainly not a logical analysis of concepts or anything like it, although some interpreters have tried their hand at making it look like it is exactly that (e.g. Strawson 1966). Jakob Fries, the founder of the philosophical tradition to which Hermann belonged, claimed it was something rather more like psychological introspection. Nelson agreed, and his 'deduction of the moral law' is a deeply introspective (and largely causal) account of the moral struggle.

In any case, what is the purpose of the exercise? Nelson's 'deduction of the moral law' is an extended meditation on who we are as human beings. As a meditation, it is as deeply personal as the original Cartesian meditations. Descartes wrote his book as an outline of the kind of meditation that one ought to do *once in one's life*. By doing the meditation oneself (as opposed to merely reading it), one would apprehend once and for all who one was—and by the sheer force of such an apprehension one's life would change forever. My point is that Nelson pursued the same goal. His 'deduction' was a guideline to a personal meditation, a *portrait* of how you have to go inside yourself and try to 'find the law', viz the moral law that binds your actions. By finding it—by 'knowing what the right thing is', like Kingsley's character—it would be easy to do it. The meditating person who went through this meditation would know once and for all what his or her duty was and how he or she would have to live his or her life from now on. Whether he or she was to be a moral person no longer would be a matter of chance *after the meditation*. In fact, the kind of chance that makes some of us good and some of us less good or outright bad would be removed—overcome forever.

So what is at stake in the 'deduction' of the moral law is far from being merely a matter of theoretical philosophy. Socrates, the ancient forerunner of Hermann's brand of Kantian philosophy, said that when he asked questions of his Athenian fellow citizens he was not just examining what they *knew* or *were ignorant of* as a matter of theory. Rather, he was putting their very *lives* to the test. Nelson's 'deduction' had the very same seriousness of purpose. Grete Hermann knew that and dealt with it accordingly.

2.5 Complementarity in Ethics?

Grete Hermann did not agree with Nelson's 'deduction'. And although she kept thinking about these issues until her death, there is an interim report in which she suggests that Bohr's idea of complementarity might help us to see what is wrong with Nelson's 'deduction'. It was her contribution to Nelson's memorial volume published in 1953 (Henry-Hermann 1953; reprinted in Henry-Hermann 1985a) whose English translation, originally published in 1991, I shall refer to from now on. I call this very long paper, entitled 'Conquering Chance', an interim report because Hermann continued thinking about these issues, as is shown by her short paper on ethology (Henry 1973a) and her extended notes (Henry-Hermann 1985b), as well as her unpublished correspondence.

For the purposes of the following discussion, let us assume that complementarity means or implies that for a given event or phenomenon there are two opposite conceptions of it—two conceptions that are both true, each within its own limits, and incompatible with each other. Hermann thought that every cognitive operation (never mind whether it is purely theoretical and oriented to knowing what is the case, or whether it is practical and oriented toward acting in the world) can be understood in two complementary ways: either in terms of causes and effects, or in terms of reasons.

It is a fact that we can and do account for somebody's actions by asking either for the causes or for the reasons of that action; and it is a well-known problem within analytic philosophy of the last half century or so that, in view of that fact, we need to explain what the relation between these two accounts is.[18] Three main purported solutions to this problem have been given so far: (1) only the causal account is valid, whereas the account in terms of reasons is a will-o'-the-wisp induced by bad, unscientific habits of thinking; (2) trying to account for an action in terms of causes implies perverting the very idea of an action, replacing it with something else—say, a behaviour or a mere bodily movement—so that an action can only be accounted for in terms of reasons; (3) there is no important philosophical difference between the two accounts, in that reasons *are* causes, the apparent problem arising from the fact that we do not know, or not yet, of any exact laws for human behaviour. As usual in philosophy, each one of these main solutions yields a bewildering proliferation of variations and sub-variations, which periodically fill the pages of the professional journals. Moreover, a whole branch of analytic philosophy, called 'the philosophy of action' or 'the theory of action', centres around them.

[18]The problem is in fact much older than twentieth century analytic philosophy. It makes its first clear appearance in a celebrated passage of Plato's *Phaedo* (97B–99D) and has been resurfacing ever since in the works of all major philosophers, although it is not always completely explicit. It underlies the whole phenomenological and existentialist movement of 'Continental' philosophy, often artificially opposed to the analytic tradition.

Solution (1) is mainly associated with early logical empiricism (or positivism), with behaviourism, and indeed with the naive philosophical position of most experimental scientists—physiologists and psychologists—as well as with contemporary adherents to naturalism, materialism and physicalism. Solution (2) is more commonly upheld by rationalistic philosophers, by followers of Wittgenstein and of Sellars, and it underlies the work of most social scientists.[19] Solution (3) was invented by Donald Davidson and independently developed in a much more scientifically sophisticated manner by Daniel Dennett (see Davidson 1963; Dennett 1984, 1987, 2003). Solutions (1) and (3) assume that a causal account is the only real account we can give, whereas solution (2) involves a distinction between actions, which can be given a rational account, and behaviour, which should be given a causal account.

Grete Hermann's proposal presupposes that distinction as well. Nonetheless, she lived and worked in complete devotion to the political and educational ideals of the Nelsonian movement. In consequence, she was completely detached from, and unaware of, the debates which started to rage shortly after her paper in the circles and journals of professional philosophers about this problem, its purported solutions and their variations, continuing unabated to this day. From the standpoint of contemporary analytic philosophy, her contribution would almost certainly be seen as just one variation of solution (2). However, it is in my opinion much more than that: it also implies a departure from Nelson, as we will see.[20]

To understand the novelty of Hermann's approach, consider the old anti-psychologistic objections raised by authors like Frege and Husserl (see Kusch 1995). One of Hermann's examples is the process of solving a simple multiplication problem (Henry-Hermann 1991, pp. 68–78). It is on the one hand clear that we could make a causal inquiry into how a given person (e.g. a child) actually goes about solving the problem. Such a causal, psychological inquiry would have to study processes of memory, attention and perhaps certain cognitive illusions to which the human mind is prone. It is on the other hand equally clear that one can make a non-causal enquiry as to the steps that logically lead to a solution. There is a whole branch of

[19]Wittgenstein and Sellars proposed solution (2) at about the same time as Hermann, but their arguments are dispersed in different papers and books. In the case of Sellars we now have an appropriate selection of essays (Scharp and Brandom 2007). Perhaps because he was less lucidly aware of the consequences of his view, nothing like that is yet available for Wittgenstein (but see Schroeder (2010) for an attempt at synthesis). Philosophically-inclined historians and social scientists (e.g. Collingwood 1999; Elster 2007; Boudon 2010) have ably defended solution (2). The practical work of researchers who work with statistics and mathematical models often write as though they prefer solution (1), but they do not usually bother to spell out their arguments for so thinking. It is clearly not an easy problem; in fact it is probably the central logical, epistemological and methodological problem of historical and social scientific research.

[20]Kant clearly follows solution (2), but all Friesians, because of their allegiance to a psychological approach to the critical questions ('how is X possible?', where X stands for ordinary human experience, scientific knowledge, prudential action, moral action, aesthetic judgment, and teleological judgment in biology) tend to hover between solution (2) and solutions (1) and (3). Hermann was, like Kant, firmly within solution (2) while yet maintaining the psychological approach characteristic of Friesians.

mathematics—indeed several—that deal with this question, and ever since the advent of the techniques of symbolic logic the field has been tremendously successful.[21]

By comparing these two kinds of enquiry—which we could call for short the logical and the psychological—it would seem (i) that it is a tremendous fallacy to confuse the two, which is why the famous critique of psychologism is generally considered definitive, (ii) that there is no third kind of inquiry different from these two, and (iii) that the two inquiries are equally legitimate. It would seem to follow that complementarity means here that what one knows by means of the causal inquiry is incompatible with what one knows by means of the logical inquiry, yet both grasp something that is true.

It may seem so, but this is not at all what Hermann was up to. In fact, if this was all she meant, then her recourse to Bohr's idea of complementarity would be less interesting than I think it is.[22] We must go deeper.

2.6 The Shifting of Perspectives in Decision-Making

Let us face it: to say that logic and psychology are complementary theories of *the same thing* (say, the solution of a simple multiplication problem) would be barely more than an uninteresting play on words. But this was not at all Hermann's point. In fact, she was not in the least interested in the *logical* aspect of thinking as such. Rather, she argued that the very process of thinking as it has actually happened in the mind of, say, a child who has been trying to solve a multiplication problem—therefore as a *real* process (not an ideal process to be studied by logicians)—has two sides that are complementary. One side is psychological in the ordinary sense of the word; I have already described it. Yet the other side of the process we may call

[21] For a recent survey of the first psychological kind of research see Campbell (2003). As for the second, Hilbert referred to the formal logical study of mathematics as *Metamathematik*. Although this name is still used, it would be clearer today to distinguish, say, between proof theory, model theory and category theory, perhaps the most important distinct branches of mathematics dealing with the logical kind of question described above.

[22] The reader beware: when Hermann (in Henry-Hermann 1953) tries to use the idea of complementarity for ethics, she does not speak of a metaphor or an analogy; she says rather that this idea will make the relation between causes and reasons clear. Related applications of the idea of complementarity had already been expressed by Bernays (1948) and indeed by Bohr himself (cf. Bohr 1933, 1937), both of whom certainly speak of an 'analogy', and there has been some debate about the extent to which we should take it seriously (Favrholdt 1999). Hermann does not refer to any of them, which—given her careful scholarship—implies that hers was an independent discovery, and in fact not intended by her as a metaphor or analogy. For the idea of incompatible conceptions that limit each other actually belongs to the Friesian tradition quite independently of complementarity (see e.g. Nelson 1949c, pp. 44–61; German original in Nelson 1970–1977, vol. III, pp. 283–303). That idea, in its turn, is very old and goes back at least to the medieval doctrine of the 'two truths': the one revealed by God in the Scriptures and the other accessible to human reason. It is important not to confuse this doctrine with the distinction, indeed separation, between 'appearance' and 'reality' originating in Parmenides' *Poem* (see Popper 1998; Mourelatos 2008). The latter distinction is monistic in character, and so more akin to solutions (1) and (3) outlined above.

psychological as well, but we should be clear that it is subtly different from the first. The first side is so to speak *external*, its form of enquiry observational: the enquirer is trying to *explain* what the child did. The other side is *internal*, for here we do not observe the process from the outside, as it were, but from the inside—we try to carry it out ourselves *in the same way* it was carried out by the child: we as enquirers are trying to *understand* the reasons the child had for proceeding the way he or she did.

In each case we are dealing with the very same real, psychological, cognitive process. And yet what we see is completely different. Both inquiries find something true and significant, but what they find is not at all the same. In short, the mind of the child behaves causally and it behaves rationally at the same time—rational in a sense that goes beyond any purely logical account. Now *that* is an interesting idea. Let us return to the case of a moral decision and consider how we go about it. When any of us is comparing the options and weighing the pros and cons of the impending choice, it is impossible for us to examine ourselves from a causal point of view. All we see is reasons and we cannot see causes. If on the contrary we concentrate on our psychological or physical state of mind—the fact that we are tired, sleepy, angry, eager, indignant, or whatnot—all consideration of what is right or wrong disappears. All we see is causes and no reasons. We cannot grasp both aspects of the same process at the same time. Hermann even added that sometimes the choice before us is so difficult that we have to shift or switch from one point of view to the other.[23] Say I become aware of the struggle to 'do the right thing', so I switch to a causal viewpoint and try to see whether my anger or my spite is blinding my thinking, and this might help with my decision when I return again to a consideration of the reasons to act in a certain way.

By the way, this switching between complementary viewpoints is not considered by Hermann to be exhausted by the opposition between causes and reasons. She explains that we can also switch between different causes—say, between divergent interests causally struggling within us (Henry-Hermann 1985a, p. 89; Henry-Hermann 1991, p. 75). What is strikingly novel about this idea—whether we want to talk in this context of complementarity or not—is that it delineates an empirical research programme for the study of moral decision making, and in fact of all decision making. This brings Grete Hermann's contribution (originally part of an extended critique of Leonard Nelson's moral psychology) quite close to contemporary research in cognitive science on rationality.[24]

---

[23]According to Hermann, this process would be analogous to Heisenberg's concept of a 'cut' ('Schnitt', mistranslated by Peter Winch as 'section'; see Henry-Hermann (1991, p. 76)). I have no technical competence to understand what is implied by this analogy or whether it has any cash value, and I defer to readers who are better informed about Heisenberg's 'cut'. The fact remains that the 'cut' (as opposed to 'complementarity') was explicitly intended by Hermann as an analogy.

[24]In fact, Hermann's argument also touches contemporary research on *causality*. In a nutshell, causality does not make sense independently of human reason and human ends. The reader might usefully compare this approach with Hermann's comments on Laplace's demon (see Sect. 11 of Chap. 15). The main exponent of the new thinking about causality is doubtless Judea Pearl (2009, see especially the Epilogue), although a more psychologically-oriented text is Sloman (2005).

Starting in the 1950s and '60s but first really gathering speed in the 1970s and '80s, experimental psychologists, more recently joined by economists and anthropologists, have converged on the idea that the older debates on the fundamental rationality or irrationality of human beings must be replaced by a 'dual process' theory. According to this theory, we have and use two systems or procedures in order to solve problems and make decisions. The primary system (or 'System 1') was evolved over thousands and thousands of years to allow us to process information fast and to act quickly. Our uncanny abilities to recognise patterns (voices, faces, movements), which have so far eluded the dedicated efforts of artificial intelligence researchers, belong to this system. In fact, we think so fast that we are not even aware that we have actually done any *thinking*. In popular parlance we often speak here of 'intuition'. On the other hand, the secondary system (or 'System 2') is much more recent in origin and is tied up with our invention of law, morality, technology, mathematics, science, and more generally with specific, codified procedures. It is slow, careful, painstaking. Both systems are very good at what they do, but they sometimes interfere with each other, producing a recognisable pattern of errors in performance.[25]

It is important to realise two things. One is that before this kind of research we had no proper theory of error.[26] The other is that in order to thrive—and indeed to survive—we need both systems. In fact, we *switch* between them according to occasion, type of problem and urgency of the matter at hand. The reader might object that these two systems both act causally, so what this kind of research has achieved is a merely *causal* account of human action or—what amounts to the same thing—of human behaviour. This would be foreign to Hermann's proposal, which—like those of Wittgenstein and of Sellars—reserves a place for *reason* in accounting for human action as opposed to human behaviour.

That objection would also be superficial, for the modern study of rationality includes the consideration of reasons in the shape of norms of thinking (taken from the full gamut of normative disciplines). Yet we are not just back to the old opposition between logic and psychology which I mentioned earlier. The most interesting approaches have a meliorative aspect to them—the idea that, by reflecting on the best norms there are as compared to the imperfect norms we follow, we might improve our problem-solving and decision-making abilities. Thus we will probably never achieve the absolute ideal, but we will make better choices and design a better world.[27]

---

[25] A recent and very engaging presentation of the main findings of the field, written by one of the pioneers and winner of the Nobel Prize in Economics, is Kahneman (2011).

[26] For thousands of years of philosophical and scientific thinking, error and its pendant irrationality were profoundly enigmatic. 'There is only one way to be right but infinite ways to be wrong' was a popular saying. Thanks to dual process thinking, we are now well on our way toward understanding and explaining in how many and which ways we can and do make mistakes, and why. Anybody who is acquainted with the Kantian project of a 'transcendental dialectic' will see the resemblance.

[27] The relevant literature is immense, but a recent book that clearly captures the meliorative spirit of the field is Stanovich (2010). Grete Hermann, ever the scientist (see Henry 1973a), would have welcomed it with open arms if she had not kept herself, alas, so far away from the mainstream.

And that was exactly, after all the minute arguments, the main point for Grete Hermann the philosopher: the human condition is such that there is no way we can have perfect access to a moral law and thus permanently transform ourselves into ideal moral beings. So she concluded that the absolutist claim characteristic of the Kantian and Friesian tradition in general and the Nelsonian project in particular does not hold. She came to that conclusion after a careful and personally painful analysis of concepts and facts. She was able to come to that conclusion not in spite of but rather precisely because of the logical consistency and psychological insights of Fries and Nelson. Nevertheless, she retained the urgent sense of the task indicated by those traditions. To fulfill that task, she concluded, one has to abandon the absolutist claim in favour of a continuous and empirically-grounded search for worthy moral goals to aspire to and appropriate (if fallible) means to try to realise them.[28]

References

Apelt EF (1849) Johannes Kepler's astronomische Weltansicht. Weigel, Leipzig
Apelt EF (1851) Die Epochen der Geschichte der Menschheit. 2 volumes, Mauke, Jena
Apelt EF (1854) Die Theorie der Induction. Engelmann, Leipzig
Apelt EF (1857) Metaphysik. Engelmann, Leipzig
Bernays P (1914) Über die Bedenklichkeiten der neueren Relativitätstheorie. Abhandlugen der Fries'schen Schule (Neue Folge) 4(3):457–482
Bernays P (1948) Über die Ausdehnung des Begriffes der Komplementarität auf die Philosophie. Synthese 7(1–2):66–70
Bohr N (1933) Light and life. Nature 131(421–423):457–459
Bohr N (1937) Causality and complementarity. Philos Sci 4(3):289–298
Boudon R (2010) The cognitive approach to morality. In: Hitlin S, Vaisey S (eds) Handbook of the sociology of morality. Springer, New York, pp 15–33
Campbell JID (ed) (2003) Handbook of mathematical cognition. Psychology Press, New York/Hove
Campbell JK, O'Rourke M, Shier D (eds) (2004) Freedom and determinism. The MIT Press, Cambridge, MA
Collingwood R (1999) The principles of history and other writings in philosophy of history. Oxford University Press, Oxford
Conrat F (1903) Helmholtz' Verhältnis zur Psychologie. Ehrhardt Karras, Halle an der Saale
Copp D (ed) (2006) The Oxford handbook of ethical theory. Oxford University Press, New York
Dahms H-J (2006) Karl Poppers erste Schritte in die Philosophie: Leonard Nelsons Paradoxien der Souveränität und Nelsons sowie Poppers Lösungsversuche. In: Jarvie I, Milford K, Miller D (eds) Karl Popper: a centenary assessment, vol I. Ashgate, Aldershot (UK), pp 83–98
Davidson D (1963) Actions, reasons and causes. J Philos 60(23):685–700

[28]I repeat: although I have mainly referred here to Hermann's main paper in moral philosophy, 'Conquering Chance', this is also the gist of Henry (1973a), a short German article Hermann wrote for the philosophical journal *Ratio*. By the way, this journal was launched by Stephan Körner with the help of Karl Popper. Popper had been introduced to Nelson's work in the mid-1920s and had come to admire it, although he recoiled from Nelson's politics (Dahms 2006, p. 86; see also Milkov 2012). When Popper and Körner launched *Ratio*, they explicitly wanted to continue Nelson's *Abhandlungen der Fries'schen Schule*, the journal in which Hermann's papers on modern physics first appeared.

Dennett DC (1984) Elbow room: the varieties of free will worth wanting. Oxford University Press, New York

Dennett DC (1987) The intentional stance. The MIT Press, Cambridge, MA

Dennett DC (2003) Freedom evolves. Viking, New York

Elster J (2007) Explaining social behavior: more nuts and bolts for the social sciences. Cambridge University Press, Cambridge

Favrholdt D (ed) (1999) Complementarity beyond physics (1928–1962), Niels Bohr's collected works, vol 10. Elsevier, Amsterdam

Friedman M (2013) Kant's construction of nature: a reading of the metaphysical foundations of natural science. Cambridge University Press, Cambridge

Fries JF (1822/1979) Die mathematische Naturphilosophie nach philosophischer Methode bearbeitet. In: Sämtliche Schriften, vol 13

Fries JF (1824/1970) System der Metaphysik. In: Sämtliche Schriften, vol 8

Fries JF (1826/1973) Lehrbuch der Naturlehre. Erster Theil: Experimentalphysik. In: Sämtliche Schriften, vol 15

Fries JF (1828/1968) Neue oder anthropologische Kritik der Vernunft. In: Sämtliche Schriften, vol 4, pp 31–478

Fries JF (1833/1973) Populäre Vorlesungen über die Sternkunde. In: Sämtliche Schriften, vol 16

Fries JF (1837–1839) Handbuch der psychischen Anthropologie oder der Lehre von der Natur des menschlichen Geistes, 2nd edn. 2 volumes. Croker, Jena

Fries JF (1837/1967) System der Logik. Ein Handbuch für Lehrer und zum Selbstgebrauch. In: Sämtliche Schriften, vol 7

Fries JF (1839/1975) Über den optischen Mittelpunkt im menschlichen Auge nebst allgemeinen Bemerkungen über die Theorie des Sehens. In: Sämtliche Schriften, vol 17, pp 411–491

Fries JF (1842/1996) Versuch einer Kritik der Principien der Wahrscheinlichkeit. In: Sämtliche Schriften, vol 14, pp 11–254

Fries JF (1967–2011) Sämtliche Schriften. 30 vols, Scientia, Aalen. Edited by G. König and L. Geldsetzer

Gilder L (2008) The age of entanglement: when quantum physics was reborn. Vintage Books, New York

Heilbroner RL (1999) The worldly philosophers: the lives, times, and ideas of the great economic thinkers. Simon & Schuster, New York

Heisenberg W (1969) Der Teil und das Ganze. Gespräche im Umkreis der Atomphysik. Piper Verlag, München

von Helmholtz H (1903) Vorträge und Reden, vol I, 5th edn. Vieweg, Braunschweig

Henry G (1973a) Die Bedeutung der Verhaltensforschung für die Kritik der Vernunft. Ratio (German edition) 15(2):197–209

Henry G (1973b) The significance of behaviour study for the critique of reason. Ratio (English edition) 15(2):206–220. Translation of Henry (1973a) by 'G. S.'

Henry-Hermann G (1953) Die Überwindung des Zufalls: Kritische Betrachtungen zu Leonard Nelsons Begründung der Ethik als Wissenschaft. In: Specht M, Eichler W (eds) Leonard Nelson zum Gedächtnis. Verlag Öffentliches Leben, Frankfurt am Main/ Göttingen, pp 25–111

Henry-Hermann G (1985a) Die Überwindung des Zufalls. Meiner, Hamburg

Henry-Hermann G (1985b) Im Gespräch mit Leonard Nelson und mit Kameraden der Akademie, 1978–1984. In: Die Überwindung des Zufalls, pp 99–178

Henry-Hermann G (1991) Conquering chance: critical reflections on Leonard Nelson's establishment of ethics as a science. Philos Invest (1):1–80. Translation of Henry-Hermann (1953) by Winch P

Hermann G (1935) Die naturphilosophischen Grundlagen der Quantenmechanik. Abhandlugen der Fries'schen Schule 6(2):75–152

Hermann G (1937a) Die Bedeutung der Modernen Physik für die Theorie der Erkenntnis. In: Drei mit dem Richard Avenarius-Preis ausgezeichnete Arbeiten von Dr G. Hermann, Dr E. May, Dr Th. Vogel. S. Hirzel, Leipzig, pp 3–44

Hermann G (1937b) Über die Grundlagen physikalischer Aussagen in den älteren und den modernen Theorien. Abhandlugen der Fries'schen Schule (Neue Folge) 6(3):309–398
Hermann G (1945) Politik und Ethik. Renaissance Publishing Co., London
Hermann G (1947) Politics and ethics. International Publishing Company, Bombay. Translation of Hermann (1945)
Hilbert D (1918) Axiomatisches Denken. Mathematische Annalen 78:405–415
Kahneman D (2011) Thinking, fast and slow. Farrar, Straus & Giroux, New York
Kane R (ed) (2002) The Oxford handbook of free will. Oxford University Press, New York
Kane R (2005) A contemporary introduction to free will. Oxford University Press, New York
Kant I (1781) Kritik der reinen Vernunft. In: Kants gesammelte Schriften, vols 3 and 4. De Gruyter, Berlin
Kant I (1793) Über den Gemeinspruch: Das mag in der Theorie richtig sein, taugt aber nicht für die Praxis. Berlinische Monatsschrift, pp 201–284
Kant I (1795) Zum ewigen Frieden. Ein philosophischer Entwurf. Friedrich Nicolovius, Königsberg
Kant I (1797) Die Metaphysik der Sitten in zwei Teilen. Friedrich Nicolovius, Königsberg
Kuhn TS (1977) The essential tension: selected studies in scientific tradition and change. Chicago University Press, Chicago
Kusch M (1995) Psychologism: a case study in the sociology of philosophical knowledge. Routledge, London
Leal F (2013) Leading a philosophical life in dark times: the case of Leonard Nelson and his followers. In: Chase M, Clark SRL, McGhee M (eds) Philosophy as a way of life: ancients and moderns-essays in honour of Pierre Hadot. Wiley-Blackwell, Chichester, pp 184–209
Leoni B (1961) Freedom and the law. Van Nostrand, Princeton, NJ
Milkov N (2012) Karl Popper's debt to Leonard Nelson. Grazer Philosophische Studien 86:137–156
Mourelatos APD (2008) The route of Parmenides, revised and expanded edn. Parmenides Publishing, Las Vegas
Nelson L (1949a) Socratic method and critical philosophy: selected essays. Yale University Press, New Haven, CT. Translated by Brown TK III
Nelson L (1949b) Critical philosophy and mathematical axiomatics. In: Nelson (1949a), pp 158–184
Nelson L (1949c) The scientific and the esthetic conception of nature. In: Nelson (1949a), pp 44–61
Nelson L (1970–1977) Gesammelte Schriften. 9 volumes. Meiner, Hamburg
Nelson L (1971) Progress and regress in philosophy: from Hume and Kant to Hegel and Fries. 2 volumes. Blackwell, Oxford. Translation by H. Palmer
Nelson L (2004) Kritische Naturphilosophie. In: Schroth J (ed) Herrmann K. Winter, Heidelberg
Nelson L (2011) Typische Denkfehler in der Philosophie. Meiner, Hamburg
Nelson L (2016) A theory of philosophical fallacies. Springer International Publishing, Switzerland. Translation of Nelson L (2011) by Leal F, Carus DG
Pearl J (2009) Causality: models, reasoning, and inference, 2nd edn. Cambridge University Press, New York
Peckhaus V (1991) Hilbertprogramm und Kritische Philosophie: Das Göttinger Modell interdisziplinärer Zusammenarbeit zwischen Mathematik und Philosophie. Studia Logica 50(2):351–354
Peckhaus V (1998) Axiomatische Ethik. In: Neisser B (ed) Zwischen Kant und Hare: eine Evaluation der Ethik Leonard Nelsons. Dipa-Verlag, Frankfurt am Main, pp 37–61
Peckhaus V (2001) 'Mein Glaubensbekenntnis': Leonards Nelsons Brief an David Hilbert. In: Toepell M (ed) Mathematik im Wandel: Anregungen zu einem fächerübergreifenden Mathematikunterricht, vol 2. Franzbecker, Hildesheim, pp 335–346
Pink T (2004) Free will: a very short introduction. Oxford University Press, New York
Popper KR (1998) The world of Parmenides: essays on the Presocratic Enlightenment. Routledge, London
Riehl A (1904) Hermann von Helmholtz in seinem Verhältnis zu Kant. Reuther & Reichard, Berlin
Scharp K, Brandom RB (eds) (2007) In the space of reasons: selected essays of Wilfred Sellars. Harvard University Press, Cambridge, MA

Schleiden MJ (1842) Grundzüge der wissenschaftlichen Botanik, nebst einer methodologischen Einleitung als Anleitung zum Studium der Pflanze. Engelmann, Leipzig

Schleiden MJ (1849) Principles of scientific botany: or botany as an inductive science. Longmans, London. Translation of Schleiden (1842) by Lankester E

Schleiden MJ (1861) Zur Theorie des Erkennens durch den Gesichtssinn. Engelmann, Leipzig

Schroeder S (2010) Wittgenstein. In: O'Connor T, Sandis C (eds) A companion to the philosophy of action. Wiley-Blackwell, Chichester, pp 554–561

Schwertschläger J (1883) Kant und Helmholtz erkenntnistheoretisch verglichen. Herder, Freiburg im Breisgau

Sloman S (2005) Causal models: how people think about the world and its alternatives. Oxford University Press, New York

Stanovich KE (2010) What intelligence tests miss: the psychology of rational thought. Yale University Press, New Haven

Strawson PF (1966) The bounds of sense: an essay on Kant's Critique of Pure Reason. Methuen, London

Warnock M (1960) Ethics since 1900. Oxford University Press, Oxford

Wilson M (2010) Back to 'back to Kant'. In: Dixon M, Domski M (eds) Discourse on the new method: reinvigorating the marriage of history and philosophy of science. Open Court, Chicago, pp 553–568

Chapter 3
Understanding Hermann's Philosophy of Nature

Giulia Paparo

> Ein philosophischer Kopf ist also derjenige, in dem das Selbstdenken leicht geweckt wird, und der dabey des Enthusiasmus für Ideen schnell empfänglich ist
> Jakob Friedrich Fries[1]

3.1 Introduction

Grete Hermann's novel and provocative ideas may be seen as arising from the confluence of her extraordinary 'philosophical mind' with her background in the philosophy of nature; an examination of this relatively unexplored background will show how it inspired Hermann, and hopefully serve in turn as a source for novel and provocative ideas in science and philosophy. In this paper I provide the philosophical context necessary to fully comprehend and appreciate Hermann's philosophical position and method. In order to do so I will delve into relatively unexplored protagonists of the history of philosophy, and establish their relationship to Hermann's thought. This is a first critical analysis of Hermann's philosophical background, and is intended both

[1] 'A philosophical mind is therefore the one that goes easily to independent reasoning, and in this process is quickly receptive of the enthusiasm for ideas' (Fries 1804/1968, p. 30). All translations, if not stated otherwise, are mine.

This paper is a development of the research for my Master's thesis at Utrecht University on *Grete Hermann: Mathematician, Philosopher, and Physicist*, under the supervision of Prof. Dr Dennis Dieks and Prof. Dr Paul Ziche. I am very grateful for their guidance. In addition, I would like to thank Dr Michiel Seevinck and Dr Guido Bacciagaluppi for encouragement and support. A special thanks for important comments and revisions goes also to Marco Santi and Charles Driver.

G. Paparo (✉)
Stiftung Deutsches Technikmuseum Berlin, Berlin, Germany
e-mail: giulia.paparo@gmail.com

E. Crull and G. Bacciagaluppi (eds.), *Grete Hermann - Between Physics and Philosophy*, Studies in History and Philosophy of Science 42,
DOI 10.1007/978-94-024-0970-3_3

as a starting point for the study of Grete Hermann's philosophical ideas, as well as providing new views on the history of philosophy and science.

Hermann stated in the introduction of her most important essay, 'Die naturphilosophischen Grundlagen der Quantenmechanik', that she aims to analyse the basic principles of quantum mechanics, specifically, the consequences for Kant's causality principle; not from the view of physics, but from a specific philosophical perspective, that of philosophy of nature (cf. Chap. 15, pp. 239–241). Naturphilosophie, or philosophy of nature, is for Hermann the only venue capable of providing answers to the philosophical problems arising from the developments of quantum mechanics. Although she states that she is grounding her work on philosophy of nature, no explicit definition of the latter term can be found in her essay on the foundations of quantum mechanics (nor in any of her other published works). Thus, it is first necessary to understand what Hermann means by 'philosophy of nature' in order to fully comprehend her work and the philosophical context in which it developed. This paper begins by taking up this question. Once explained, it will lead us further into the study of two, largely neglected, philosophers of nature; namely Friedrich Jakob Fries and Leonard Nelson. This study of Fries and Nelson will further shed light on the historiography of philosophy as it pertains to philosophy of nature, as well as on Hermann's position in it. Consequently, an understanding of Hermann's use of philosophy of nature adds to the comprehension of both the importance of her work, and its place in the history of philosophy.

When speaking about Naturphilosophie, Grete Hermann is referring to what Carl Siegel called 'critical philosophy of nature' (Siegel 1913) and in particular to Jakob Friedrich Fries, whom she had come to know throughout the work of his latter-day follower, Leonard Nelson. From her first meeting with Nelson, Hermann was immediately fascinated by the way Nelson was carrying out philosophical discussions, leading her to take the philosophy of nature elaborated by Fries as the privileged perspective to analyse natural science—in particular the philosophical problems concerning the understanding of causality in quantum mechanics. As detailed later in this paper, Hermann takes many elements of Fries' and Nelson's philosophy of nature and creates original developments of her own within the field. However, to show the original contribution of Hermann, it is first necessary to outline the main features of Fries' and Nelson's philosophies, with a focus on the aspects relevant to Hermann's work. After this first brief review of these authors, both scarcely considered in the secondary literature,[2] Hermann's position on philosophy of nature and the influence Friesian philosophy had in the development of her ideas will be examined, highlighting the philosophical and historical value of her proposal. At the same time, this study contributes to the history of philosophy in proposing a re-evaluation of Friesian philosophy and its importance in the development of science in the 19th century.

[2]Nelson and Fries are omitted by most general histories of philosophy, and there are very few critical studies on either of the two authors. In recent years the tendency has been changing, especially towards Fries' philosophy, and new critical studies have appeared such as Pulte (1999), Geldsetzer (1999) and Gregory (2006).

3.2 Fries' Philosophy of Nature

Jakob Friedrich Fries was born the son of a cleric of ancient nobility in Moravia,[3] and after some years at the seminary he enrolled in the law faculty in Leipzig. Soon, his interest in philosophy prevailed and he gave up his law studies. By 1796, he was living in Jena, where he was attending Fichte's lectures on philosophy, but soon he became attracted to the natural sciences, in particular chemistry. However, it was the precision and clarity of Kant's philosophical system that largely inspired his philosophical spirit, impressing Fries in his early years of study at the theological seminary, and now attracting all his devotion. The reading of Kant was one reason why after working as a tutor in Switzerland he returned to Jena, to pursue a career as a philosopher and university professor.[4]

Most of his life Fries searched for a philosophical argument that was comparable in clarity and precision to the proofs of geometry, and he thought to have found one such in the work of Immanuel Kant. Fries found in Kant a challenge and a source of inspiration. From the first encounter with Kant's writings, Fries' philosophical work can be seen as a psychological revision of Kant's thought. In the *Neue oder anthropologische Kritik der Vernunft* (*New or Anthropological Critique of Reason*), Fries explains that he is trying to carry out an extension of Kant's ideas in a programme that he calls 'propaedeutic of general psychology' (Fries 1828/1968, 1935).

Fries' reworking of Kant starts by criticising Kant's transcendental deduction; the attempt to prove the very 'possibility of experience' is for Fries mistaken, as this cannot be proved, but only exhibited (aufgewiesen). Therefore, for Fries the way to the a priori possibilities of knowledge goes only through internal analysis and observation.

> The truth about which humans argue, with respect to which they can err and doubt, is never this transcendental truth of the agreement of idea [Vorstellung] and object. That is rather the empirical truth of consciousness which requires the correct comparison of mediated representations with the immediate. (Fries 1828/1968, pp. XXVII–XXVIII; translation in Gregory 2006)

Thus, for Fries it is through reflection and inner observation that we can (re)gain knowledge of the existence of a priori laws. What Fries carries out has been called 'epistemological psychologism' (cf. Sachs-Hombach 1999), since the objects of observation are contents of our mind and it is in our mind that the starting point of any study of knowledge is to be found. This also leads Fries to introduce the concept of fallibility of knowledge; our judgements, since they involve a distance in time and space from the immediate knowledge of the senses, are for Fries subject to error. For this reason, the natural sciences have to undergo a continuous revision and re-examination, and the fallibility of our knowledge has to be accepted (cf. Gregory 2006).

From Fries' revision of Kant's critique of pure reason stem many consequences, such as a new concept for science and organisms. Kant had divided the organic from

[3]On the relevance of the Moravian past in Fries' philosophy see Gregory (2006).

[4]This biographical sketch is mainly based on Gregory (2006), Gregory (1983), and Henke (1867).

the inorganic world; he understood the knowledge of organisms as being accidental, which means that it could not be exhaustively explained by the natural laws of mechanics and that thus the regulative intervention of reason was necessary (cf. Kant 1790, § 70 and § 78). This created a decisive separation between the organic and inorganic world and what we can know about both domains. Fries, in line with the romantic Naturphilosophie of Schelling,[5] believed in the unity of nature, and thus overcame this division by considering both the organic and inorganic as being empirically based and intelligible without the intervention of reason.[6] He recognised two types of processes in nature: indifference and cyclic process (Kreislauf). Matter in contact was acting by indifference, while the cycle was the typical process of organisms. Both the organic and the inorganic world were therefore explainable by empirical laws by looking at them in terms of their processes.

Furthermore, Fries 'dynamised' and expanded Kant's concept of science (cf. Pulte 1999). Actually, Kant accepted as 'science proper' only those sciences that respected the three necessary (and independent) conditions of mathematicity, apodicticity and systematicity (cf. Kant 1781). Consequently, physics and mathematics were accepted in the realm of science, while other subjects were excluded, because, as in the case of biology and psychology, they lacked the proper degree of mathematisation (cf. Kant 1786). In contrast, Fries' anthropological revision of Kant developed a new methodology of science which softened Kant's definition in an 'empirical' direction, expanding the realm of science to new areas of research such as chemistry and biology.

Similarly to the re-evaluation of the general definition of science, Fries widened Kant's perspective in mathematics as well. Fries studied the new developments in the foundations of mathematical physics, and was the first German philosopher to speak about a 'philosophy of mathematics' (cf. Pulte 1999). He elaborated a meta-theory of mathematics which considered the problem of the origin and foundation of mathematical knowledge.

For these reasons, Fries' proposal has been considered a 'scientifically adequate' continuation of Kant, opening up possibilities for science that Kant had dismissed, such as the new developments in chemistry and biology, while focusing significant attention on the mathematical foundations of physics (cf. Pulte 2006). Fries' concept of science, and corresponding openness to different disciplines, constituted the grounds on which Hermann considered the relationship between philosophy, mathematics and natural science.

However, Fries' elaboration on Kant, and the possibilities it offered for both philosophy and science, have been largely neglected or misunderstood.[7] Siegel stated that 'there is only one thinker who can claim to have pursued Kant's philosophy of

[5]More on the relation between Fries' and Schelling's philosophical systems is to be found in Gregory (1983).

[6]For Fries, not only what is alive, but more broadly everything that is changing in cyclical form would count as an organism. The living organism, therefore, was seen as only a specific part of the organic world (cf. Gregory 1997).

[7]Geldsetzer (1999) stresses that in the historiography of philosophy, Fries has been generally referred to only with restriction to three aspects: (1) the alleged psychologism, (2) Kantianism and (3) his

nature in the most rigorous way, and as well with precise examination: this thinker is J. F. Fries' (Siegel 1913, p. 118). But with the exception of Siegel and a few others, Fries' work received little attention in the history of philosophy and was too quickly accused of psychologism. In recent years there has been a reversal in the attitude towards Fries. Historians and philosophers of science tried to look at the importance of Fries' ideas in their context, re-evaluating the novel aspects of Friesian philosophy of nature.[8] As Fries' work provided the grounds for Grete Hermann's ideas in philosophy of nature, an analysis of Hermann's work in turn contributes to the reappraisal of Fries as well. Fries' dynamising of Kant, his parallel work on philosophy and science, and his expansion of the concept of science played indeed a central role in Grete Hermann's work, as will be shown later.

3.3 Nelson's Case

Grete Hermann was exposed to Friesian Philosophy thanks to her encounter with the German philosopher Leonard Nelson. Although Nelson was an earnest follower of Fries' ideas, his work is not a pedantic reproduction of them. Contrary to what was asserted in the refusal of his first dissertation proposal,[9] he actively revisited Fries' work by amplifying and criticising it. Nelson studied Fries' works for many years and extended Fries' natural philosophy in several respects, in particular on the relationship between critical philosophy and the contemporary developments of mathematics, and its ethical and political consequences. In a way, Nelson's relationship to Fries reflects Fries' relationship to Kant.[10] They both endeavour to carry out a renewal of their mentors by pursuing an in-depth understanding of their works, and at the same time confronting the mentor's philosophy with contemporary advancements in science. It is by trying to follow Fries' understanding of philosophy of nature (science and its development as the object of study) that Nelson engages in a long-term dialogue with David Hilbert. The discussion carried out between Nelson and Hilbert on the relationship between philosophy and mathematics is an important indicator of the intellectual background against which Hermann elaborates her ideas. Nelson's life

(Footnote 7 continued)
antisemitism. Bianco (1980) points out that the studies of Fries have been mainly apologetic or polemic.

[8]Cf. for example Pulte (2006), Gregory (2006), and Geldsetzer (1999).

[9]His first dissertation thesis was refused because it was considered not to be an independent work, but mainly deputising Fries' thoughts ('keine selbständige Arbeit [...] sondern vor allem Fries'sche Gedanken vertrete'), as Peckhaus reports in Peckhaus (1991).

[10]Although Kant was a famous philosopher in Fries' time, Fries was almost unknown when Nelson 'discovered' him. In the words of Kraft, 'just as Felix Mendelssohn-Bartholdy (to whom Nelson was related through his mother's family) rediscovered Bach's forgotten masterpiece, "The Passion According to St. Matthew", so Nelson rediscovered the forgotten writing of a forgotten philosopher, J. F. Fries (1773–1843), whose work had fallen into oblivion by a coincidence of adverse cultural and political circumstances, namely, the crushing effect of post-Kantian philosophical mysticism as cultivated by Fichte, Hegel, Schelling and the police state of Metternich' (Kraft 1949, p. XI).

and his encounter with Hilbert, therefore, play a key role in Hermann's subsequent study of philosophy of nature.

Nelson was born in 1882 in Berlin to a family of Jewish lawyers, and could boast having Felix Mendelssohn-Bartholdy and Du Bois-Reymond as his relatives. He first studied philosophy, psychology and theoretical physics in Heidelberg and Berlin, then he moved to Göttingen where he worked until his early death in 1927.[11] In Göttingen, Nelson was surrounded by both a climate fertile for the development and spread of his ideas and a hostile academic establishment, which obstinately opposed his professorship. For instance, in 1921, when Hermann first attended Nelson's seminars, he was still busy trying to secure his position at Göttingen University and was not allowed to examine her for the Staatsexamen.[12]

It was only thanks to the personal intervention of David Hilbert that in 1916–1917 Nelson could finally obtain his Habilitation. Some time before, in 1896, during his early reading of Fries, Nelson had understood that his own mission in philosophy was not only the development and defense of his own system, but also the diffusion of his ideas, through which he could put into practice his philosophical convictions. This pedagogical drive led Nelson to found schools and political movements like the Internationaler Jugendbund (the IJB, founded in 1918, which in 1925 became the ISK, Internationaler sozialistischer Kampfbund), and the school in Walkemühle, both following the same political and educational socialist ideals. In 1903, shortly after he had started studying in Göttingen, he founded together with the philosophers Alexander Rüstow, Carl Brinkmann and Heinrich Goesch, the Neue Fries'sche Schule (New Friesian School). The mathematicians Gerhard Hessenberg, Otto Meyerhof and Kurt Grelling soon also joined the New Friesian School.[13] Nelson, of not yet twenty years—who had already the charisma that would later characterise him—convinced the older and more influential mathematician Hessenberg and the physiologist Kaiser to support the publication of Nelson's early project, the *Abhandlungen der Fries'schen Schule*.[14] The journal aimed at propagating Fries' interpretation of Kant as the epistemological ground for a 'Philosophy of Natural Science'. Notwithstanding the initial personal and political success of Nelson's ideas, his academic carrier had a 'sluggish progression' (schleppender Verlauf), since Husserl and the majority of the professors of the philosophy department opposed him, and his Habilitation proposal was repeatedly rejected. Only in 1919 did he receive the professorship for the Extraordinariat für systematische Philosophie, and from that moment

[11]For the biography I am mainly following Peckhaus (1991) and Hieronimus (1964).

[12]The Staatsexamen was the final exam in the philosophy course of studies. Hermann graduated in 1925 with a thesis on transcendental idealism. After many difficulties, Nelson was allowed to examine her, but only under the control of Ach (cf. Hermann 1928/1985, and above, Chap. 1).

[13]On the individual members see (Peckhaus 1991, p. 132 ff.).

[14]'Mit einem ungewöhnlichen philosophischen Unternehmen trat ein kaum zwanzigjähriger Göttinger Student, Leonhard Nelson, an den Verlag heran: einer neuen Folge der "Abhandlungen der Fries'schen Schule" [...] Es war nicht möglich, ihm den abenteuerlich erscheinenden Plan auszureden, und da er zweifellos ein außergewöhnlich fähiger Mensch war und auch zwei schon ältere Gelehrte, den Mathematiker G. Hessenberg und den Physiologen K. Kaiser gewonnen hatte, wurde das Unternehmen 1904 begonnen, noch ehe der geistige Vater das Doktorexamen bestanden hatte'

on, until his early death in 1927, he was engaged in propagating and elaborating Fries' ideas.

From the very first meeting, Nelson and Hermann did not have an easy relationship, and Hermann's feelings towards Nelson were a mixture of both admiration and fear. In her personal memoirs she described in detail her first encounter with such an important figure in her life (cf. Hermann 1928/1985, and above, Chap. 1). Hermann's brother had taken Nelson's class and had been annoyed by Nelson's dogmatism; nevertheless, he suggested Nelson to Grete. In 1921 she followed Nelson's seminar on 'Typical Mistakes in Philosophy' ('Typische Denkfehler in der Philosophie'). Although initially fascinated by his way of thinking, she was sceptical towards his arrogant (eingebildet), authoritative attitude and the numerous rules which were imposed on the class.[15] Nonetheless, in the winter of 1921/1922 she was sitting again in the benches of his class on 'Exercises in the Philosophy of Religion' ('Übungen über Religionsphilosophie'). She admired Nelson's critical thinking and especially his *method*, although at the same time she was scared by the *results* this could bring. A strange fear captured her, when confronted with the dilemma of having to choose between Nelson's philosophical method, in which she believed, and the security and hope of religion. In her words, the dilemma was

> [...] either I had to give up the hope in a worldview consistent with the religious life and consequently draw disagreeable ethical consequences or betray the method of philosophy. As far as I at the time knew about it, I was convinced of the certainty and necessity of this method. (Hermann 1928/1985, p. 182)

Overcoming this fear thanks to Nelson's philosophy was a fundamental step in Hermann's life, and most of her philosophical work can be seen as an elaboration of this single step.[16] However, it took her some time to undertake this important step. First she avoided Nelson's classes and then she moved for a year to Freiburg University, and only in 1924 did she return to Nelson's seminars. As she deepened her

(Footnote 14 continued)
['With an unusual philosophical project, a scarcely twenty-year-old Göttingen student, Leonhard Nelson, approached the publishing house: a new series of the *Abhandlungen der Fries'schen Schule* [...] It was impossible to talk him out of the adventurously-looking plan, and since he was undoubtedly an uncommonly capable person and had also won [for the project] two already senior scholars, the mathematician G. Hessenberg and the physiologist K. Kaiser, the project was started in 1904, even before its spiritual father had passed his doctoral exam (*eds.*)]—thus the publisher Ruprecht describes the first publication of the journal, in Peckhaus (1991, p. 151).

[15]Nelson's rules were punctuality, regular participation, and the fact that the discussion had to go on until late at night. Grete Hermann voiced her criticism of these rules, but also expressed a complete approval of the fourth rule, namely 'the demand to never refrain from answering because of fear of disgrace' ('die Aufforderung, sich nie aus Furcht vor Blamage vom Antworten abhalten zu lassen'), cf. Henry-Hermann (1953, p. 180).

[16]From Hermann (1928/1985, p. 182): 'Durch Nelsons Herausforderungen habe ich es allmählich gelernt, mir Schritt für Schritt den Mut zur Wahrheit zu erkämpfen, der dazu gehört, sich einer als zwingend anerkannten Denkmethode nun auch rücksichtslos im eigenen Denken anzuvertrauen.' ['Through Nelson's challenges I have gradually learned, step by step, to carve out that courage for the truth that is required in order to trust utterly within one's own thinking a method of thinking that one has recognised as cogent' (*eds.*).]

study of the work of Friedrich Fries, Nelson's philosophical mentor, her fear faded away; Nelson supported her and after many vicissitudes[17] she passed her final exam under his supervision. The collaboration between the two could now begin. Nelson asked Hermann to supervise the critical edition of his works[18] and she accepted, yet not without the original hesitation. The critical study of Nelson's work kept her busy until the last years of her life, yet even in the final agreement with Nelson's ideas, she never lost her initial critical attitude toward the Friesian philosopher.

As described by her own words, what immediately captured her attention, and served as the primary basis for her support of Nelson's school, was his 'method of philosophy, of whose certainty and necessity I was convinced'.[19] For the reader, the question is left open whether she is here referring to his general way of reasoning in philosophy, to his effective way of conducting philosophy classes (his stress on punctuality, critical thinking and on many little rituals, such as having discussions until deep in the night), or to what he called 'the Socratic method'. In fact, these three aspects cannot be entirely separated from each other in Nelson's work and they likely all played a role in rousing Hermann's interest in *the method of philosophy*.

In his lecture on *The Socratic Method*, Nelson clearly explains what he understands philosophy and its method to be:

> The function to be performed by the philosophical method is nothing other than making secure the contemplated regress to principles, for without the guidance of method, such regress would be merely a leap in the dark and would leave us where we were before—prey to arbitrariness. (Nelson 1965, p. 9)

The philosophical method consists of the regression to the principles; it works regressively from the consequences back to the reason and discards all other unnecessary characteristics from the original judgement. This process does not bring new knowledge (since it is deductive) but causes a transformation; through reflection, vague, confused judgements are transformed into clear concepts. Philosophy, therefore, consists of the application of this method, and it will have as a result the sum of all the universal rational truths discovered with this reflective method. However, Nelson asks himself: how is it possible to teach this method? He points out that only the history of philosophy can be effectively communicated by instruction, whereas the art of philosophising must be acquired through practice—nevertheless, he does believe that such practice can benefit from guidance. This guided practice is what he calls 'the Socratic Method', and it provides the examples on how to perform the regression to principles. According to Nelson, 'The Socratic Method consists of freeing instruction from dogmatism; in other words, in excluding all didactic judgements from instruction' (Nelson 1965, p. 10). Essentially, the student of philosophy can only learn how to ascertain principles if he or she is standing on his or her own two feet and not being limited by any imposed dogmatic judgements.

---

[17]As already noted, Nelson was not allowed to examine at the Staatsexamen.

[18]Nelson's works were collected in Nelson (1975).

[19]'Methode des Philosophierens, von deren Sicherheit und Notwendigkeit ich überzeugt war' (Hermann 1928/1985, p. 182).

This method is faced by what Nelson calls the general problem of education: how is it possible to teach a method which is itself opposing any authority? He solves this by following Socrates' example, wherein the teacher does not provide answers, but only helps students in formulating their questions, and in not being afraid to doubt. Even if students get scared when confronted with all their doubts, or are in a 'benumbed' state,[20] the Socratic teacher does not help the students, but lets them find their way through reasoning back to the first principles. This does not involve proposing a solution or answer, but only indicating the way to go—what Socrates called the ars maieutica, the art of midwifery.[21]

In theoretical reasoning, as well as in his political activities, we see the importance Nelson, in contrast to Fries, places on education. His example concerning the importance of pedagogic methods was followed by Hermann, both in its theoretical reasoning and practical implications. This can be seen in her pedagogical engagement in the Walkemühle school, and in her critical reasoning about Nelson's work. The Socratic Method is for Nelson something peculiar to philosophy. He underlines the fact that in mathematics, the basic principles are grasped more easily and are not wrapped in obscurity—unlike in philosophy—and therefore the regression is not even necessary. In this manner, he explains his opinion on mathematics and its relationship to philosophy:

> The brilliant development of the science of mathematics and its universally acknowledged advance are explained by the fact that its principles—ignoring for the moment the problems of axiomatics—are easily grasped by the consciousness. They are intuitively clear and thus completely evident, so evident that, as Hilbert recently remarked on this same platform, mathematical comprehension can be forced on everyone. (Nelson 1965, p. 7)

In giving a description of how mathematics works its way to the first principles, Nelson mentions David Hilbert's programme of a new axiomatisation of mathematics as *the* paradigmatic example for the discipline. Indeed, Nelson understood his own work as the philosophical foundation for Hilbert's mathematics, and as an alternative to the logicism of Frege and the conventionalism of Poincaré (cf. Peckhaus 1991). Nelson's position was an elaboration of Fries' critical philosophy, and from Fries he inherited two key positions on the relationship between philosophy and mathematics. First, he wanted to pursue the Friesian dream of the construction of a philosophy based on a rigourous scientific method, such as the one used by mathematics and the natural sciences.[22] Second, Nelson elaborated on Fries' concept of kritische Mathematik

[20] 'I consider', says Meno to his teacher Socrates, in the dialogue bearing his name, 'that both in appearance and in other respects you are extremely like the flat torpedo fish; for it benumbs anyone who approaches and touches it [...] For in truth I feel my soul and my tongue quite benumbed and I am at a loss what answer to give you' (Plato [1967], 80a–80b).

[21] 'My art of midwifery is in general like theirs; the only difference is that my patients are men, not women, and my concern is not with the body but with the soul that is in travail of birth,' explains Socrates in Plato's *Theaetetus* (cf. Plato [1997], 149a–151d).

[22] The aim to ground philosophy on as rigorous a scientific method as mathematics and natural science ('[...] unsere Philosophie auf ebenso strenger wissenschaftlicher Methode beruht wie die Mathematik und die Naturwissenschaften') was declared in the programme of *Die Abhandlungen Fries'schen Schule* (cf. Peckhaus 1991, p. 151).

(critical mathematics) as a philosophically grounded method for mathematics. By kritische Mathematik, Fries understood the study of mathematics as using the method of the regression to principles and the analysis of concepts. Critical mathematics had both the task of indicating, and of questioning, the validity of basic principles (or axioms).[23]

This is not the place to delve further into the relationship between Hilbert and Nelson,[24] but it will suffice to note that even if Hilbert and Nelson were looking at philosophy and mathematics from different angles, they agreed that the collaboration between mathematics and philosophy would lead to the advancement of both subjects. It is from this mutual relationship between mathematics and philosophy that Hermann's ideas developed. Initially educated as a mathematician at the same university in which Hilbert was carrying out his programme, she directly encountered Leonard Nelson—the main living endorser of Friesianism and supporter of critical mathematics. Thus, Nelson's philosophy, his views on politics and education, and the mutual relationship between philosophy and mathematics as pursued in the collaboration with Hilbert, have to be regarded as the intellectual stimuli on which Hermann's ideas grew.

3.4 The Friesian Hermann

After the above description of Fries' and Nelson's critical philosophy, it is clear what Hermann means when speaking about natural philosophy: she is referring to the natural philosophy as elaborated by Fries and Nelson. Fries based his philosophy on a revision of Kant; however, by assigning a different role to experience and to organisms, he elaborated a new concept of science, which included disciplines that were left out by Kant. Nelson carried out a revision of Fries' work, with a particular focus on the relationship between mathematics and philosophy. This allowed him to engage in a close and mutually beneficial collaboration with David Hilbert. In addition, Nelson's philosophy went hand in hand with his political and educational ideas. It was while trying to propagate his ideas that he met Hermann and initiated her to the Friesian school. However, the precise influence of Friesian philosophy of nature on Hermann's work remains uncertain. The following section addresses this question, and in so doing completes the picture of Hermann's use of, and position within, natural philosophy.

It is my contention that the influence of the philosophy of nature of Fries and Nelson is evident in three main aspects throughout Hermann's work, particularly in her 1935 essay on the foundations of quantum mechanics (Hermann 1935, translated as Chap. 15 of this volume). These three areas are (1) her aim and choice of subjects to investigate, (2) the method she used for pursuing her research, and (3) her general understanding of the meaning of philosophy and its relationship to science.

---

[23]A critical study of Nelson's position on critical mathematics is Bernays (1928).

[24]For further readings see Peckhaus (1991) and Peckhaus (2001).

The influence of Friesian philosophy on Hermann's aims and subject matter is evident in her choice to look at quantum mechanics and its relation to transcendental idealism. This is, of course, a topic of chief interest for any follower of Kant's ideas, including Fries and Nelson. Hermann is among the first to tackle the controversial problem of the relationship between quantum mechanics and Kantian philosophy (cf. below, Chap. 4). She describes how '[i]n its bold and successful advance, modern physics' has forced 'from the Olympus of the a priori' concepts which have long been accepted as the foundation of the knowledge of nature, and how in the same vein quantum mechanics seems to contradict the law of causality (Chap. 15, pp. 239–240). As stated by the author herself, her aim is not a critique of the physical theory. In fact, although for Hermann a revision of physics is possible, this would not solve the problem of causality, for which only philosophy of nature can adequately provide answers (ibid.). Hermann's first declared aim is therefore 'to scrutinise the revision of the law of causality announced by the theory' [of quantum mechanics] (Chap. 15, p. 240). Through a careful philosophical examination of the physical theory and its interpretation Hermann shows that quantum mechanics does not contradict at all Kant's (and Fries') views on causality; on the contrary, she will conclude it supports them. The third chapter of 'Die naturphilosophischen Grundlagen' is then dedicated to illustrate the parallel between Fries' and Nelson's critical philosophy and quantum mechanics. As described, Fries has underlined the fallible character of knowledge; as such, the Kantian categories should not be understood as absolute patterns for the order of our experience, but more as an arbitrary attempt to order and limit the immensity of nature. Hermann claims that we can derive the same lesson from quantum mechanics: what we can know about nature is only a part, a relative view; we cannot have an absolute knowledge of the situation, but only the experience and understanding relative to the context of our observation. Therefore, she concludes, not only has the new experimental knowledge of modern physics not contradicted the principle of causality as understood by Fries, but also a fundamental parallel between quantum mechanics and Friesian philosophy exists.

> Rather, closer examination reveals that despite all prima facie discrepancies with the apparent conclusions also of the critical philosophy, the crucial discoveries of quantum mechanics fit consistently together with the principles of that philosophy, and through these their significance for the knowledge of nature becomes intelligible. (Chap. 15, p. 278)

Hermann takes up the challenge posed to Friesian philosophy by the development of physics, and successfully defends and supports Friesian thought throughout her work.

The second aspect of Hermann's thought that is also characteristic of Friesian and Nelsonian philosophy is the method—and the importance assigned to it—she uses to carry out her research on the philosophical foundations of quantum mechanics. Fries claimed that 'the two conditions of the art of philosophising are to think precisely, or to know the rules of the correct use of reason [...], and secondly, to think purposefully',[25] and Nelson had elaborated Fries' methodology further into his

[25] 'Beym Philosophiren ist die Methode die Hauptsache, das andere muss sich dann von selbst ergeben, denn es soll unser eigenes Werk sein [...] Die beyden Bedingungen der Kunst zu

Socratic Method. Similarly, Hermann's methodology is characterised by following a strict logic of thinking, a careful analysis of language and a step-by-step reasoning through deduction back to the basic principles. She starts investigating causality by clarifying the definition of the term and by distinguishing between two definitions (the principle of causality and the law of causation), which she claims have been erroneously confused.

> Seen in this light, the difficulties into which the advocate of the law of causation is plunged by the discoveries of quantum mechanics are rooted in the fact that various principles have been merged together [...] This assumption has in fact not been introduced explicitly, but has crept into the criterion of the applicability of the principle of causality as an undisclosed and seemingly self-evident premise. (Chap. 15, p. 264)

The philosophical method used by Hermann consists in the 'clarification of terminology' (Klärung der Terminologie) and in accord with Nelson's and Fries' ideas is the fundamental and necessary tool that Hermann uses for answering her initial question of the status of causality in quantum mechanics.

As already shown for her method and aim, overall Hermann analyses the developments of the natural sciences (specifically contemporary physics: in one case analysing quantum mechanics, in another essay relativity theory) as the Friesian school advocated. The overall aspect of influence is that like Fries, Hermann understands as a duty of philosophy the critical study of the natural sciences; philosophy illuminates and assigns meaning to the fundamental and problematic aspects of science, while at the same time benefitting from the challenges posed by scientific development. In Hermann's words:

> [...] the scientific progress that has been obtained in these theories precisely through the willingness to abandon or revise old familiar concepts provides the guarantee that new and fruitful points of view have been introduced here into research. (Chap. 15, p. 240)

This argumentation is based upon Fries' and Nelson's view of the relationship between natural science and philosophy; Nelson claimed that 'every philosophical statement that is in accord with the exact sciences *can* be true, but every philosophical statement that is contradiction with the exact sciences *has* to be wrong'.[26] Therefore, also the way Hermann understands the focus of philosophy and its relationship to science—as looking at contemporary developments of natural science as a source of challenge and inspiration, and at the same time working on the clarification of its principles—is defined by her Friesian background.

(Footnote 25 continued)
philosophiren sind bestimmt denken, oder die Regeln des richtigen Verstandes-Gebrauches (nicht nur im Schließen, sondern vorzüglich im Gebrauche der Begriffe) zu kennen, und zweytens zweckmässig denken, d.h. jene Regeln gehörig anwenden wissen' ['When philosophising, method is the main thing—the rest must follow by itself, for it must be our own work [...] The two conditions of the art of philosophising are to think precisely, or to know the rules of the correct use of reason (not only for inferring, but especially in the use of concepts), and secondly, to think purposefully, i.e. to know how to apply those rules appropriately' (*eds*.)] (Fries 1804/1968, §§ 9–10; pp. 32–33).

[26]'Jedes Philosophieren, das mit den exakten Wissenschaften widerstreitet, *muß* notwendig falsch sein' (Nelson 1973, p. 10).

3.5 Hermann's Philosophy of Nature

We have seen from the influence on the aims, method and general understanding of philosophy that Hermann's work on the foundations of quantum mechanics is grounded in Friesian philosophy of nature. However, Hermann's work is not a mere replication of Fries' and Nelson's ideas: she goes further than her teachers in a number of ways, widening Friesian philosophy by *extending its realm of study* and by *softening its absolutism* (especially regarding Nelson's philosophy).

First of all, the choice Hermann makes to look into quantum mechanics and its consequences for the Kantian category of causality, while in line with Friesian philosophy, also adds new aspects to it. Since Friesian philosophy is based on Kant's philosophical system, the problem of validity of Kant's categories is an argument of chief importance for Friesian philosophy. An analysis of the newest developments in quantum mechanics and its implications for Kant's category of causality, however, reveals that Hermann carried out a widening of the interests and objects of study of Friesian philosophy. As described above, Friesian natural philosophy, up until Hermann, had been mainly concerned with mathematics, and its relation to philosophy or critical mathematics. Fries had widened Kant's conception of 'science proper' to include chemistry, but had still chosen mathematics as the paradigmatic methodology, and consequently elaborated a first meta-theory of pure mathematics. Similarly, Nelson had dedicated himself to the study of mathematics and to the solution of the antinomies. While Fries had devoted some little attention to the new developments in physics, as carried out by Euler and Lagrange, his follower Nelson disdained the changes in physics. Nelson believed, like an old-school Kantian, that Newton's mechanics was the only possible physical theory and that the problems of quantum mechanics and the theory of relativity would end up explained and formulated as classical mechanical problems (cf. Heckmann 1985). On the contrary, Hermann, although coming from a mathematical background, sees in the new physical theories an important contribution to epistemology and philosophy of nature. She looks into the consequences of quantum mechanics (and of physical theories in general) for the philosophy of nature, and in this way she broadens the Friesian perspective. Friesian philosophy, under Hermann, is now not solely looking at mathematics, but includes also a new perspective, entailing the incorporation of the newest developments in physics. The relationship between mathematics, philosophy and physics is central to Hermann's reflections, and thus as a consequence of her work, Friesian philosophy of nature and its realm of study are expanded.

The other original development of Hermann's philosophy with regard to Friesianism is her softening of the absolutist aspects in Nelson's philosophy. In a letter to her friend and political colleague Gustav Heckmann she writes: 'I want to understand how to free Nelson's philosophy from its misleading absolutist demands, and through which modifications to the kernel of his philosophy this can be achieved and in turn made valid'.[27] This sentence condenses the position of Hermann towards Nelson's philosophy, with which she has been involved since 1925, when Nelson asked her to collaborate in the edition of his lectures on ethics (Nelson 1975).

Fries distinguished between theoretical and practical natural philosophy. Theoretical philosophy investigates the existence of things, while practical philosophy of nature is concerned with their purposes. After Fries, Nelson understood theoretical philosophy as the method of axiomatics, and took Hilbert's system as the example to follow for 'exact science', and at the same time identified practical philosophy with ethics. In both realms, Nelson wants to find the way back to fundamental principles (Axiomen) that once discovered must be accepted by everyone. Nelson believed in the universal validity of reason (Vernunft) and understood his mission in leaving nothing, in both physics and ethics, to chance (vom Zufall entziehen). For Nelson, anyone confronted with an answer that is indicated by reason must see it as the only possible solution. This position tends to absolutism and dogmatism, which is exactly what his philosophical method endeavoured to liberate us from. Hermann is aware of these tendencies in Nelson's philosophy, and thus, looks to modify his theory. In both realms Hermann manages to free Nelson's philosophy from these absolutist colours, without falling into relativism yet preserving the role of rationality (Vernunft). In 'Die Überwindung des Zufalls' ('Conquering Chance') she shows how, by using the same methods as Nelson, we should get to a different conclusion, namely that reason and sense experience (Vernunft und Sinnlichkeit) always work together in every ethical judgement. Although rationality still plays a major role, justice is given to experience and to the possibility of different ethical and political judgements (cf. Henry-Hermann 1953). Similarly, the absolutist aspects are left out of the other part of Friesian philosophy, the theoretical philosophy of nature. Here Hermann accepts the indeterministic aspects of physical theories, without having to dismiss the Kantian categories of understanding:

> This necessary self-limitation of the physical study of nature remains by all means compatible with the commitment to the guiding methodological principles of physical research as such, according to which wherever the causes of an observed process are not known, it is a physically meaningful problem to look for them.[28] (Hermann 1948, p. 382)

Quantum mechanics may not provide precise predictions, yet it still maintains a rational causal structure. Both in physics and in ethics, and without negating the role of reason, Hermann restores an element of indeterminacy that Friesian and Nelsonian philosophy lacked.

---

[27] 'Ich möchte verstehen, durch welche Modifikationen der Wahrheitskern der Nelsonschen Philosophie von irreführenden Absolutheitsansprüchen befreit werden und sinngemäss geltend gemacht werden kann' (Heckmann 1985, p. XI).

[28] 'Diese notwendige Selbstbescheidung physikalischer Naturerkenntnis bleibt dabei durchaus vereinbar mit dem Festhalten an dem methodischen Leitsatz jeder physikalischer Forschung, wonach es überall da, wo wir die Ursachen eines beobachteten Vorgangs nicht kennen, ein physikalisch sinnvolles Problem ist, nach ihnen zu suchen'.

3.6 Conclusion

After a brief overview of the philosophy of nature, and in particular of the philosophy of nature of Fries and Nelson, Hermann's ideas now have been grounded in their broader context. This allows both an understanding as to the development of her ideas, as well as a critical examination of the previously little explored, yet seemingly fruitful, school of thinking.

My first focus was on explicating the meaning that philosophy of nature, as stemming from the critical philosophy of Jakob Friedrich Fries and his follower Leonard Nelson, has in Hermann's philosophy. Hermann used the Friesian philosophy of nature as a privileged perspective for her study of modern physics, and this stands out in three main aspects of her work—her aim, her understanding of philosophy, and the method of pursuing her research. She looks at the meaning of the principle of causality as determined by quantum mechanics, because of the challenge quantum mechanics poses to critical philosophy. In doing so, Hermann applies the method of the analysis of concepts and the regression to the principles as she had learned it from Nelson and Fries. I have consequently shown that her general understanding of the meaning of philosophy and its relation to science is of Friesian origin. Next, I addressed Hermann's novel adaptations of Friesian philosophy. Although she chose to specifically follow this school of thought, Grete Hermann's philosophical ideas were not a mere reproduction of Friesian philosophy of nature, but made valuable and original contributions to it. For instance, I noted that Grete Hermann pursued a widening of the Friesian perspective on an interdisciplinary level, and at the same time lessened some of its absolutist tendencies. In addition, the study of Hermann's use of philosophy of nature, and her contributions to it, supplements and hopefully stimulates the recent studies of philosophy of nature, in re-evaluating the long-neglected Friesian school. I hope this deeper reading and understanding of Hermann's philosophical background and reasoning will provide a fruitful source of inspiration in philosophy and science.

References

Bernays P (1928) Über Nelsons Stellungnahme in der Philosophie der Mathematik. Die Naturwissenschaften 16:142–145

Bianco B (1980) J. F. Fries, rassegna storica degli studi (1803–1978). Bibliopolis, Napoli

Fries JF (1804/1968) System der Philosophie als evidente Wissenschaft aufgestellt. In: Sämtliche Schriften, vol 3, pp 7–410

Fries JF (1828/1968) Neue oder anthropologische Kritik der Vernunft. In: Sämtliche Schriften, vol 4, pp 31–478

Fries JF (1935) Neue oder anthropologische Kritik der Vernunft. Verlag Öffentliches Leben, Berlin

Fries JF (1967–2011) Sämtliche Schriften, 30 vols. Scientia, Aalen. Edited by König G and Geldsetzer L

Geldsetzer L (1999) Jakob Friedrich Fries' Stellung in der Philosophiegeschichte. In: Hogrebe W, Herrmann K (eds) Jakob Friedrich Fries: Philosoph, Naturwissenschaftler und Mathematiker, Lang, Frankfurt am Main, pp 13–56

Gregory F (1983) Die Kritik von J. F. Fries an Schellings Naturphilosophie. Sudhoffs Archiv 6:146–157

Gregory F (1997) The Newtonian vitalism of J. F. Fries. In: Cimino G, Duchesneau F (eds) Vitalisms from Haller to the cell theory. Florence, Olschki, pp 149–151

Gregory F (2006) Extending Kant: the origins and nature of Jakob Friedrich Fries's philosophy of science. In: Friedman M, Nordmann A (eds) The Kantian legacy in nineteenth-century science. The MIT Press, Cambridge, MA, pp 101–121

Heckmann G (1985) Vorwort. In: Henry-Hermann G, Die Überwindung des Zufalls: Kritische Betrachtungen zu Leonard Nelsons Begründung der Ethik als Wissenschaft. Verlag Öffentliches Leben, Frankfurt, pp xi–xx

Henke ELT (1867) Jakob Friedrich Fries: aus seinem handschriftlichen Nachlasse dargestellt. Brockhaus, Leipzig

Henry-Hermann G (1953) Die Überwindung des Zufalls: Kritische Betrachtungen zu Leonard Nelsons Begründung der Ethik als Wissenschaft. In: Specht M, Eichler W (eds) Leonard Nelson zum Gedächtnis. Verlag Öffentliches Leben, Frankfurt am Main/ Göttingen, pp 25–111

Hermann G (1928/1985) Erinnerungen an Leonard Nelson. In: Henry-Hermann G, Die Überwindung des Zufalls: Kritische Betrachtungen zu Leonard Nelsons Begründung der Ethik als Wissenschaft. Verlag Öffentliches Leben, Frankfurt, pp 179–210

Hermann G (1935) Die naturphilosophischen Grundlagen der Quantenmechanik. Abhandlugen der Fries'schen Schule 6(2):75–152

Hermann G (1948) Die Kausalität in der Physik. Studium Generale 1(6):375–383

Hieronimus E (1964) Theodor Lessing, Otto Meyerhof, Leonard Nelson: bedeutende Juden in Niedersachsen. Niedersächsische Landeszentrale für Politische Bildung, Hannover

Kant I (1781) Kritik der reinen Vernunft. In: Kants gesammelte Schriften, vols 3 and 4. De Gruyter, Berlin

Kant I (1786) Metaphysische Anfangsgründe der Naturwissenschaft. In: Kants gesammelte Schriften, vol 4. De Gruyter, Berlin, pp 465–565

Kant I (1790) Kritik der Urteilskraft. In: Kants gesammelte Schriften, vol 5. De Gruyter, Berlin, pp 165–485

Kraft J (1949) Introduction. In: Nelson L, (ed) Socratic method and critical philosophy: selected essays by Leonard Nelson. Dover, New York 1965, pp ix–xxii

Nelson L (1965) Socratic method and critical philosophy: selected essays by Leonard Nelson. Dover, New York

Nelson L (1973) Geschichte und Kritik der Erkenntnistheorie. In: Henry-Hermann G (ed) Leonard Nelson: Gesammelte Schriften in neun Bänden, vol 2. Meiner, Hamburg

Nelson L (1975) Vom Selbstvertrauen der Vernunf: Schriften zur kritischen Philosophie und ihrer Ethik. Meiner, Hamburg

Peckhaus V (1991) Hilbertprogramm und Kritische Philosophie: Das Göttinger Modell interdisziplinärer Zusammenarbeit zwischen Mathematik und Philosophie. Studia Logica 50(2):351–354

Peckhaus V (2001) 'Mein Glaubensbekenntnis': Leonards Nelsons Brief an David Hilbert. In: Toepell M (ed) Mathematik im Wandel: Anregungen zu einem fächerübergreifenden Mathematikunterricht, vol 2. Franzbecker, Hildesheim, pp 335–346

Plato [1967] Meno. In: Plato in twelve volumes, vol 5. Harvard University Press, Cambridge, MA

Plato [1997] Theaetetus. In: Cooper J (ed) Plato: complete works. Hackett, Indianapolis, pp 137–234

Pulte H (1999) '...sondern Empirismus und Speculation sich verbinden sollen.' Historiographische Überlegungen zur bisherigen Rezeption des wissenschaftstheoretischen und naturphilosophischen Werkes von J. F. Fries und einige Gründe für dessen Neubewertung. In: Hogrebe W (ed) Jakob Friedrich Fries: Philosoph, Naturwissenschaftler und Mathematiker, Lang, Frankfurt am Main, pp 57–94

Pulte H (2006) Kant, Fries and the expanding universe of science. In: Friedman M, Nordmann A (eds) The Kantian legacy in nineteenth-century science. The MIT Press, Cambridge, MA, pp 101–121

Sachs-Hombach K (1999) Ist Fries' Erkenntnistheorie psychologistisch? In: Hogrebe W, Herrmann K (eds) Jakob Friedrich Fries: Philosoph, Naturwissenschaftler und Mathematiker, Lang, Frankfurt am Main, pp 57–94

Siegel C (1913) Geschichte der deutschen Naturphilosophie. Akademische Verlagsgesellschaft, Leipzig

Part II
Hermann on Quantum Mechanics

Chapter 4
The Convergence of Transcendental Philosophy and Quantum Physics: Grete Henry-Hermann's 1935 Pioneering Proposal

Léna Soler

4.1 Introduction

In the 1930s, Grete Hermann (1901–1984) was a young woman trained in physics, mathematics and philosophy. She was deeply convinced that transcendental philosophy is the best framework to articulate both ethical and epistemological issues. At the time, it was increasingly said in philosophical and scientific circles that the just-born quantum mechanics refuted Kantian philosophy, especially the Kantian table of categories and its concept of causality. Having heard about this, Hermann decided to tackle the problem. She worked on it for more than a year and submitted

I came to hear about Hermann's writings during my DEA study (the university year just before a PhD) devoted to the relationship between Kantian philosophy and quantum mechanics. It was Michel Bitbol, then my DEA supervisor, who drew my attention to this writing and encouraged me to work on it. I am happy to have the occasion to thank him warmly here, for this, indeed, but more fundamentally for the inspiration I always found in his way of doing philosophy and for his friendly support. I initiated the translation of Hermann's most important text into French and edited it with an introduction and a long critical postface in Hermann (1935/1996). The present paper is based on my analyses in that book and is a slightly edited version of the paper published as Soler (2009). The latter is itself derived from a conference speech, delivered 2 March 2001 in Bremen (in German, thanks to the collaboration of Alexander Schell), at the invitation of the Philosophisch-Politische Akademie on the occasion of the Hermann Centenary Celebration. A French version of this lecture has been translated into English by Dr Edmund Jephcott (A & G Translations) at the request of the Society for the Furtherance of Critical Philosophy (SFCP). This English translation has been published by the SFCP (Soler 2004). Interested readers can acquire a copy of this volume from Keith Martin, SFCP Treasurer, http://sfcp.org.uk/contact/).

L. Soler (✉)
Laboratoire d'Histoire des Sciences et de Philosophie—Archives Henri Poincaré, Nancy, France
e-mail: lena.soler@univ-lorraine.fr

E. Crull and G. Bacciagaluppi (eds.), *Grete Hermann - Between Physics and Philosophy*, Studies in History and Philosophy of Science 42,
DOI 10.1007/978-94-024-0970-3_4

her proposals to Werner Heisenberg and Carl Friedrich von Weizsäcker. At the end she published arguments that aim to show, first, that the Kantian category of causality still constitutes a necessary condition of possibility of quantum physics, and second, that—more fundamentally—quantum physics and critical philosophy manifest an essential structural convergence that reinforces the strength of transcendental philosophy.

This chapter will present, and sketch the significance, of Hermann's interpretation of the relationships between transcendental philosophy and quantum physics.[1] Hermann's contributions to the philosophy of physics, although almost unknown, especially outside Germany, will appear very important in a historical perspective, as well as of great philosophical interest from a contemporary point of view, especially for transcendentally-oriented philosophers.

4.2 The Philosophy of Science in Hermann's Work: Motivation and Situation

Hermann was a pupil and great admirer of Leonard Nelson, and she followed her teacher in concerning herself primarily with ethics and political philosophy. Comparatively, her works on the philosophy of science occupy a relatively marginal position. What caused Hermann to take an interest in physics?

Hermann, following Nelson, considered Kant's philosophy—or more precisely, its reinterpretation by Fries—to be the basis on which twentieth-century philosophy should run. Yet in the 1930s, a number of participants in the debates concerning quantum mechanics—still a very young science at that time—took the view that the new physics called into question, or indeed definitively refuted, some fundamental aspects of the critical philosophy inaugurated by Kant. Given the significance of transcendental philosophy for Hermann, it was of crucial importance to her to investigate whether twentieth-century physics did or did not effectively refute the fundamental principles of such a philosophy.

The principal published texts in which Hermann sets out her reflections on this subject are relatively few in number and are confined to a limited period: they correspond to three relatively short essays produced during the years 1934–1937:

1. 'Die naturphilosophischen Grundlagen der Quantenmechanik' ('The Natural-Philosophical Foundations of Quantum Mechanics') in *Abhandlungen der Fries'schen Schule* (Hermann 1935b),[2]

[1]For a more detailed account, see my introduction and postface in Hermann (1935/1996).

[2]English translation in Chap. 15. Leonard Nelson edited this journal from 1904 (reviving the enterprise begun by two followers of Fries, Apelt and Schleiden, and which was continued from 1847 to 1849 until it was interrupted during the 1848 Revolution due to political disagreements between the editors). An abridged version of the 1935 essay by Grete Hermann was also published in *Die Naturwissenschaften* as Hermann (1935c).

2. 'Die Bedeutung der modernen Physik für die Theorie der Erkenntnis' ('The Significance of Modern Physics for the Theory of Knowledge') in Hermann (1937a), and
3. 'Über die Grundlagen physikalischer Aussagen in den älteren und den modernen Theorien' ('On the Foundations of Physical Statements in Earlier and Modern Theories') also in 1937 (Hermann 1937c).

The truly major text containing Hermann's fundamental theses and arguments is that of 1935, which deals with quantum mechanics. The later texts either revisit the developments of 1935 or apply analytical principles similar to those of 1935 to the theory of relativity, and arrive at similar conclusions.[3] For this reason, the essay of 1935 (translated into French in Hermann 1935/1996) will be the primary focus of this chapter.

4.3 Physics and Causality in the 1930s

As we saw, Hermann's first concern is the apparent incompatibility between transcendental philosophy and twentieth-century physics. In the 1935 essay, the reflection is more especially focused on the relationship between causality and quantum physics.

Hermann characterised the problem in the following manner:

- Kant listed the conditions of possibility of knowledge, and thus in particular, the conditions of possibility of *any future physics*.
- These conditions include the category of causality, which seems to imply that the predictions of any proper (according to Kant) science must be strictly deterministic in the following sense: there must be a *univocal (one-to-one) connection between cause and effect* (between initial and final conditions); one and the same cause can produce only *one single*, well-defined effect.

[3]Hermann's contributions to the philosophy of science comprise four other articles: the first (Hermann 1935a) is a review of Karl Popper's *The Logic of Scientific Discovery*, which was written in 1934 and later became famous (Popper 1934). The second (Hermann 1936) is the text of a short talk given in Copenhagen in June 1936, at the Second International Congress for the Unity of Science, 'Zum Vortrag Schlicks'. This was a reply to a paper by Moritz Schlick regarding the causal problem, in which Hermann brought to bear the theses of 1935. The third (Hermann 1937b), 'Die Naturphilosophische Bedeutung des Übergangs von der klassischen zur modernen Physik' ('The Philosophical Significance of the Transition from Classical to Modern Physics'), represents Hermann's contribution to the 9th International Congress of Philosophy, 'Congrès Descartes', held in Paris in 1937; in it the conclusions from the essay of 1935 are again summarised. Finally, an article of about ten pages from 1948 entitled 'Die Kausaliltät in der Physik' ('Causality in Physics', Hermann 1948), provides an extremely clear synthesis of Hermann's previous works. [To these published works should be added the manuscript from 1933 on 'Determinismus und Quantenmechanik' ('Determinism and Quantum Mechanics', Hermann 1933), to appear in Herrmann (2017), and translated here as Chap. 14 (*eds.*).].

- Now, the predictions of quantum mechanics are *statistical*: one and the same initial state can be followed by *several different* final states, and scientists know in advance (before any actual experiment) *only the probability* of each possible result.
- Should we therefore conclude that the Kantian category of causality is refuted by the new physics? That strict causality is not, after all, a necessary condition of any physics?

Hermann was not the first to frame the problem in this way. The general strategy underlying such a framing was to call Kant before the tribunal of history, according to the following reasoning: in listing the conditions determining the possibility of any future physics, Kant had drawn his inspiration from the physics of his time (that of Newton, retrospectively called 'classical'); do his propositions still hold good for physics after Kant? I.e., can the Kantian conditions of sensibility and understanding be maintained in particular in light of twentieth-century physics, which seemed to break so radically with the fundamental principles of classical physics?

A number of philosophers and physicists had already applied a similar line of reasoning to non-Euclidean geometries and to the theory of relativity.[4] Specifically regarding the latter, they had enquired whether relativity theory refuted those conditions of sensibility set out by Kant which depended upon Euclidean space and absolute time. With the emergence of quantum physics, it was the pure concepts of understanding that became threatened: at least at first glance, it was the specific concept of causality that appeared refuted.

These were the questions on the side of Kantian philosophers. However, philosophers were not the only ones to be concerned by causality in the 1930s. Causality was also preoccupying physicists themselves. On their side the debate essentially concerned so-called 'hidden variables', and the corresponding problem was framed by the following alternatives:

- Should it be understood that quantum physics and its statistical predictions express no more than a deficiency of human knowledge? Stated differently: should certain variables, as yet unknown to physicists, be allowed to exist which univocally determine all measurement results? Are there 'hidden variables' or parameters that, if known, would place a given cause in a one-to-one correspondence with a single determinate effect?
- Or should the statistical character of predictions, that is, the association of a plurality of possible effects with a given cause, be recognised as determinate because it expresses a fundamental aspect of phenomena or of the relationship between human beings and the physical world?

Hermann addressed the questions of the philosophers as well as those of the physicists. Here are the conclusions she finally reached, briefly summarised:

[4]See, for example, Schlick (1974, especially Sects. 37–40).

- Quantum physics does not refute the category of causality.
- There are no hidden variables; quantum physics is complete and its predictions will remain statistical.

Considered in detail, Hermann's conclusions and the arguments supporting them are complex and subtle. A full understanding of them would require an extensive discussion drawing simultaneously on philosophy, physics and history (cf. Hermann 1935/1996, postface). In this chapter, attention will be confined to the following:

- Some points that make Hermann's 1935 essay particularly noteworthy (Sect. 4.4),
- Indications as to what constitutes the central originality of her thesis concerning quantum physics (Sect. 4.5),
- Some remarks concerning the strengths and weaknesses of Hermann's interpretation (Sect. 4.6), and
- Some of Hermann's general conclusions concerning the relationship between critical philosophy and quantum physics (Sect. 4.7).

4.4 Noteworthy Aspects of Hermann's Work in Philosophy of Science

1. Hermann's attempt to investigate quantum physics from the perspective of Kantian philosophy is, chronologically, one of the first. Philosophers of science are often accused of lagging behind the advancement of science. For once, this accusation does not apply. Hermann was truly a pioneer in developing a philosophical interpretation of quantum physics.[5]

Indeed, in 1934 quantum physics had been a physical theory worthy of the name for only a few years. Since the Solvay Congress of 1927, the term 'quantum mechanics' had referred to something fairly definite and stable: a formalism which was a synthesis of Heisenberg's matrix mechanics, Schrödinger's wave mechanics and Dirac's theory of transformations—together with a more unified interpretation of this formalism proposed by Bohr and Heisenberg and later called the 'Copenhagen interpretation' or the 'orthodox interpretation'. Hermann was to adopt numerous ingredients of this interpretation in her own account: the complementarity of wave and particle representations and of the conjugate variables of position and velocity; the idea that the results obtained have validity only in relation to a particular experimental arrangement; etc.

2. Hermann had two-fold training, scientific (physics and mathematics) and philosophical. This is uncommon enough to deserve mention, and clearly is not without

[5] Along with a few others, among them Kojève (1932/1990), Bachelard (1934), and Cassirer (1937) (English translation in Cassirer 1956).

relevance to her contributions to the philosophy of science. Indeed, Hermann had sufficient mastery of physics to be able to study in depth the theory of quantum mechanics and its formalism, and to engage in high-level dialogue with scientists.

3. Hermann's text of 1935 is the outcome of discussions she held during a year in Leipzig with Heisenberg and a group of major physicists who were among the originators of quantum physics.

Leipzig was one of the centres, along with Göttingen and Copenhagen, that contributed significantly not only to the development of quantum theory but to the clarification of its philosophical foundations. Werner Heisenberg, the famous pioneer in this venture, organised a seminar in Leipzig that brought together a considerable number of eminent scientists, such as the Swiss Félix Bloch, the Soviet Lev Landau, the Hungarian Edward Teller, and from Germany, Rudolf Peierls, Friedrich Hund and Carl Friedrich von Weizsäcker, who was still very young at that time. The latter took a passionate interest in the philosophical questions raised by the new physics, and for this reason he played a leading role in the dialogue with Hermann.

In the course of 1934, Hermann went to Leipzig in order to participate in Heisenberg's seminar. It seems that her decision resulted from the following events, according to von Weizsäcker (1963). Hermann wrote a manuscript dealing with causality in quantum physics (Chap. 14 below) which she sent to Bohr and Heisenberg. Bohr asked von Weizsäcker to read the manuscript and possibly to respond to Hermann. Von Weizsäcker did so, indicating in his letter in what way the theses of the article seemed erroneous to him. Having received a letter with very similar content from Heisenberg, Hermann decided to travel to Leipzig to discuss the matter with the two physicists in person.

It was at the end of a year of debate with these prestigious figures that Hermann wrote her essay of 1935. By the end of these discussions it seems that she had succeeded in convincing Heisenberg of her point of view.[6] In his scientific autobiography, *Physics and Beyond*, Heisenberg devoted an entire chapter to the discussions between Hermann, von Weizsäcker and himself. This chapter, entitled 'Quantum Mechanics and Kantian Philosophy', presents the content of the arguments and their progression, together with the compromise that emerged from them. Heisenberg's tone at the end is quite positive: 'We had the feeling that we had all learned a good deal about the relationship between Kant's philosophy and modern science' (Heisenberg 1971 p. 124).

For his part, von Weizsäcker reviewed Hermann's essay in highly eulogistic terms in an article published as von Weizsäcker (1936), presenting Hermann's essay as the first 'positive and indisputable contribution to elucidating the implications of quantum mechanics for the theory of knowledge', and adding that 'a fruitful debate on this subject could hardly be opened in a clearer or more objective manner'. Moreover, in his book *The World View of Physics* (von Weizsäcker 1952), von Weizsäcker develops

[6] To begin with, at least; he seems later to have changed his mind under the influence of Bohr.

a relationship between critical philosophy and quantum physics that in many respects is akin to that of Hermann.[7]

4. In addition to the original ideas she developed on the relationship between the Kantian category of causality and quantum physics, to which I shall return below, Hermann's essay of 1935 contains the first critique of von Neumann's argument demonstrating the impossibility of completing quantum physics by means of hidden parameters.[8]

In 1932, von Neumann claimed to prove that the statistical character of quantum physics was not due to a deficiency in human knowledge, and that it was therefore pointless to hope to discover hidden variables. Quantum mechanics is complete, he claimed, in the sense that the physicist already knows everything there is to be known. According to him, the predictions of quantum physics are statistical because quantum phenomena themselves are indeterministic (in the sense that given identical initial conditions, different final conditions can indeed follow).

Historically, von Neumann's proof played a very important role. Indeed, between 1932 and 1964, its existence and assumed validity deterred physicists from trying to develop theories of hidden variables. But the situation changed in 1964 when John Bell attacked Neumann's 'proof', providing what has since been considered the definitive refutation of Neumann's argument (Bell 1966). Yet in 1935—*thirty years before Bell*—Hermann had produced a refutation of von Neumann based on arguments very similar to those of Bell.

Indeed, Hermann identified as problematic the very premise in von Neumann's reasoning on which Bell based his famous refutation. This premise—the condition of additivity—stipulates that the expectation value of the sum of two physical quantities is equal to the sum of each of their expectation values. Such a property, while trivial for variables capable of being measured simultaneously (i.e., variables of classical physics or non-conjugate variables of quantum physics), must be proven in the case of those conjugate quantities used in quantum physics. Hermann and Bell both insist on this point. This is really striking when their statements are compared side by side. For instance, Hermann writes

[7]When addressing the question of the relationship between physics and philosophy in his 1963 interview with T.S. Kuhn (von Weizsäcker 1963), von Weizsäcker emphasised that around 1933–1934, the Leipzig group formed a unified front defending the new ideas associated with physics against attacks by philosophers. He then went on to speak spontaneously of Hermann, emphasising above all, in the brief account he gave of her, her two-fold training in mathematics and philosophy. Von Weizsäcker referred to her as an extremely intelligent person and remarked that her great clarity of mind made discussion with her easy. He added that Hermann was probably right in maintaining that Kantian philosophy, when correctly interpreted, was in no way placed into difficulty by modern physics, itself also correctly interpreted. Then he alluded to Hermann's manuscript dealing with causality in quantum physics.

[8]Max Jammer seems to be the first to notice this point: cf. Jammer (1974, p. 272).

> [T]he sum of two such [conjugate] quantities is not immediately defined at all: since a sharp measurement of one of them excludes that of the other, so that the two quantities cannot simultaneously assume sharp values, the usual definition of the sum of two quantities is not applicable. Only by the detour over certain mathematical operators assigned to these quantities does the formalism introduce the concept of a sum also for such quantities. (this volume, Chap. 15, p. 252)

Compare with Bell:

> A measurement of a sum of noncommuting observables cannot be made by combining trivially the results of separate observations on the two terms—it requires a quite distinct experiment [...] The additivity of expectation values [...] is a quite peculiar property of quantum mechanical states, not to be expected a priori. (Bell 1987, p. 4)

Or again, consider Hermann regarding Neumann's argument:

> [A] detailed assessment shows here, too, that this mathematically otherwise faultless argumentation introduces into its formal assumptions, without justification, a statement equivalent to the thesis to be proven [...] In words: *the expectation value of a sum of physical quantities is equal to the sum of the expectation values of the two quantities*. Neumann's proof stands or falls with this assumption. (Chap. 15, pp. 251–252)

And in the words of Bell,

> Von Neumann's essential assumption is: Any real, linear combination of any two Hermitian operators represents an observable, and the same linear combination of expectation values is the expectation value of the combination. (Ibid., p. 4)

Now, despite this similarity between Hermann's and Bell's arguments, and despite the fact that Bell's paper quickly convinced all physicists after its publication, Hermann's refutation had no impact. In fact, it remained entirely unknown—and this is highly surprising if one bears in mind that physicists such as Heisenberg and von Weizsäcker must have known of it.

Whatever the reasons may be, this fact has important historical implications. Indeed, if Hermann's refutation of Neumann had not been a 'dead letter', the history of interpretations of quantum physics would certainly have been very different. Theories involving hidden variables, which have proliferated since Bell's paper in the 1960s, would probably have flourished much earlier, and the Copenhagen interpretation—so long regarded as one of the only acceptable available interpretations—would perhaps have enjoyed a less exclusive monopoly.[9]

[9]Even the habits underlying the theoretical practices of physicists, including judgements of simplicity, could perhaps have been substantially transformed. This is important, since the main argument today against one the most prominent theories of hidden variables—namely the Bohmian interpretation of quantum physics—is its alleged lack of simplicity: it is said to be less simple than the standard interpretation of quantum mechanics. I develop this point in relation to the question of the contingency of our history of science in Soler (2006). See also Cushing (1994). [For more on von Neumann's proof, see this volume, Chap. 8 (*eds.*).]

4.5 The Core of Hermann's Original Interpretation

Let us now turn to Hermann's conception of the link between quantum physics and Kantian philosophy. With her refutation of von Neumann's proof Hermann had reopened the door to the possibility of discovering hidden variables. Thus one might believe that she would proceed to engage in an attempt to save the Kantian category of causality by invoking, as others had done, the existence of hidden variables which determine the unique cause of each effect. But that is not the case. Hermann set out on another original path: that of retaining the universal validity of the pure concept of causality while accepting, with Bohr and Heisenberg, the definitive character of statistical predictions.

The core of her original interpretation is, in essence, the following. The results of measurements actually obtained for quantum objects cannot be univocally predicted with certainty. However, *after* having performed a quantum measurement, and *after* having gained knowledge of its result (previously not predictable with certainty), it is possible, by working backwards, to reconstitute, retrospectively and completely, the causal chain which has necessarily produced such a result.

To properly grasp what is meant by these causal chains reconstructed a posteriori, it is necessary to emphasise that the causal chains under consideration connect:

- On the one hand, a phenomenon resulting from measurement (for example, a spot on a screen)
- And on the other hand, the value of some theoretical variable (for example, a value for the momentum).

The effect is the phenomenon resulting from the measurement. The cause is the value of the variable.

This type of causal chain already plays a part in classical physics. For example, when you measure an object's weight, the observed phenomenon equated with the effect is the needle movement and its stopping on the scale at a certain graduation mark. The value of the theoretical variable equated with the cause of such a phenomenon is the determinate weight of the object weighed. The link between the two is a causal scenario of the following kind: the weight causes the vertical displacement of the scale pan, which in turn causes, through a series of specifiable mechanical actions, the deflection of the needle.

Hermann carried this classical theory of measurement over to cases of quantum measurements. For example, take the case of measuring the momentum of an electron by illuminating this electron under a microscope. Because the electron is illuminated, there is an interaction between the electron and the incident light. This light is then captured in the microscope, and from it information about the electron is derived. The effect—the phenomenon resulting from the experiment—is a discrete impact on the photographic plate. The cause is the momentum of the electron at the instant of interaction with the incident light.

There are two differences between the quantum and classical cases:

1. Contrary to the situation in classical physics, in quantum physics the causal scenario cannot be anticipated; it is known only once the measurement has actually been performed and only once the relevant phenomenon has actually been observed.
2. In quantum physics, although the causal chain connecting the cause to the effect makes use, at the same time and with regard to the same object, of classical concepts such as waves or particles, it involves representations that are contradictory according to a classical account. Here Hermann reverts to Bohr's idea of complementarity. The same physical system, depending on the measurement, is treated now as a wave, now as a particle. For instance, in our example, the light that interacts with the electron is treated firstly as corpuscular (the situation represented by an electron–photon collision) and secondly as undulatory (after the interaction, the light is seen as a flat wave which passes through the lenses of the microscope and converges at a point on the photographic plate).

For every measurement carried out on a quantum object, it is possible, according to Hermann, to reconstitute a posteriori its causal chain. In addition, Hermann proposes a verification procedure, which she calls an 'indirect' procedure for her a posteriori causal reconstitutions. By this procedure she believed she had proved that such causal scenarios were not only possible, but also necessary.[10]

From the above elements taken all together, Hermann draws the following conclusions:

1. Because the causes of any phenomenon resulting from a quantum measurement can always be univocally determined (albeit only a posteriori) and because a single causal scenario connects the phenomena resulting from measurements to theoretical variables also in the case of quantum physics, then the Kantian category of causality remains a necessary condition for quantum physics.
2. Because one is in possession of all the causes that determined any result of measurement, the hypothesis of hidden variables loses all credibility: seeking additional parameters that are supposed to put an end to the statistical character of the quantum description becomes, in principle, pointless.

4.6 Strengths and Weaknesses of Hermann's Interpretation

The main strength of Hermann's interpretation is the essential point she establishes: that in order to exist, both classical physics and quantum physics require that the physicist be able to establish one-to-one connections between:

[10]This proof is discussed at length in the introduction and postface to Hermann (1935/1996).

- The great diversity of phenomena that constitute the results of measurements (impacts on photographic plates, deflections of needles, etc.) and
- The values of a limited number of variables involved in the theory (position, velocity, quantity of movement, etc.).

Judiciously, Hermann places emphasis on the *only one-to-one connection necessary for the existence of quantum physics*. If the physicist were unable univocally to interpret a given phenomenon resulting from measurement as the definite value of a specific theoretical variable, the phenomenon would lose all meaning, all connection with our theories. In this, Hermann emphasises something crucial.

The weak point of Hermann's thesis is that she is not content to assert merely that the one-to-one character of the connection under consideration is a condition for the possibility of physics. She goes much farther in her interpretation of this one-to-one connection by asserting:

- that this connection is *causal* in type,
- that the concept of causality involved is *essentially equivalent* to the Kantian concept of causality, and
- that the a posteriori causal scenario is *necessary*.

Each of these three points is open to question. Take the last one. One can readily imagine other scenarios than those proposed by Hermann for linking a cause to an effect. And since there are no means of deciding among alternative scenarios, this undermines the assumed necessity of Hermann's causal scenarios.

Turning to the two other points, one might stress that Hermann's causal scenarios connect the phenomena resulting from measurement to *only one* of two *conjugate variables*, and that it remains impossible *to bring together, at the same time, two of these causal chains*, one of which would culminate in one of the variables (for instance the position) and the other in the conjugate variable (for instance the momentum). Hence Hermann's interpretation in no way allows the conjugate variables to be simultaneously measured, and therefore in no way allows the reconstitution of the continuous trajectory of an object. It is precisely on the basis of the possibility of gaining access to such continuous trajectories that classical physicists conceived of causality. For them, causal behaviour meant that the values of two conjugate variables of an object at a given time (position *and* momentum equated with *the* cause) univocally determined the subsequent trajectory (position *and* momentum at a later time equated with *the* effect). Here one can readily attack Hermann's conclusions by claiming that the concept of causality involved is very different from (or at least cannot be identified with) the classical, Kantian concept of causality.

4.7 A General Comparison Between Transcendental Philosophy and Quantum Physics

Having shown that the category of causality—the Kantian category that seemed most threatened by the advent of quantum physics—nevertheless continued to constitute a condition of the possibility of quantum physics, Hermann went on to consider, at the most general level, the relationship between quantum physics and transcendental (or critical) philosophy. Her conclusion is that quantum physics and transcendental philosophy converge with regard to essentials, at least if 'transcendental philosophy' is understood to mean, as Hermann understood it, not the Kantian system taken in its precisely literal form, but Kantian philosophy as re-read, clarified and reinterpreted by Fries.

Hermann compares the principal assertions of critical philosophy and quantum mechanics on three points:

(a) For critical philosophy, the Kantian categories 'provide the theoretical schema for interpreting sensation' (Chap. 15, p. 274). Now, in order to interpret the results of measurement, quantum physics must necessarily make use of classical concepts. Thus, in classical physics, as in quantum mechanics, the same fundamental classical concepts mediate the transition from the diverse material of sense data to knowledge of nature, although in the second case their applicability is limited.[11] Quantum physics and classical physics therefore rest, once and for all, on the same conditions of possibility. The a priori forms listed by Kant are not specifically threatened by the advent of quantum physics.

(b) The above item also implies that quantum physics does not specifically call into question the assertion of critical philosophy that the table of Kantian categories is complete, i.e., that Kant's twelve pure concepts are sufficient to order the flux of sensations for knowledge. The advent of quantum theory obliges us neither to add a pure concept to the table, nor to remove or modify one.

(c) If one subscribes to the clarification—carried out by Fries and then by Nelson—of the true implications of the Kantian theses, critical philosophy also shows that the application of the categories to the diversity of phenomena remains limited in the sense that the pure concepts are only ideal models which, 'merely

[11]Of course, to conclude from this that the Kantian categories continue to constitute the conditions of possibility of quantum physics, it would also be necessary to have demonstrated that the entire edifice of classical physics actually does rest on such categories. According to Hermann, such a demonstration is yet to be made, and extends beyond the framework of her essay. Kant apparently believed he had provided such a proof, at least with regard to the physics of his time. In Kant (1786, published in English in Kant 1970), he sets out to demonstrate that the twelve categories listed in the *Critique of Pure Reason* do indeed constitute the necessary conditions of the possibility of physics. When examined, the demonstration appears to posit as the foundation of physics laws that are essentially similar to the fundamental principles of Newtonian physics (conservation of matter, principle of inertia and law of action and reaction). Now, a modern epistemology could hardly have recourse to such a demonstration to prove that the Kantian categories continue to constitute the conditions of the possibility of post-Kantian physics, and this already holds for post-Kantian classical physics (for example, electromagnetism).

as *analogies*', provide 'the guide to the interpretation of sensation' (Chap. 15, p. 274). This means that description extends only to the structures of connections, but properly speaking, never isolates absolute substances, causes or effects. Description therefore remains relative. Nevertheless, the structures of connections represent spatio-temporal relationships that are objective and unequivocally determined.

Quantum mechanics confirms the limits of the application of the fundamental concepts that make knowledge possible: classical concepts, like the categories, are no more than analogies, which should not be understood literally. Classical physics is concerned only with differential equations within which nothing refers (properly speaking) to substances, causes or effects, although such concepts remain indispensable in guiding research and organising the diversity of perception into a knowledge of macroscopic objects. In the same way, quantum physics does not allow the systems it describes to be identified with waves or particles properly speaking, although it cannot do without such concepts in organising the diverse material of perception into a knowledge of atomic phenomena.

In fact, quantum physics goes still further than transcendental philosophy. It confirms that physics has access only to structures of connections, and shows in addition that these structures of connections are in each case relative to the experimental situation through which the experimenter gains knowledge of them. This, according to Hermann, is the major philosophical lesson of the new physics: quantum mechanics, far from contradicting the fundamental principles of transcendental philosophy, radicalises them still further.

Finally, quantum mechanics, to an even greater degree than transcendental philosophy, forces us to abandon the dream of a universal science capable of embracing all aspects of reality within a single description. Indeed, not only is knowledge divided into different types of description (psychology, physics, ethics, etc.) that constitute several perspectives on the world—as Kant's analysis had already shown—but in addition, as is shown by quantum theory, the disintegration of truth into a multitude of perspectives infiltrates the very heart of physics itself:

> [T]he natural-philosophical novelty of quantum mechanics is describable thus: the splitting of truth goes deeper than philosophy and natural science had previously assumed. It penetrates into the physical knowledge of nature itself; instead of merely delimiting its scope against other possibilities for grasping reality [e.g. axiological, ethical, aesthetic, etc.], it separates various equally legitimate representations within the physical description that cannot be unified into a single picture of nature. (Chap. 15, p. 277)

Hermann insists, however, that these convergences between quantum physics and critical philosophy should not mask the independence of the paths followed by each of these two kinds of discourse: the quantum description is 'explicitly grounded in guidance through experience' and is 'independent of philosophical speculations' (Chap. 15, p. 277), while the critical system—the antinomies in particular—'thoroughly depend on mathematical and philosophical considerations' (Chap. 15, p. 277).

But according to Hermann, this observation confers still greater value on transcendental philosophy. For her, the fact that these convergences arise from wholly independent approaches based on distinct principles underlines the credibility of the fundamental principles of transcendental philosophy. Such convergence, she wrote, 'represents no justification' of this philosophy, yet it 'represents an empirical confirmation, which is all the more significant [...]' (Chap. 15, p. 275). In short, the prestige attaching to the exact sciences is to an extent reflected back on critical philosophy. Hermann concludes her fundamental essay of 1935, which was to provide the inspiration for all her later writings on the philosophy of science, with these words:

> Therefore, even if it remains the undeniable merit of physical research to have advanced the understanding of the natural-philosophical foundations of our knowledge of nature by a decisive step, this advance means just as little a break with the prior philosophical development as quantum mechanics represents a break with classical physics. Rather, closer examination reveals that despite all prima facie discrepancies with the apparent conclusions also of the critical philosophy, the crucial discoveries of quantum mechanics fit consistently together with the principles of that philosophy, and through these their significance for the knowledge of nature becomes intelligible. (Chap. 15, p. 278)

4.8 Conclusion

This paper is confined to presenting, as faithfully as possible, the central elements of Hermann's pioneering contributions to the philosophy of quantum physics. Of course, her conceptions, like any philosophical analysis, are open to criticism from various directions. These critiques, which have been barely sketched here, show the fecundity of Hermann's position, and they provide an excellent springboard for subtle analyses of the relationship between causality and quantum physics.

All in all, Hermann provides one of the first contributions to the philosophy of quantum physics—and an original one—as well as interesting analysis of the relationship between modern physics and Kantian-inspired philosophy. One can only regret that her writings have not been, and are not, better known.

References

Bachelard G (1934) Le nouvel esprit scientifique. Presses Universitaires de France, Paris

Bell JS (1966) On the problem of hidden variables in quantum mechanics. Rev Mod Phys 38:447–452

Bell JS (1987) Speakable and unspeakable in quantum mechanics. Cambridge University Press, Cambridge

Cassirer E (1937) Determinismus und Indeterminismus in der modernen Physik. Göteborgs Högskolas Arss-skrift XLII, Gothenburg

Cassirer E (1956) Determinism and indeterminism in modern physics: historical and systematic studies in the problems of causality. Yale University Press, New Haven

Cushing JT (1994) Quantum mechanics: historical contingency and the Copenhagen hegemony. University of Chicago Press, Chicago and London

Heisenberg W (1969) Der Teil und das Ganze. Gespräche im Umkreis der Atomphysik. Piper Verlag, München
Heisenberg W (1971) Physics and beyond: encounters and conversations. Harper & Row, New York. Translation of Heisenberg (1969)
Hermann G (1933) Determinismus und Quantenmechanik, manuscript. Dirac Archives DRAC 3/11
Hermann G (1935a) Besprechung: K. Popper. Logik der Forschung. Physikalische Zeitschrift 36:481–482
Hermann G (1935b) Die naturphilosophischen Grundlagen der Quantenmechanik. Abhandlugen der Fries'schen Schule 6(2):75–152
Hermann G (1935c) Die naturphilosophischen Grundlagen der Quantenmechanik. Die Naturwissenschaften 23(42):718–721
Hermann G (1935/1996) Les fondements philosophiques de la mécanique quantique. Vrin, Paris. Translation of Hermann (1935b) by Schnell A. Edited, with introduction and postface by Soler L
Hermann G (1936) Zum Vortrag Schlicks. Erkenntnis 6:342–343
Hermann G (1937a) Die Bedeutung der Modernen Physik für die Theorie der Erkenntnis. In: Drei mit dem Richard Avenarius-Preis ausgezeichnete Arbeiten von Dr G. Hermann, Dr E. May, Dr Th. Vogel. S. Hirzel, Leipzig, pp 3–44
Hermann G (1937b) Die naturphilosophischen Bedeutung des Übergangs von der klassischen zur modernen Physik. In: Bayer, R (ed) Travaux du IX. Congrès international de philosophie, Paris, 1–6 Août 1937. Hermann, Paris, pp 99–101
Hermann G (1937c) Über die Grundlagen physikalischer Aussagen in den älteren und den modernen Theorien. Abhandlugen der Fries'schen Schule (Neue Folge) 6(3):309–398
Hermann G (1948) Die Kausalität in der Physik. Studium Generale 1(6):375–383
Herrmann K (ed) (2017) Grete Henry-Hermann: Philosophie–Mathematik–Quantenmechanik. Springer, Berlin
Jammer M (1974) The philosophy of quantum mechanics: the interpretations of quantum mechanics in historical perspective. John Wiley, Hoboken, NJ
Kant I (1786) Metaphysische Anfangsgründe der Naturwissenschaft. In: Kants gesammelte Schriften, vol 4. De Gruyter, Berlin, pp 465–565
Kant I (1970) Metaphysical foundations of natural science. Bobbs Merill, London
Kojève A (1932/1990) L'idée du déterminisme dans la physique classique et dans la physique moderne. Le Livre de Poche, Paris, published posthumously in 1990
Popper KR (1934) Logik der Forschung. Springer, Vienna
Schlick M (1974) General theory of knowledge. Springer, New York
Soler L (2004) The contributions of Hermann to the philosophy of physics. In: Shipley P, Mason H (eds) Occasional Working Papers in Ethics and the Critical Philosophy, vol 3, pp 23–31
Soler L (2006) Philosophie de la physique: dialogue à plusieurs voix autour de controverses contemporaines et classiques, Epistémologie et philosophie des sciences, vol 41. L'Harmattan, Paris
Soler L (2009) The convergence of transcendental philosophy and quantum physics: Grete Henry-Hermann's 1935 pioneering proposal. In: Bitbol M, Kerszberg P, Petitot J (eds) Constituting objectivity: transcendental perspectives on modern physics, Western Ontario Series in Philosophy of Science, vol. 79. Kluwer, Dordrecht, pp 329–344
von Weizsäcker CF (1936) Review of Hermann, 'Die naturphilosophischen Grundlagen der Quantenmechanik'. Physikalische Zeitschrift 37(14):527–528
von Weizsäcker CF (1952) The world view of physics. Routledge and Kegan Paul, London
von Weizsäcker CF (1963) Interview with T. S. Kuhn and J. L. Heibron, Archive for the History of Quantum Physics, https://www.aip.org/history-programs/niels-bohr-library/oral-histories/4947

Chapter 5
Carl Friedrich von Weizsäcker's 'Ortsbestimmung eines Elektrons' and its Influence on Grete Hermann

Thomas Filk

5.1 Introduction

Most likely Grete Hermann's contribution to the historical development of quantum theory would be almost forgotten today if Werner Heisenberg had not dedicated a whole chapter to her in his book *Der Teil und das Ganze* (Heisenberg 1969; the following citations are taken from the English translation, *Physics and Beyond*, Heisenberg 1971, pp. 117–118 and 124). Referring to the years 1934/35 when he was in Leipzig, he writes in the opening sentences of this chapter:

> We were offered a special occasion for philosophical discussions [...] when the young philosopher Grete Hermann came to Leipzig. [...] Grete Hermann believed she could prove that the causal law—in the form Kant had given it—was unshakable. Now the new quantum mechanics seemed to be challenging the Kantian conception, and she had accordingly decided to fight the matter out with us.

The mere fact that Heisenberg devotes a chapter of his book to the philosophical discussions with Hermann can be taken as evidence for a positive and lasting impression; however, referring to her twice as 'junge Philosophin' (the 'young philosopher') does not really give credit to the fact that Hermann was about nine months older than Heisenberg and had a PhD in mathematics. The chapter ends with the remarks:

> 'Science progresses not only because it helps to explain newly discovered facts, but also because it teaches us over and over again what the word "understanding" may mean.'

I greatly acknowledge the kind hospitality as well as many interesting discussions during the Grete Hermann meeting in Aberdeen, 5–6 May 2012.

T. Filk (✉)
Albert-Ludwigs-Universität Freiburg, Freiburg im Breisgau, and Parmenides Center, Munich, Germany
e-mail: thomas.filk@t-online.de

E. Crull and G. Bacciagaluppi (eds.), *Grete Hermann - Between Physics and Philosophy*, Studies in History and Philosophy of Science 42,
DOI 10.1007/978-94-024-0970-3_5

> This reply, based partly on Bohr's teachings, seemed to satisfy Grete Hermann to some extent,[1] and we had the feeling that we had all learned a good deal about the relationship between Kant's philosophy and modern science.

Towards the end of her time in Leipzig, Hermann wrote an article entitled 'Die naturphilosophischen Grundlagen der Quantenmechanik'. A short version of this article leaving out all the interesting technical details appeared in *Die Naturwissenschaften* (Hermann 1935a); the full version was published in *Abhandlungen der Fries'schen Schule* (Hermann 1935b), which was 'hardly the place where the devotees of von Neumann's defective proof were likely to ever discover it' (Gilder 2008, p. 158). In this quotation, Louisa Gilder refers to a proof of John von Neumann (von Neumann 1932), according to which the indeterminism of quantum theory cannot be explained by an extension of quantum theory by additional, hitherto unobserved variables. In her article Hermann points out that von Neumann's proof is based on an assumption that is physically unjustified and that, according to her assessment, is circular by putting what von Neumann wanted to prove already into the assumptions. (For more details see Chap. 7 by Michiel Seevinck in this volume.)

Whether Grete Hermann was really convinced of the 'Copenhagen Credo' that served as a philosophical interpretation of the quantum formalism remains open. The first eight sections of Hermann's article leave the reader with the impression that she is arguing in favour of additional variables not included in the formalism of quantum theory that could save the principle of causality. After a very detailed and rigorous analysis (of which the refutation of von Neumann's assumptions is a part) she comes to the conclusion that additional variables are not ruled out according to current knowledge or the status of quantum theory. Such additional variables would indeed have been an explanation of the observed indeterminism in full agreement with Kantian ideas of causality. Yet, at the end of her Section 8, one senses an abrupt change in style and argumentation which gives the article a twist in a different direction. She argues that quantum mechanics does not need a completion by such variables because '*one already knows these causes*' (Chap. 15, p. 254), and essentially the same statement is repeated several times, often emphasised in italics. However, her argumentation in favour of a causally complete quantum formalism is by far not as convincing as her previous refutation of all arguments against hidden variables. Heisenberg's very cautious expression that Grete Hermann was 'wie uns schien, einigermaßen' satisfied adds to the impression that this 'Bohrian' style of argumentation may not really have been her full conviction. Perhaps historians of science will discover the reasons for her surprising change of mind [cf. also this volume, Chap. 8 (*eds.*)].

In the present article, my primary concern is with the influence of an article of Weizsäcker on Grete Hermann's argumentation. Already in 1931, at the age of nineteen and having been in the group of Werner Heisenberg for only about a year, Weizsäcker published the results of a theoretical investigation of the Heisenberg microscope (von Weizsäcker 1931). Heisenberg had suggested that he should

[1] The expression Heisenberg uses in the original German edition—'[sie war], wie uns schien, einigermaßen zufrieden'—seems to express even more uncertainty about this.

mathematically analyse the measurement of the location of an electron by the scattering of a single photon that then passes through an optical lens and is finally registered on a screen in the image plane of that lens. The question was whether a quantum mechanical (even quantum field theoretical) analysis of this situation would lead to the same results as the classical treatment in the context of wave optics or even geometrical optics. In Sect. 5.2, I will review the main arguments of this article.

Grete Hermann devotes a whole section of her 1935 article to the analysis by Weizsäcker and uses his results for her argumentation according to which quantum theory in its present form is already causally complete. In Sect. 5.3, I will analyse her arguments and the way she interprets Weizsäcker's article. For me, her reasoning does not sound convincing—maybe except for one argument, which is not even explicitly mentioned in her text—and I doubt whether Hermann was herself convinced by it. Finally, Sect. 5.4 addresses the question to which extent the ideas of Einstein, Podolsky and Rosen were already contained in the articles of Weizsäcker and Hermann. I will finish with a few concluding and summarising remarks.

5.2 Weizsäcker's Analysis of the Heisenberg Microscope

Weizsäcker had met Werner Heisenberg in Copenhagen (where his father was a diplomat) when he was fourteen, and this meeting greatly influenced his decision to study physics. Around 1930, at the age of eighteen, he joined Heisenberg's group, and in April 1931 he submitted the results of an investigation of the Heisenberg microscope to *Zeitschrift für Physik* entitled 'Ortsbestimmung eines Elektrons durch ein Mikroskop' (von Weizsäcker 1931; 'Determination of the position of an electron by a microscope'). He opens the article by writing: 'In the following I will discuss a particular thought experiment for the determination of the location of an electron, namely the imaging of an electron that is illuminated with light of a sufficiently short wavelength, by a microscope.'

Figure 5.1 shows the basic set-up of the Heisenberg microscope. An electron e is allowed to move freely within a plane L. The electron is hit by a single photon for which the wave vector $\mathbf{k}$ is known. We now assume that the photon is scattered by the electron and its quantum state is described by a spherical wave emanating from the location of the electron at the moment of scattering. The elaborate calculations of Weizsäcker, based on the quantum field-theoretic formalism developed by Heisenberg and Pauli, show that one can now essentially use classical wave optics to deduce that the wave function of the single photon propagates through the optical lens in just the same way as a classical spherical electromagnetic wave. He discusses the limitations of such an approach, but the essential results remain the same. In particular, this wave is focused in a small region in the image plane behind the lens, and if we put a photographic plate into this plane it will register the photon at point P.

The location of P, even if it is produced by just a single photon, allows us to deduce the location of the centre of the spherical wave function and, thereby, the location of the electron at the moment of scattering.

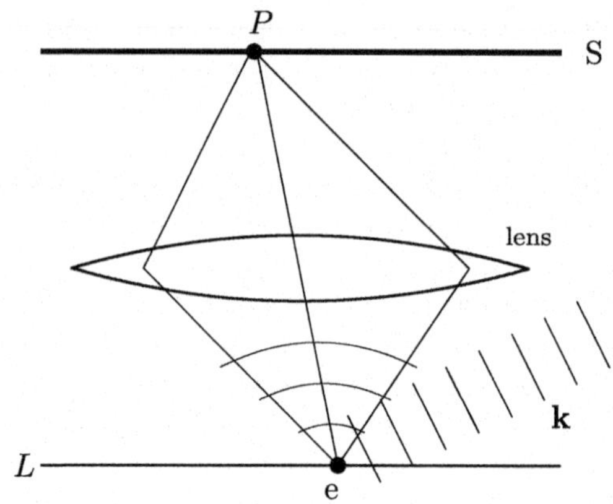

Fig. 5.1 An electron e is located in a plane L and allowed to move freely within this plane. A single incoming photon, described by a planar wave vector **k**, is scattered by the electron in the form of a spherical wave. The wave is guided through an optical lens according to the classical laws of optics, and the photon hits the screen S at point P, which is located in the image plane of the electron

On the other hand, we can also put the photographic plate into the focal plane of the optical lens (Fig. 5.2). From the location of point P' where the photon hits this plate we now obtain the information about the direction from which the photon entered the lens.

The process can now be described as follows: the photon is scattered by the electron as a planar wave with wave vector **p**. This implies that there is no particular scattering centre but the state of the electron has to be thought of as distributed over the 'whole plane' and the electron is described by a momentum eigenstate. The information about the scattering centre of the photon and electron is now lost, instead the difference between the initial wave vector **k** and the wave vector **p** of the

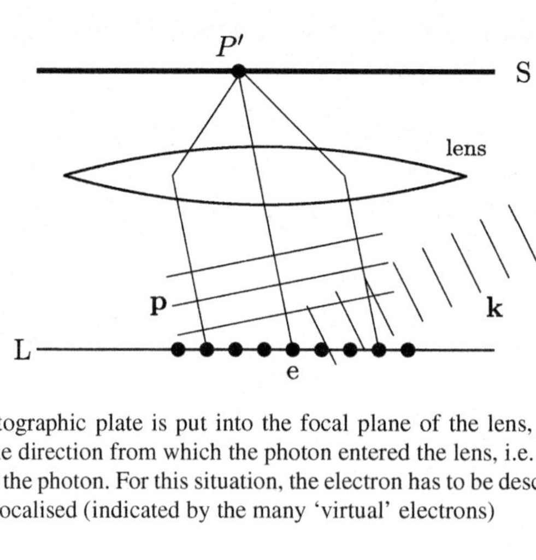

Fig. 5.2 If the photographic plate is put into the focal plane of the lens, point P' contains the information about the direction from which the photon entered the lens, i.e. about the wave vector **p** (or momentum) of the photon. For this situation, the electron has to be described by a momentum eigenstate that is delocalised (indicated by the many 'virtual' electrons)

scattered wave is equal to the momentum transfer from the photon to the electron. If we know the initial momentum of the electron, we can infer from P' the momentum of the electron after the scattering process.

Expressed using today's terminology, after the scattering the electron and the photon are to be described by an entangled state. We can expand this entangled state in two different bases: (1) a momentum basis for the electron as well as for the photon (which, after the scattering, is described by a planar wave with a fixed wave vector), and (2) a position basis where the electron at the moment of scattering is located at a point x and the scattered photon propagates away from this centre x in the form of a spherical wave. Very formally this means for the state of the total system after the scattering took place (and before the photon is absorbed by the photographic plate):

$$|\Psi_f\rangle \simeq \int \mathrm{d}p \; a(p) \; |p\rangle_{\mathrm{e}}|k-p\rangle_\gamma \simeq \int \mathrm{d}x \; b(x)|x\rangle_{\mathrm{e}}|\mathrm{s.w.}(x)\rangle_\gamma \tag{5.1}$$

where $|\mathrm{s.w.}(x)\rangle$ refers to the state of a spherical wave with centre x, and $a(p)$ and $b(x)$ are certain expansion coefficients. The integration over x and p extends over the possible values these quantities can assume under the restriction that the electron is bounded to the plane L.[2] It is important to notice that both expansions describe the same total state of both particles, but with respect to different bases.

Depending on whether the photon is measured in the image plane or the focal plane, one either measures the centre x of the spherical wave of the photon or the momentum $\mathbf{p}$ of the planar wave of the photon. Due to the entanglement, this indirectly implies a measurement of the location x of the electron or a measurement of its momentum $\mathbf{k} - \mathbf{p}$ (actually, the momentum transfer from the photon to the electron). The situation is very similar to the one that is used by Einstein, Podolsky and Rosen four years later in their famous EPR paper (Einstein et al. 1935), and I will come back to this point in Sect. 5.4.

5.3 Weizsäcker's Impact on Grete Hermann's Argumentation

The tenth section in Grete Hermann's article is devoted to an analysis of Weizsäcker's article.[3]

Hermann first describes the set-up of the Heisenberg microscope and the two cases described above: (1) putting the photographic plate in the image plane yields

[2] The entanglement may not be maximal due to the size of the microscope and other influences; however, this does not change the basic structure of the argument.

[3] Interestingly enough, now it is she who simply writes about 'a student of Heisenberg', mentioning his name only in the footnote with a reference to the publication. The mutual high scientific respect between Weizsäcker and Grete Hermann is well known; however, taking into account the political situation in 1934 Germany, it remains an open question to what degree this esteem also extended into the private domain between a conservative aristocrat from a diplomatic family and a member of the Internationaler Sozialistischer Kampfbund (see discussion of this organisation in Chap. 11).

information about the location of the electron, and (2) putting the plate in the focal plane yields information about the momentum transfer from photon to electron. She also mentions a third case:

> Finally, if one sets up no photographic plate at all, but allows the light quantum to pursue its path without detecting it, then one obtains yet a third—though not in the same way intuitive—description of the state after the collision. In this case, the physical system composed of the light quantum and the electron is assigned a wave function that describes a linear combination: each of its terms is the product of one wave function describing the electron and one describing the light quantum. Through this linear combination the light quantum and the electron are thus not described each by itself, but only in their relation to each other. Each state of the one is associated with one of the other. (Chap. 15, p. 258)

This is an absolutely clear and precise description of what later became known as 'entanglement' (this term was coined by Erwin Schrödinger in Schrödinger (1935)). Hermann continues by arguing that

> [...] depending on how one procures one's knowledge of the observed system, or, as we can say for this, depending on the relevant observational context, one can obtain different wave functions for the same system and for the same instant—namely for the electron at the time immediately after the collision with the light quantum. Thus the quantum-mechanical characterisation, unlike the classical one, does not pertain to the physical system still somehow 'in itself', and this means here: independently of which observation one uses to procure one's knowledge of it. (Chap. 15, pp. 258–259)

In my view it is not really clear to which instant exactly Grete Hermann refers: the entangled state of photon and electron immediately after the collision, or the state of the electron immediately after the registration of the photon in one of the planes—the focal plane or the image plane. At least from the modern perspective, the total (entangled) state of the electron and the photon immediately after the collision is independent of the observational context; however, it can be expanded in terms of different bases as in Eq. (5.1). The electron alone does not have a definite state immediately after the collision, but only 'relative states' with respect to the basis one chooses for the description of the photon. This state is conveniently chosen with reference to the type of measurement one intends to make (the 'observational context'), and therefore the 'relative states' of the electron also depend on this context.

Hermann now proceeds by arguing that in retrospect, after the observer knows the outcome of the measurement of the photon one can, with 'sufficient reason', reconstruct the events which led to this outcome. If, e.g., the photon has been measured in the focal plane, one can reconstruct the momentum of the electron. Here, she gives up predictability as a necessary condition for causality—which she defends in the opening sections of her article as 'indispensable'—and replaces it essentially by Leibniz's 'principle of sufficient reason' (Leibniz 1714/1890). This principle of sufficient reason only requires that *after* something has happened it should be possible to figure out the sufficient reasons why it was so and not otherwise, and thus replaces the requirement of predictability as a characterisation of causality. In Section 9 (before she discusses Weizsäcker's example) she explains:

> [...] if one makes such a measurement [...] then one obtains for this new state not only a quantum mechanical description that assigns this quantity a sharp value, but moreover in the context of this mode of description one can also find causes for precisely this unpredictable value having had to result. (Chap. 15, p. 256)

And after she described Weizsäcker's example, she emphasises:

> At the same time, the example shows that the quantum mechanical formalism cuts off the question about new features to be discovered, upon which the outcomes of arbitrary measurements depend, by itself providing sufficient reasons for these outcomes, and that it nevertheless affords no clues for the prediction of all measurement outcomes. So, in the case under consideration, it is in principle impossible to predict at which point the light quantum will darken the photographic plate set up, say, in the focal plane of the system of lenses. Nevertheless, the inference from the observation of this point to the momentum imparted to the electron during the collision, allows one to identify precisely in this exchange of momentum the cause for the light quantum being found exactly at this point on the plate. (Chap. 15, p. 259)

Regarding this aspect of Hermann's thesis, Max Jammer writes:

> It seems, however, that Hermann's claim of retrodictive causality is unwarranted. In the author's opinion she did not prove, as she claimed, that a retroactive conceptual reconstruction of the measuring process provides a *full* explanation of the particular result obtained. Although such a reconstruction may prove the *possibility* of the result obtained, it does not prove its *necessity*. Thus in the Weizsäcker–Heisenberg experiment her reconstruction, starting from the observation, accounts for the fact that the photon *can* impinge on the photographic plate where it impinges, but not that it *must* impinge there. (Jammer 1974, p. 209)

In my opinion this assessment is absolutely correct, but for me it is hard to believe that Grete Hermann, who proved to be such a sharp analyst in her rejection of the arguments against hidden variables, had overlooked this obvious objection. So why does she insist that quantum mechanics already gives a complete description of the causal chain of events leading to a particular outcome? Her arguments, even though repeated several times, are not really convincing and often circular. The following is just an attempt to figure out what she might have had in mind and to express this in more contemporary language.

Grete Hermann uses the expression 'observational context' ('Beobachtungszusammenhang'), which in a very general way refers to the experimental set-up which in turn determines the physical quantity one wants to measure. Given a state of a quantum system $|\psi\rangle$ and a measuring device that has a pointer basis $\{|\varphi_i\rangle\}$ with an initial state $|\varphi_0\rangle$, we may express the initial state of the total system before the interaction takes place as

$$|\Psi_{\text{init}}\rangle = |\psi\rangle \otimes |\varphi_0\rangle \,, \tag{5.2}$$

and after the interaction between both systems (but before a reading of the measuring device) by

$$|\Psi_1\rangle = \sum_i a_i |s_i\rangle |\varphi_i\rangle \,, \tag{5.3}$$

where the state $|\psi\rangle$ has been expanded according to the eigenstates of the observable that is represented by the measuring device. Hermann now acknowledges that a different measuring device (representing a different observable) leads to a different expansion:

$$|\Psi_2\rangle = \sum_j b_j |s'_j\rangle|\varphi'_i\rangle . \tag{5.4}$$

Note that $|\Psi_1\rangle$ and $|\Psi_2\rangle$ need not be the same states because the measuring devices are different. However, if one considers the photon the 'measuring device' for the electron (and Hermann explicitly emphasises that almost any system can act as a measuring device in certain situations), there is no 'pointer basis' distinguished and the two states are the same.

After the interaction between the quantum system and the measuring device has taken place, there is no 'state of the quantum system' by itself, but its states can only be defined *relative* to a state of the measuring device, and different measuring devices lead to different 'relative states' for the system. This is the 'observational context'. If the outcome of the measurement is known, we can deduce the state of the quantum system that is correlated to this outcome:

- for measuring device (1) and result k we deduce the state of the quantum system to be $|s_k\rangle$,
- for measuring device (2) and result l we deduce the state of the quantum system to be $|s'_l\rangle$.

Unfortunately, Hermann is not very explicit about the next point. Some passages like the ones cited above can be interpreted in the sense that in her opinion, the characteristic features described by $|s_k\rangle$ or $|s'_l\rangle$ were already present *before* the interaction between the measuring device and the system took place, but that due to the restrictions of the quantum formalism this state can never be known to the observer in advance. Under this assumption, everything that happened during the measuring process followed a deterministic causal chain.

On the one hand, she is well aware that the uncertainty relations do not express a lack of knowledge on the side of the physicist (she explicitly mentions interference experiments, which cannot be explained by assuming that we simply do not *know* through which slit a particle passes). What she does not seem to consider is that these states (depending on the measuring device) might have been *created* during the measuring process, and that this 'creation' (what is today known as the collapse process) is not causal.

On the other hand, she must have been aware of such ideas: she criticises Schrödinger for the opinion expressed in his 1934 paper that a classical form of causality may be maintained if one gives up outdated classical concepts like 'location' or 'spatial geometry' (Schrödinger 1934). In this article, Schrödinger satirically attacks the notion of 'quantum measurement' and proposes to replace it by 'Prokrustie', referring to the giant in Greek mythology who forces guests into his bed by stretching or compressing them. In this context, Schrödinger remarks: 'I know that the experimenter cannot choose the value [of the result of a measurement]; but nevertheless he forces his victim into *one* of his beds while it fits into none' (ibid., p. 519).[4]

A possible reason for Hermann's opinion that quantum theory is already causally complete could have been related to an aspect of the Heisenberg microscope (and, more generally, the type of entanglement involved in this situation) which, however, is never explicitly emphasised, neither in Weizsäcker's article nor in Hermann's article: during the interaction between the electron and the photon, the complete information about the state of the electron is transferred to the state of the photon. It carries the full information about both the location of the electron as well as the momentum of the electron. The decision of the experimentalist to put the photographic plate into the image or the focal plane allows him to extract either one of these two complementary bits of information about the electron from the photon. (As the experimentalist can, in principle, make this decision after the interaction between electron and photon has taken place and the electron is gone, this is a particular form of 'delayed choice experiment'; cf. Wheeler 1978).

This complete information transfer from electron to photon might have contributed to Grete Hermann's opinion that the quantum formalism does not need an extension by hidden variables because it is already causally complete.

5.4 EPR Anticipations?

In a letter from 1967, Jammer points out to Weizsäcker that the situation of the Heisenberg microscope is analogous to the one described by EPR. In his answer on 13 November 1967, Weizsäcker writes:

> The problem which led to this paper was certainly closely related to that raised by Einstein, Rosen and Podolsky. Except that Heisenberg, who suggested it to me, and I as well regarded this state of affairs not as a paradox, as conceived by the three authors [...] (Jammer 1974, p. 179)

After quoting the letter from Weizsäcker, Jammer comments as follows:

> It may well be that Heisenberg and Weizsäcker were fully aware of the situation without regarding it as a problem. But as happens so often in the history of science, a slight critical turn may open a new vista with far-reaching consequences. As the biochemist Albert Szent-Györgyi once said: 'Research is to see what everybody has seen and to think what nobody has thought'. (ibid., p. 180)

As I will argue in the following, there was indeed a 'slight critical turn' distinguishing the argumentation of EPR in their 1935 article from similar situations that

[4] As a side remark, in the same article Schrödinger mentions already the possibility of a 'measurement without interaction' by remarking that the non-detection of a particle by a detector which surrounds a decaying atom completely except for a small hole gives very precise information about the trajectory of this particle. The same situation was later emphasised in a famous article by Renninger (1960).

had been discussed before. (This new type of strategy on the side of Einstein is also emphasised in Jammer (1974).) Already many times before, Einstein's criticisms of quantum theory employed examples that implicitly relied on entanglement. A famous example is the light box thought experiment from 1930 (cf. Bohr 1949, and Jammer 1974). However, until about 1930 his strategy was to prove that quantum mechanics was wrong. In all these cases, Einstein tried to construct thought experiments that seemed to violate the uncertainty relations, and in all cases Bohr's reply was that this violation of the uncertainty relations cannot be experimentally verified.

Applied to the case of the Heisenberg microscope, Einstein's old type of argumentation might have been that one can use the photon to measure the *location* of the electron while simultaneously one can measure directly the *momentum* of the electron. This would mean that both the position as well as the momentum of the electron are known, which would violate the uncertainty relations. Bohr's answer to this hypothetical situation may have been that Einstein could not test whether his knowledge about the position of the electron was indeed correct. Measuring the position of the electron *after* the momentum measurement will, in general, yield a different result compared to the one obtained from the measurement of the first photon. Measuring this position *before* the momentum measurement (in which case it will agree with the photon measurement) may destroy any information about the momentum such that an additional momentum measurement is irrelevant.

The new type of attack against quantum mechanics that EPR use in their 1935 article does not refer to simultaneous measurements of complementary variables or untestable statements. Applied to the microscope, the new type of argument is: we can freely choose to measure the photon in the image plane or in the focal plane. Now we can predict ('with probability equal to one' and without 'disturbing the electron in any way'—these expressions appear in Einstein et al. (1935)) the result of the corresponding measurement (location or momentum) performed on the electron. Their conclusion is that both values must be an 'element of reality', which they are not in the formalism of quantum theory: this is the 'slight turn' of view. I should remark that already from 1931 onwards, Einstein used this new type of argumentation in several articles in order to point out the incompleteness of quantum theory (see, e.g., Jammer 1974, Chap. 6, especially p. 170 ff.). However, the EPR paper seems to be the one that provoked the strongest reactions. In a letter to Heisenberg on 15 June 1935 (Pauli 1985, pp. 402–405), Pauli most clearly contrasts the old and the new type of Einstein's strategy:

> He now has understood that much that two quantities which correspond to non-commuting operators cannot be measured at the same time and that one cannot assign numerical values to them simultaneously.

(This refers to the old type of attack of Einstein against quantum theory.)

> Now comes the 'deep feeling' and he proceeds: 'Because measurements of system 2 cannot disturb particle 1, there must be something called "the physical reality", which is the state of particle 1 in itself, independent of which measurements have been performed at system 2.'

(This sharply characterises the new strategy.)

The old type of argument—seeking possibilities to violate the uncertainty relations—is indeed addressed in Weizsäcker's article. He writes that in order to control the position measurement performed with the photon one has to use the scattering of a second photon, and then he proceeds:

> Now it is obvious that in this case a later measurement of the momentum [of the electron] does not allow any conclusion about the direction into which the two light quanta have been scattered, because the momentum of the electron between the first and the second scattering process has its own undetermined value; the same holds obviously for any other control of the position measurement. If, however, one performs a momentum measurement of the electron *before* one has checked the position measurement, the question whether the microscope has determined the position of the electron correctly loses its meaning due to the loss of knowledge about the position induced by the momentum measurement.

This is exactly the Bohr-type rejection of Einstein's old strategy mentioned above.

Referring to the resemblance between his 1931 article and the EPR article, Weizsäcker writes (von Weizsäcker 2002): 'I do no longer know whether I became aware of it in 1935 on the occasion of the work of Einstein, Podolsky and Rosen [...]', but it seems very likely that he did not realise the relation to EPR, neither in 1935 nor later, until it was pointed out by Jammer. The philosopher Walter Schindler, the long-time assistant (and, in questions of Kantian philosophy, often a kind of personal consultant) of Weizsäcker, remarks that 'during the 1960s we often discussed the EPR article, but he [Weizsäcker] never mentioned his work from 1931. In fact, I didn't know about this article until recently' (Schindler 2012). Interestingly enough, Weizsäcker had given Schindler a copy of Grete Hermann's article (the long version) during the 1960s.

So, how close is Grete Hermann to an EPR-type argument? Despite her precise characterisation of entangled states, she does not seem to realise the point that EPR later make, even though she comes quite close. She explains that from the location where the photon hits the photographic plate (put into the focal plane) one can retrospectively use the causal relationships to determine the momentum transfer from the photon to the electron. She writes (Chap. 15, p. 259): '[...] the posited causal claim can, indeed, be used indirectly for the prediction of an observation result, and be checked by carrying out this observation: it is sufficient for reaching a conclusion about the change of momentum of the electron, which for its part can be checked.' Here she correctly argues that the measurement of the momentum of the photon leads to a prediction for the momentum of the electron, and she seems to assume (like EPR) that the momentum of the electron has a definite value after the interaction (and the momentum transfer) between electron and photon has taken place.

It is most likely that she was also aware of the same situation for position. We can measure the location of the photon on the photographic plate, now placed into the image plane, and from this deduce the location of the electron (at the moment of scattering). However, she does not conclude from this situation that both location and momentum of the electron must be 'elements of reality'. It seems that for her only the observational context (where we put the photographic plate) renders, in retrospect, one of these properties a fact.

5.5 Conclusion

Some of the arguments that Hermann invokes in order to show that the existing quantum formalism does not need an extension by hidden variables because it is already causally complete, still remain mysterious or unclear. Her style of argumentation in this context is in complete contrast to the more mathematical and accurate refutation of all objections against the possibility of hidden variables. It is difficult to believe that the sudden change in style and argumentation at the end of Section 8 of her article has no external reason. Whether this reason may be found in a certain social 'pressure' from her discussion partners in physics, who presumably tried to avoid even the mere thought of hidden variables, will presumably never be completely uncovered. In fact, many of them had their own no-hidden-variables arguments (see, e.g., Bacciagaluppi and Crull 2009), and for this reason the refutation of von Neumann's assumptions may not have left such a deep impression on them.

Concerning her arguments in favour of the causal completeness of the existing quantum formalism, the main point that remains unclear from her text is whether she considers the electron and the photon immediately after their interaction as entangled or not. If yes—and her clear characterisation of 'entanglement' in the case where the photon is not registered supports this possibility—then the total state of both particles (Eq. (5.1)) does not depend on the observational context; only its expansion with respect to a particular basis does. In this case, the instant that cannot be causally explained is the reduction of this total entangled quantum state to a separable state when the photon hits the photographic plate (wherever it is placed behind the lens). Today, this reduction of the quantum state—or wave collapse—is considered to be the most critical 'postulate' in the formalism of quantum theory, but in 1935 the collapse problem did not seem to be the main issue. Only after the photon has been registered is it possible to assign a definite value to the momentum or the location of the electron, depending on the location of the plate.

The other alternative—that already from the moment of interaction between electron and photon, the state of the two particles factorises—leads to a problem: the factorised state does indeed depend on the observational context, i.e., on which property of the photon is (later) measured. It is very likely that Hermann was not thinking of a 'delayed choice' scenario, because the actual time scales involved in such an experiment make delayed choice impossible, in particular if one takes into account the experimental capacities of that time. This means that the observational context is already defined when the scattering between electron and photon takes place. This could also be the reason why Hermann misses the EPR argument: as the observational context is given at the moment of scattering, only one of the quantities—position or momentum—is relevant for causal completeness and becomes an element of reality.

On the other hand, in Section 9 of her article she explicitly remarks that any process in physics may become part of a measurement process, and that the momentum measurement of a particle may, e.g. after a scattering process, also be considered as a measurement of the momentum of the scattered particle. Taking into account her discussion of Weizsäcker's article and the Heisenberg microscope, the same should also hold for position measurements. In this latter case, however, I cannot see how she is able to avoid the conclusion of EPR that both momentum and position are 'elements of reality'. Yet, this is in contrast to earlier arguments of Hermann according to which the classical variables—momentum and location—cannot be 'hidden variables', as this would contradict interference experiments.

It is difficult to believe that Grete Hermann did not see these question marks behind her arguments, which brings us back to the question of whether she was really convinced by her own conclusions, or only 'wie uns schien, einigermaßen'—satisfied by them.

References

Bacciagaluppi G, Crull E (2009) Heisenberg (and Schrödinger, and Pauli) on hidden variables. Stud Hist Philos Mod Phys 40(4):374–382

Bohr N (1949) Discussion with Einstein on epistemological problems in atomic physics. In: Schilpp PA (ed) Albert Einstein: philosopher-scientist. The Library of Living Philosophers, Evanston, IL, pp 199–241

Einstein A, Podolsky B, Rosen N (1935) Can quantum-mechanical description of physical reality be considered complete? Phys Rev 47:777–780

Gilder L (2008) The age of entanglement: when quantum physics was reborn. Vintage Books, New York

Heisenberg W (1969) Der Teil und das Ganze. Gespräche im Umkreis der Atomphysik. Piper Verlag, München

Heisenberg W (1971) Physics and beyond: encounters and conversations. Harper and Row, New York. Translation of Heisenberg (1969)

Hermann G (1935a) Die naturphilosophischen Grundlagen der Quantenmechanik. Die Naturwissenschaften 23(42):718–721

Hermann G (1935b) Die naturphilosophischen Grundlagen der Quantenmechanik. Abhandlugen der Fries'schen Schule 6(2):75–152

Jammer M (1974) The philosophy of quantum mechanics: the interpretations of quantum mechanics in historical perspective. Wiley, Hoboken, NJ

Leibniz G (1714/1890) The principles of nature and of grace. In: The philosophical works of Leibniz, translated from the Latin and French by G.M. Duncan. Tuttle, Morehouse and Taylor, New Haven, pp 209–217

von Neumann J (1932) Mathematische Grundlagen der Quantenmechanik. Springer, Berlin

Pauli W (1985) Scientific correspondence with Bohr, Einstein, Heisenberg a.o., vol II, 1930–1939. Springer, Berlin. Edited by Meyenn K von

Renninger M (1960) Messung ohne Störung des Messobjekts. Zeitschrift für Physik 158:417–421

Schindler W (2012) Private communication with author

Schrödinger E (1934) Über die Unanwendbarkeit der Geometrie im Kleinen. Die Naturwissenschaften 22(31):518–521

Schrödinger E (1935) Die gegenwärtige Situation in der Quantenmechanik, I–III. Die Naturwissenschaften 23:808–812; 823–828; 844–849

von Weizsäcker CF (1931) Ortsbestimmung eines Elektrons durch ein Mikroskop. Zeitschrift für Physik 70(1–2):114–130

von Weizsäcker CF (2002) Der Aufbau der Physik, 4th edn. dtv, Munich

Wheeler J (1978) The 'past' and the 'delayed-choice' double-slit experiment. In: Marlow R (ed) Mathematical Foundations of Quantum Theory. Academic Press, New York, pp 9–48

Chapter 6
'In the No-Man's-Land Between Physics and Logic': On the Dialectical Role of the Microscope Experiment

Mélanie Frappier

6.1 Introduction

In the 1959 appendix on thought experiments to the English edition of *The Logic of Scientific Discovery*, Popper returned to the criticism of Heisenberg's interpretation of quantum mechanics he had published twenty-five years earlier in the original 1934 German edition of the book.[1] Popper was especially harsh towards Heisenberg's analysis of the microscope thought experiment where one attempts to determine the state of an electron at a specific moment by observing the particle's position through a microscope, illuminating it with a single photon. From this imaginary set-up, Popper argued, Heisenberg had wrongly concluded to the existence of insuperable limits—the indeterminacy relations—that precluded the simultaneous determination of the position and momentum of quantum particles, thus invalidating the principle of causality in the quantum domain (Popper 1959, p. 451). Popper was adamant: not only was Heisenberg's argument flawed, but even as a mere illustration of the indeterminacy relations, the microscope experiment was simply 'a bad illustration [...] quite inadequate as a basis for interpreting these formulae—let alone the whole quantum theory' (ibid., p. 452).

[1]Popper notes the differences between the two editions in the preface, appendices and, most importantly, footnotes of the English edition (Popper 1959).

I want to thank Elise Crull and Guido Bacciagaluppi for their generous and unrelenting support. I am indebted to them and to all the other participants of the 'Grete Hermann: between physics and philosophy' workshop (Aberdeen 5–6 May, 2012) for stimulating discussions. I also wish to thank William Demopoulos, Gordon McOuat, and Robert DiSalle for their comments and constant encouragements. Many thanks to Carol A. Leadenham, from the Hoover Institution, for her invaluable assistance. Part of this research was made possible by the Canadian Social Sciences and Humanities Research Council, whose financial support I gratefully acknowledge.

M. Frappier (✉)
University of King's College, Halifax, Canada
e-mail: melanie.frappier@ukings.ca

E. Crull and G. Bacciagaluppi (eds.), *Grete Hermann - Between Physics and Philosophy*, Studies in History and Philosophy of Science 42,
DOI 10.1007/978-94-024-0970-3_6

Despite Popper's uncompromising critique of 'Heisenberg's programme,' the microscope experiment had been rapidly adopted by the physics community of the 1920s, and it apparently remained at the core of the interpretation of the indeterminacy relations thirty years later. 'Few imaginary experiments,' Popper lamented, 'have exerted a greater influence on thought about physics than this one' (Popper 1959, p. 451). The only way to explain the experiment's incredible success, Popper reasoned, was to assume that physicists had simply failed to take it 'seriously' (ibid., p. 452).

Popper's remark is at first disconcerting, given the barrage of criticisms Heisenberg's paper had faced upon publication. Both Bohr and Dirac had been quick to point out that the indeterminacy relations did not arise from the photon's disturbance of the electron, but from the microscope's aperture (cf. Cassidy 1991 and Frappier 2004). Rapidly, Kennard (1927), Robertson (1929), and Schrödinger (1930) had improved the mathematical definition of 'indeterminacy' and had illuminated its relation to the electron's wave function.

Popper was, it is true, generally wary of the apologetic use of such imaginary experiments. Although he readily admitted that thought experiments could reliably be given 'critical' and 'heuristic' roles in science, he remained circumspect of the apologetic use of thought experiments. This new argumentative role, Popper argued, had only been given to thought experiments at the turn of the twentieth century in the relativistic analysis of the behaviour of clocks and rods before being co-opted—with much less success—by quantum physicists. 'In this development,' Popper noted mournfully, 'an important part was played by Heisenberg's imaginary microscope' (Popper 1959, p. 443).

Popper remained silent as to exactly how the microscope experiment had gained—or rather failed to gain—any evidential significance in the debate surrounding the validity and interpretation on the indeterminacy relations. His comments remain perplexing until we remember that Popper's understanding of the development of scientific knowledge in terms of 'conjectures' and 'refutations' grew out of his reflections upon Leonard Nelson's Socratic method (Milkov 2012, p. 150), an approach in which thought experiments find a positive, but delicate role (cf. Sect. 6.2).

Popper, I argue here, rightly understood Heisenberg's 1927 microscope paper as the very kind of multi-faceted, critical and dialectical engagement that Nelson associated with rational inquiry. As we shall see in Sect. 6.4, under such a reading, the microscope thought experiment is neither the mere rhetorical argument many have taken it to be (see Beller 1999), nor an inductive argument in favour of the indeterminacy relations (a position I partly defended in the past in Frappier 2004, following Norton 1996), nor a device giving us a priori, intuitive access to the fundamental principles of nature (as Brown 1991 might suggest). It instead appears as the core of a broad dialectic argument exploring the source of the dispute between those who, following Schrödinger, argued that quantum systems were waves following continuous paths and those agreeing with Heisenberg that quantum objects must be particles jumping instantaneously between distant locations in space. Using parallel anschaulich and theoretical analyses of the microscope experiment, Heisenberg concluded—in a move typical of Nelson's theory of philosophical argumentation (Leal 2016, p. 3)—that neither position was correct: both rested on the unsupported assumption that a quantum object has a definite position and momentum at all times. Abandoning the principle, Heisenberg argued, revealed an unexpected, but plausible redefinition of kinematic concepts that naturally explained the codependence of quantum conjugate variables as well as quantum mechanics' failure to offer a complete causal explanation of quantum phenomena.

As we will see in Sect. 6.5, Popper forcefully argued that Heisenberg's attack against the principle of causality ultimately failed, because Heisenberg had equivocated, under the term 'measurement', both measurements proper and state preparation procedures.[2] The microscope experiment was not, as Heisenberg thought, a case of joint measurements, but a procedure where a single measurement is also acting as a state preparation procedure. Consequently, Popper reasoned, the indeterminacy relations could not be interpreted as formulae describing the state of a quantum object at a specific time, but rather as limits imposed on our ability to manipulate quantum objects into groups of similarly prepared quantum systems. This statistical interpretation of the indeterminacy relations, Popper concluded, required neither the revision of our kinematic concepts, nor the abandonment of the principle of causality. It simply showed the incompleteness of quantum mechanics.

From Popper's perspective, the continued popularity of the microscope experiment was disheartening. He readily attributed the cold reception his statistical interpretation had received from Heisenberg and his students to his failure to produce an EPR-like thought experiment demonstrating that the indeterminacy relations were not the logically unsurpassable limits Heisenberg had made them to be (Popper 1959, p. 236).[3] Popper's focus on Heisenberg's original programme and his isolation from the physics community probably explain why he failed to realise that, in the years separating Heisenberg's original paper and the publication of the *Logik der Forschung*, quantum physicists had become at least tacitly aware that the photon and electron formed an entangled system in the microscope experiment. Section 6.6 discusses how Grete Hermann soon used this surprising fact to develop a new dialectic argument that challenged both Heisenberg's rejection of the principle of causality and Popper's statistical interpretation. Yet, despite offering the most complete account of the microscope argument yet devised, Hermann's argument nonetheless failed to convince the physics community of the need to revise the criteria used to evaluate

---

[2]On the importance of the fallacy of equivocation in dialectic arguments, see Nelson (2016). [For more on the measurement/preparation distinction, see below Chap. 9 (*eds.*).]

[3]For more on the reception of Popper's work by Heisenberg and his students, see Combourieu (1992). The Register of the Sir Karl Raimond Popper Papers, 1928–1995, preserved at the Hoover Institution, also has the letters Popper and Heisenberg exchanged in 1934–35 on questions raised in *Logik*.

the empirical adequacy of our causal claims or to convince her colleagues of the possibility to interpret quantum mechanics as a causally complete theory.

Failed thought experiments have been repeatedly used as evidence that thought experiments are, even in science, nothing but 'intuition pumps', that should never be trusted. But the history of the microscope thought experiment—a failed thought experiment if there is one—points toward a different conclusion, for the analyses of the experiment provided by Heisenberg, Popper, and Hermann suggest that failed thought experiments may ultimately be more useful to us than successful ones, at least when they are, as Popper would suggest, 'taken seriously' and properly embedded in an ongoing dialectic inquiry.

6.2 Leonard Nelson's Socratic Method

That dialectics is part of scientific practice is an old idea. Galileo's dialogues readily come to mind as wonderful examples of the role it plays in scientific debates. Until the mid-twentieth century, dialectics pervaded liberal education and many were the physicists who, like Heisenberg, had studied in their youth classic dialectical texts such as Plato's dialogues. But in the Göttingen of the 1920s, dialectics had gained a new life through the work of Leonard Nelson.[4] Keen on establishing close working relationships with scientists, Nelson had organized in 1913 the Jakob Friedrich Fries Society, a group that brought together mathematicians (notably Hilbert, Courant, and Bernays), physicists (like Max Born) and philosophers to critically assess the most recent scientific arguments and mathematical proofs (cf. Milkov 2012, p. 139, 2013, p. 12).[5]

Nelson's striking arrogance made him a difficult professor and colleague. Despite his admiration for Nelson's liberalism and rationalism, Born (Heisenberg's future advisor) found the philosopher's inability to accept open contradiction especially shocking.[6] Born was not the only member of Nelson's circle to find fault with him. As Popper pointed out, a number of the articles found in Nelson's memorial collection, notably Hermann's, were quite critical of Nelson's position. 'Dogmatic Nelsonians,

[4]See Chap. 3, for a detailed discussion of Nelson's philosophy.

[5]Milkov convincingly argues that the Fries Society was a forerunner of the Berlin Society for Empirical Philosophy and remarks that Grete Hermann gave one of the last presentations to the Berlin group in 1934 (Milkov 2013, p. 12).

[6]In his autobiography, Born recalls the public debate he organised with mathematician Theodore von Kármán to discuss Nelson's neo-Kantianism. More experienced and aware that neither Born nor von Kármán could present an alternative to his own views, Nelson used his position as chair of the debate to ridicule, in quite a Socratic manner, his two opponents. 'He made fun of us "harmless and naïve" scientists and had the laughers on his side', Born remembers, adding: 'I think he was quite right to fight for his idealistic views with all possible means, for we had nothing better to offer. I felt this in my innermost heart, and overcame my initial resentment, remaining on friendly terms with Nelson and his followers. But from that time on, our relations were of a more private character' (Born 1978, p. 94).

as far as I know, never existed', Popper asserted, adding somewhat sarcastically, 'and this is in spite of the fact that Nelson may well be described as the founder of a religion' (Popper 2008, p. 15).

This is not surprising, given that Nelson's importance to philosophy is mostly found in the resurgence of peirastic dialectics against positivism as key to the clarification of scientific concepts and principles. Without the Socratic method, Nelson urged, 'a regress [to fundamental principles] would be merely a leap in the dark and would leave us where we were before—prey to the arbitrary' (Nelson 1965, p. 9).

The aim of the Socratic method was not to produce new knowledge, but to reveal the understanding implicit in our judgements, mathematical statements, and scientific theories. The 'critical examination of scientific judgements', Nelson argued, yields the 'cognition hidden in the very form of the judgement', adding:

> If the requirement of simple and clear language is observed, it is possible, in Socratic teaching, merely by writing the theses of two mutually contradictory doctrines on the blackboard, to focus attention on the verbal definition underlying them, disclose its abuse, and thereby overthrow both doctrinal opinions. The success of such a dialectical performance is achieved—and this is its significant feature—not by flashes of inspiration but methodically, i.e., through a step-by-step search for the hidden premise at the bottom of the contradictory judgements. (Nelson 1965, p. 32)

Since even our most common assumptions and well-established scientific principles can tacitly embed prejudices, it is, Nelson argued, essential for the Socratic method to be 'un-philosophical enough to orient itself by means of examples'. The dialectical approach must start from concrete cases, slowly moving from consequences to causes, eliminating accidental facts in the process (Nelson 1965, p. 25). Here, thought experiments find a natural role. For example, Nelson suggested, rather than abstractly discussing the notion of substance, we should start such reflection by imagining the puzzled reaction the staunchest sceptics would display if, after a meeting, they did not find their coats on the pegs where they had previously left them (ibid., p. 9).

For Nelson, the strength of peirastic dialectics rested in its elimination of dogmatism. The aim of the method was not to establish conclusions beyond any doubt, but to survey the contradictory positions and question their premises in a dialogue that would reveal the limits of our language and cognition and thus would lead to a re-conceptualisation of the situation through the clarification of our fundamental concepts (Milkov 2012). The Socratic method did not yield proofs or demonstrations, only probable results, and so the only 'truth' criterion available, Nelson argued, was the conviction that our understanding had finally conformed to nature. As Milkov put it, for Nelson, 'human understanding proceeds from the conviction that immediate knowledge harmonizes with reality' (ibid., p. 148).

It is important to note that Nelson did not see this belief as springing from the dialectician's intuition, nor as the result of an arbitrary decision. The Socratic method was meant to provide a number of measures to guide dialectical inquiries. From requiring the examination of a large number of concrete cases, to its insistence on a sort of reflexive epistemology, to the confrontational behaviour it encouraged amongst its practitioners, it offered a number of different 'dialectical checks' that came into play to ensure that any argument advanced in a discussion was not simply an 'intuition pump'.

Such a view has two important consequences for the evaluation of thought experiments. The first one (and least recognised) is that as the dialectical debates evolve and the concepts, assumptions, and questions under discussion are redefined, so too are the thought experiments.[7] This in turn means that any thought experiment advanced as part of such a dialectical debate will be distorted if analysed in isolation of the Socratic framework. This, I want to argue here, is why Popper accused physicists not to have taken the microscope experiment seriously, for as we will see, from its inception, this experiment was meant to be at the core of a dialectic inquiry examining the most fundamental principles behind our interpretation of quantum phenomena.

6.3 On the Dialectic Aspects of Heisenberg's 1927 Paper

The years 1926–1927 saw Heisenberg pitted in a debate against Erwin Schrödinger on the nature of quantum discontinuities. Schrödinger, who pictured quantum objects as waves, strongly believed that it would eventually be possible to explain quantum discontinuities in terms of wave interferences, a hypothesis Heisenberg dismissed as 'too good to be true' (Heisenberg 1971, p. 72). Heisenberg, who conceived quantum systems as particles, instead argued that quantum discontinuities originated in the instantaneous jumps quantum objects performed between distant spatial locations. Despite 'making sense' of certain phenomena, such as atomic spectra, Heisenberg's corpuscular interpretation remained inconsistent with the well-established classical notions of state, path, velocity, and position. It was clear that any argument claiming to push the debate forward would have to be based solely on claims accepted as probable by most, if not all physicists. In short, the only solution to this deadlock was a dialectical inquiry, which Heisenberg would attempt to provide in 'Über den anschaulichen Inhalt der quantentheoretischen Kinematik und Mechanik', the paper where he introduced the microscope experiment (Heisenberg 1927/1983).

The dialectic sources that informed Heisenberg's reasoning remain unknown. There is no evidence of a direct influence of Nelson on Heisenberg, although the latter arrived in Göttingen in 1922, some five years before Nelson's death.[8] But it is clear that Heisenberg was steeped in the Socratic method. After a strong humanistic education, highly influenced by dialectics, Heisenberg studied at Göttingen, under Born—who for a while had shared Nelson's critical approach—and David Hilbert, whose axiomatic method—which bears the mark of Nelson's influence—deeply influenced Heisenberg's own views on the nature of scientific knowledge (Milkov 2013; Frappier 2004).

---

[7]Something Ian Hacking has strongly denied (Hacking 1993).

[8]Heisenberg was in Göttingen as a visiting student from October 1922 to May 1923 and then returned for his *Habilitation* in 1923–24 (Cassidy 1991).

Heisenberg's opening sentence—to the effect that 'We believe we understand the physical content of a theory when we can see its qualitative experimental consequences in all simple cases [...]' (Heisenberg 1927/1983, p. 62)—echoes Nelson's demand that the search for the fundamental principles of science should start with a critical analysis of a series of simple concrete cases, for every judgement thus obtained, Nelson tells us, 'comprises, in addition to the particular data supplied by observation, a cognition hidden in the very form of the judgement' (Nelson 1965, p. 9).

To develop specific examples able to dialectically solve such a dispute, it is first necessary to methodically examine all the possible reasons lurking behind the disagreement in order to identify the concepts or principles at the source of the conflict. In his introduction, Heisenberg carefully surveys the various causes that could potentially explain his disagreement with Schrödinger (Heisenberg 1927/1983, pp. 62–63). Quickly dismissing the idea that the solution to the dispute can be found in a revision of the mathematical formalism of quantum theory or in a reformulation of physical geometry, Heisenberg notes: 'When one admits that discontinuities are somewhat typical processes that take place in small regions and in short times, then a contradiction between the concepts of "position" and "velocity" is quite plausible' (ibid., p. 63). That a revision of these concepts is, indeed, needed is unsurprising, Heisenberg adds, as 'quantum mechanics arose exactly out of the attempt to break with all ordinary kinematic concepts [...]' (ibid., p. 62). Moreover, Heisenberg adds in a way reminiscent of Nelson's demand to adjust our everyday concepts to the new scientific discoveries, 'we have good reason to become suspicious every time uncritical use is made of the words "position" and "velocity" for in quantum mechanics, position (q) and momentum (p) are related by a the surprising commutation relation: $pq - qp = -i\hbar$' (ibid., p. 63).

From the outset, Heisenberg thus framed his discussion as a dialectical one. Neither logic nor experiments, he argues, could ultimately resolve these 'arguments about continuity versus discontinuity and particle versus wave' (Heisenberg 1927/1983, p. 62). The only way to dissolve the disputes about the nature of quantum objects is 'a more precise analysis of these kinematic and mechanical concepts' (ibid., p. 63), that is, a revision of the conditions that make judgements about position or momentum possible. 'When one wants to be clear about what is to be understood by the words "position of the object", for example of the electron [...]', Heisenberg wrote, 'one must specify definite experiments with whose help one plans to measure the "position of the electron"; otherwise this word has no meaning' (ibid., p. 64).

This remark has repeatedly been taken as evidence of the young Heisenberg's extreme positivism.[9] But a dialectical reading offers a richer interpretation. The Socratic method demands that scientific definitions be anchored in simple experimental situations ('The first step', Nelson wrote, 'is to [...] secure a firm footing in

[9]Cassidy, for example, believes that for Heisenberg, 'The physicist cannot know any more than what he or she can actually measure' (Cassidy 1991, p. 228). In their otherwise excellent piece on the uncertainty principle, Hilgevoord and Uffink (2014) similarly affirm that Heisenberg 'adopted an operational assumption: terms like "the position of a particle" have meaning only if one specifies a suitable experiment by which "the position of a particle" can be measured. We will call this assumption the "measurement = meaning principle"'.

experience [...]' (Nelson 1965, p. 27)). This, of course, does not mean that scientific concepts are completely defined by experimental procedures. From a dialectical perspective, Heisenberg's insistence is not a sign of a reductive operationalist stance, but evidence that he believed experience to be necessary, albeit not sufficient, to constrain the interpretation of a theory and define its fundamental concepts. He makes this clear when, speaking about Einstein's relativity, he remarks that 'the possibility of employing usual space-time concepts at cosmological distances can be justified neither by logic nor by observation' (Heisenberg 1927/1983, p. 62).

Despite being convinced that a redefinition of the kinematic notions was necessary, it is only after long discussions with Niels Bohr and a stimulating correspondence with Wolfgang Pauli that Heisenberg found the key to the problem (Frappier 2004). While a delicate procedure in practice, measuring the position of a quantum object, like an electron, is not a theoretically impossible operation. As Heisenberg remarked, by using a single photon rather than a beam of light to observe the particle, one can readily minimize the "disturbance" caused by the procedure on the electron. Because of its interaction with the electron, the lone photon will recoil away from its original trajectory and enter a microscope, finally hitting the observer's eye or a photographic plate, allowing for the determination of the electron's position at the time of the interaction (Heisenberg 1927/1983, p. 64).

The problem remains the determination of the precision with which this experimental set-up will enable us to measure the particle's position. Heisenberg's analysis is shrewd but simple, using only two well-known and widely accepted sets of experimental laws to derive a surprising 'inaccuracy relation':

1. Classical optics suggests that the highest precision attainable in the case of a microscope position measurement is of the order of the wavelength used;
2. The phenomenological law describing the Compton effect says that when a photon and an electron interact, the change in the electron's momentum will be (approximately) inversely proportional to the change in the photon's wavelength.

From this, Heisenberg continues, one can readily infer that the 'inaccuracy' or 'indeterminacy' of the measurement of an electron's position (q_1) will be inversely proportional to the 'inaccuracy' or 'indeterminacy' of our knowledge of the particle's momentum (p_1), according to the indeterminacy relation: $p_1 q_1 \approx h$.[10]

Reflecting on this and other similar experiments, Heisenberg concludes:

> We might summarize and generalize the results of the preceding section in this statement: *All concepts which can be used in classical theory for the description of a mechanical system can also be defined exactly for atomic processes in analogy to the classical concepts.* The experiments which provide such a definition themselves suffer an indeterminacy introduced purely by the observational procedures we use when we ask of them the simultaneous determination of two canonically conjugate quantities. (Heisenberg 1927/1983, p. 68; emphasis original)

This passage has repeatedly been interpreted as the proposal of an ignorance interpretation explaining away our inability to describe the motion of quantum objects

[10]Heisenberg ignores the microscope's numerical aperture in his account here. More on this later.

along continuous spatial trajectories as caused by the uncontrollable effects any measurement procedure will have on a quantum system. Such readings misrepresent the role played by the microscope experiment in Heisenberg's paper. The microscope experiment is not meant to establish the 'indeterminacy relation' as a principle of nature. It is also not acting as a reductio ad absurdum argument for it does not demonstrate the inconsistency of classical kinematic concepts. The microscope experiment simply suggests that, given the current state of the combined theoretical and empirical knowledge of the late 1920s, no empirical evidence can be given in support of the classical assumption that the state of a particle is completely defined at all times by two numbers, its position and momentum. Simply put, the microscope experiment first and foremost suggests that our current knowledge of the physical world does not demand that we ascribe specific values to the position and momentum of physical objects. On the contrary, as the microscope experiment shows, the empirical knowledge embedded in some of our best laws and theories allows for a non-classical redefinition of our kinematic concepts, one where, as in quantum mechanics, canonically conjugate variables, such as position and momentum, are not independent of one another (Heisenberg 1931). As Heisenberg puts it, the indeterminacy relation thus obtained simply 'creates room for the validity of the relations which find their most pregnant expression in the quantum-mechanical commutation relations [...]' (Heisenberg 1927/1983, p. 68).

We find here again evidence against any operationalism in Heisenberg's thinking. While the microscope experiment suggests the need to redefine kinematic concepts, Heisenberg makes clear that this redefinition can only arise from a close examination of the description of the world offered by our most recent and robust empirical theory, quantum mechanics. That the theory offers an adequate avenue for the redefinition of these concepts was suggested to Heisenberg by the fact that a formula formally similar to the indeterminacy relation derived from the previous semi-classical analysis of the microscope experiment can be obtained by approximating the case where the position of an electron is 'determined' as 'q' with an uncertainty q_1' as a Gaussian of width q_1 centered on q' (Heisenberg 1927/1983, p. 69). From this apparently innocuous mathematical representation of the electron in the quantum formalism, Heisenberg derives the equation $p_1 q_1 = \hbar$, which he believes 'corresponds [...] to the experimental fact that the value p' is measured for p and the value q' for q' (ibid.).

But this conclusion is clearly incorrect. Nowhere in the derivation from the formalism is there any representation of the electron's interaction with the photon or any appeal to the experimental set-up. While the indeterminacy relation obtained through the analysis of the microscope experiment clearly arises because of the measurement procedure, this new relation simply expresses the fact that, in the quantum formalism, the position and momentum functions are Fourier transforms of one another.

This should have puzzled Heisenberg. Yet, unaware of the discrepancy existing between the derivation of the indeterminacy relation from the microscope experiment and that obtained from the formalism, Heisenberg first suggests a new 'geometrical interpretation' where kinematic concepts are represented as tensors and goes on to apply this abstract representation to different possible experimental set-ups to show how it easily and 'naturally' leads to a correct prediction of quantum phenomena.[11] Again, Heisenberg's approach resonates with Nelson's and its claim that the only truth criterion is the perceived harmony between our knowledge and experience. The analyses of the microscope and other thought experiments, Heisenberg argues, all point to the same conclusion: our best classical theories and phenomenological laws do not empirically anchor our assumption that objects simultaneously exhibit a definite position and momentum. Once we recognise this fact, we are free to re-interpret the p- and q-matrices of quantum mechanics as representing the position and momentum of quantum objects. The shocking consequence of this redefinition, Heisenberg continues, is not only that it demands that we abandon our belief that objects follow continuous paths when moving through space, but that it leads to the 'final failure of causality': 'But what is wrong in the sharp formulation of the law of causality, "When we know the present precisely, we can predict the future"', writes Heisenberg, 'is not the conclusion but the assumption. Even in principle we cannot know the present in all detail' (Heisenberg 1927/1983, p. 83).

Even before its publication, Heisenberg's paper drew heavy criticisms. As Bohr readily pointed out the paper's analysis of the thought experiment only paid lip service to the microscope, wrongly ascribing our inability to estimate the change in the electron's momentum to the Compton effect rather than to the diffraction of the light as it passes through the microscope's lens. Heisenberg's insistence that his analysis favoured a discontinuous description of the motion of quantum objects in spacetime was untenable, Bohr added, for the microscope experiment could only be properly understood if one accepted the dual nature of the photon, which travels as a wave through the microscope, but hit the photographic plate as if a particle.

After a brutal discussion with Bohr, Heisenberg abdicated (Cassidy 1991, p. 242), adding the following 'note in proof' to his paper:

> [...] Bohr has brought to my attention that I have overlooked essential points in the course of several discussions in the paper. Above all, the uncertainty in our observation does not arise exclusively from the occurrence of discontinuities, but is tied directly to the demand that we ascribe equal validity to the quite different experiments which show up in the corpuscular theory on one hand, and in the wave theory on the other hand. (Heisenberg 1927/1983, p. 83)

Bohr, and then Heisenberg, would go on to recycle the microscope experiment as an illustration of Bohr's new Complementarity Principle (Bohr 1928; Heisenberg 1949, p. 14). The new analysis was not a dialectical argument based on universally acceptable propositions and failed to convince many. But for the purpose of our story, the important point is that Bohr's repurposing of the thought experiment as an illustration of the Complementarity Principle drew attention away from the weak analogy Heisenberg had drawn between the microscope experiment and the formal derivation of the indeterminacy relations from the quantum mechanical formalism—an analogy which, according to Popper, another heir of Nelson's Socratic method,

[11] Of note is a *second* microscope experiment where an electron is believed to be in a given excited state.

was even more problematic for Heisenberg's interpretation of quantum mechanics than his bungled analysis of the microscope experiment.

6.4 Popper on the Failure of Heisenberg's Programme

Popper never met Nelson. It was Julius Kraft, a distant German relative who had studied law under the neo-Kantian philosopher, who introduced Popper to Nelson's work (Hacohen 2000, p. 120). During their common time in Vienna in 1924–25, Popper and Kraft enjoyed endless discussions that led to the development of Popper's falsificationism (Popper 2008, p. 17; Milkov 2012, p. 150).[12]

Despite a deep appreciation for the role dialectics could play in science, Popper would always remain wary of its many pitfalls and the possibility of using it to protect rather than unearth prejudices.[13] Whereas Nelson, Heisenberg, and Hermann actively used dialectics to attempt to find common grounds between opposing positions, Popper clearly preferred to use the critical tools of dialectics to expose the contradictions plaguing specific approaches. So while Popper probably appreciated more than most Heisenberg's attempt to appeal to dialectics to redefine the fundamental concepts of quantum mechanics, he remained highly critical of the attempt. Although he would later himself adopt an indeterminist interpretation of quantum mechanics, the Popper of the *Logik* launched a scathing attack against Heisenberg's position which he read as wavering between subjectivism (the indeterminacy relations being described as limitations on our knowledge of quantum objects) and a failed attempt to offer an instrumentalist interpretation of quantum mechanics. To Popper, Heisenberg's repudiation of the principle of causality was especially disturbing, amounting to nothing less than a rejection of the two fundamental ideals of science: realism and objectivity (Howard 2012).

Ironically, from Popper's perspective, the failure of Heisenberg's programme rested on the physicist's refusal to appeal to any un-measureable variables in hope to free quantum physics from any 'metaphysical claims'. This was undeniably the fundamental insight that had guided Heisenberg in the development of matrix mechanics in 1925. If only observables were included in the new theory, Heisenberg had argued, quantum mechanics would be bound to make only observable, testable predictions. The converse was implied. If quantum mechanics did not describe a given phenomenon, for example the path followed by an electron in an atom, one could assume that no such path existed (Heisenberg 1927/1983, p. 74). The 1927 indeterminacy

[12] After World War II, Popper and Kraft would renew their collaboration in the founding of the journal *Ratio*, conceived as a successor to *Abhandlungen der Fries'schen Schule*, the journal Nelson had created in 1904 and where Hermann's essay discussing her version of the microscope experiment appeared (Milkov 2012, p. 147). Hermann sat on the editorial board of *Ratio* and contributed a few papers to it. Correspondence between Popper and Hermann dating from this period is kept at the Hoover Institution in the Register of the Sir Karl Raimond Popper Papers, 1928–1995.

[13] Popper would come to associate a certain kind of 'dialectics' with what he considered the inherent dogmatic core of Hegelianism and Marxism (see Popper 1940).

paper built on this philosophy. Interpreted in the light of the microscope experiment understood as a joint position-momentum measurement, the indeterminacy formula suggested that the present state of an electron could never be so sharply determined as to enable one to make anything but probabilistic predictions about the system's future behaviour. Our inability to know the present with precision took away, Heisenberg claimed, our ability to predict the future, and by doing so left us without any empirical justification for adopting the principle of causality (ibid., p. 83).

For Popper, there were two problems with Heisenberg's argument. For one, even accepting Heisenberg's problematic dual analysis of the indeterminacy relations and their interpretations as limits on the attainable precision of simultaneous measurements on conjugate variables, Heisenberg's anti-metaphysical programme collapsed on itself.[14] As many had noticed before Popper, while quantum mechanics is unable to *predict* with precision the path a particle will take between two measurements, it can be calculated post factum, a result at odds with Heisenberg's representation of the electron as a Gaussian suggesting that the particle was some kind of blurred object. Heisenberg had repeatedly dismissed this apparent inconsistency as of little consequence to physicists, arguing that since the past paths that could be calculated from the quantum formalism could not be used to make testable predictions, the question of whether particles really followed such continuous trajectories between measurements was but 'a matter of personal belief' (Heisenberg 1949, p. 20).

Yet, these incongruous results undermined Heisenberg's efforts to rid quantum physics of all metaphysical claims. After all, Popper teased, '*it is possible to calculate* such a "senseless" or metaphysical path in terms of the new formalism' (Popper 1959, p. 220; emphasis original). For those interpreting quantum formulae as describing the behaviour of individual systems, only two, equally doomed, options seemed open. The first one—suggested by an objective interpretation of the indeterminacy relations—was to follow Heisenberg's suggestion that electrons were some kind of 'blurred' or 'smeared' entities that could not be conceived as following well-defined paths between measurements. If this was the case, Popper argued, quantum mechanics should be revised because it described phenomena—the particles' past paths—that simply did not exist. The second option was to accept that electrons do move along well-defined, but unobservable and unpredictable, paths. Under this reading, the indeterminacy relation should be downgraded to a measure of our ignorance, telling us little of the 'real' state of the quantum objects. Somewhat cheekily, Popper concluded: 'Even from the point of view of Heisenberg's own interpretation of his theory, it does not seem that his programme has been fully carried out' (ibid., p. 218).

All of this, Popper argued, was very much a false dilemma,[15] originating in Heisenberg's inability to understand quantum mechanics as a fundamentally statistical theory that does not describe the dynamical behaviour of individual quantum objects, but offers instead statistical predictions on ensembles of similarly prepared

[14]This is not to say that Popper proved that simultaneous measurements were indeed possible. For a possible answer to such an attack, see Filk (this volume, Chap. 5).

[15]On the prevalence of false dilemma in dialectics, see Nelson (2016).

systems. Under this reading, the indeterminacy relations are not about the relations existing between pairs of measurements, but describe the random scattering observed when *ensembles* of similarly prepared particles are subjected to what Popper called *physical (or technical) selections*, that is, what we usually call 'state preparation procedures'. Measurements, Popper explained, inform us as to the states of quantum objects when they interact with the measuring apparatus. In this sense, they are truly about the *past* of the object. Physical selections, for their part, are procedures that separate out the systems displaying a certain property out of a larger ensemble of objects. For example, a screen with a narrow aperture positioned in front of a beam of particles will select out the corpuscles whose positions let them go through the aperture, while blocking all the other particles from continuing onwards. Such selections can be used to obtain information about the state of the system *after* the procedure. For example, we know that if a particle has gone through the screen, it must afterwards be located within a certain region of space. In this sense, physical selections are very much about the future.

Applied to the microscope experiment, this distinction reveals an asymmetry that sheds light on Heisenberg's apparently inconsistent interpretation of the indeterminacy relations. With respect to position, Heisenberg's original microscope experiment is a measurement procedure. It enables physicists to determine the electron's location at the time of its interaction with the photon. But with respect to momentum, the experiment is a physical selection, not a measurement. As long as the electron's initial momentum is known, physicists can use such a set-up to determine the range within which the momentum of the electron will fall *after* its interaction with the photon. But the microscope experiment does not enable us to determine the electron's momentum prior to the measurement.[16] A more careful investigation of the origin of the

---

[16]In one of the new appendices to the 1959 edition of the *Logic*, Popper adds that this distinction can be obtained from a close analysis of the microscope experiment, writing:

> [...] Heisenberg's discussion *fails to establish that measurements of position and of momentum are symmetrical*; symmetrical, that is, with respect to the disturbance of the measured object by the process of measurement. For Heisenberg *does* show with the help of his experiment that in order to measure the *position* of the electron we should have to use light of a high frequency, that is to say, high energy photons, which means that we transfer an unknown momentum to the electron and thus *disturb* it, by giving it a severe knock as it were. But Heisenberg *does not* show that the situation is analogous if we wish to measure the *momentum* of the electron, rather than its position. For in this case, Heisenberg says, we must observe it with a low frequency light—so low that *we may assume that we do not disturb the electron's momentum by our observation*. The resulting observation [...] will fail to reveal the electron's position, which will thus remain indeterminate.
>
> Now consider this last argument. There is no assertion here that we have *disturbed* (or 'smeared') the electron's position. For Heisenberg merely asserts that we have *failed to disclose it*. In fact, his argument implies that we have not disturbed the system at all [...]. Thus *the two cases—the measurement of position and that of momentum—are far from analogous or symmetrical*, according to Heisenberg's argument. This fact is veiled, by the customary talk [...] about the '*results of measurement*' whose uncertainty is admittedly symmetrical with respect to position and momentum. Yet in countless discussions of the experiment, beginning with Heisenberg's own, it is always assumed that his argument establishes the *symmetry of the disturbances*. (Popper 1959, p. 451–452; emphasis original)

indeterminacy relations, Popper argues, clearly shows that they are not a measure of the electron's blurriness, but rather a measure of the statistical distribution physicists obtain when performing a given procedure on an aggregates of electrons.[17]

In other words, for Popper, the indeterminacy relation is a limit on our ability to create an ensemble of similarly prepared objects, not a limit on our ability to know the state of a specific system. If microscopic position measurements are performed on electrons prepared with a given momentum, the particles' momenta will scatter. However, Popper continues, after the procedure it is still afterwards possible to precisely ascertain the new momentum of each electron. 'These measurements of the single momenta', Popper explains, '[…] will give in each single case results as precise as we like, and at any rate very much more precise than Δp, i.e. the mean width of the region of the scatter' (Popper 1959, p. 231, note *1). The indeterminacy relation, Popper concludes, is not an unsurpassable limit on measurements, but a 'statistical scatter relation' (ibid., p. 225).

In Popper's mind, the strength of a statistical interpretation was that the past paths of particles that could be obtained post factum now played an important role in the *confirmation* of the theory. If for Heisenberg it was a matter of taste to accept or not that electrons followed specific paths between measurements, for Popper these paths were essential to test quantum mechanics' predictions. As he explained,

> Admittedly [the position and momentum between two measurements] do not serve as initial conditions or as a basis for the derivation of predictions; but they are indispensable nevertheless: *they are needed for testing our predictions*, which are *statistical predictions*. For what our statistical scatter relations assert is that the momenta must scatter when positions are more exactly determined, and *vice versa*. (Popper 1959, p. 230–231; emphasis in the original)

Although Popper himself would come to reject this interpretation of quantum mechanics and embrace indeterminism (Howard 2012), the analysis of quantum measurements offered in the *Logik* raised important challenges to Heisenberg's position. While not proof of the incompleteness of quantum mechanics, Popper's analysis demonstrated that the microscope experiment could neither invalidate the principle of causality, nor illustrate the indeterminacy relations, at least not if interpreted as limits on joint position–momentum measurements.

The original debate about the nature of quantum objects had been reframed as a question about the completeness of quantum mechanics. On the one side, Heisenberg took our inability to precisely describe the future behaviour of quantum systems as evidence that the quantum world was intrinsically indeterministic. On the other, Popper understood the theory's capacity to apparently offer a description of the past trajectories of quantum objects as evidence for the existence of, yet hidden, deterministic laws of nature. The microscope thought experiment strangely stood between the two camps as a procedure that could be used to determine either the

---

[17]Popper later corrected this claim, admitting he should have spoken of 'an aggregate—or of a sequence—of repetitions of an experiment undertaken with *one* particle (or *one* system of particles)' (Popper 1959, p. 225 note *1; emphasis original)

past position of an electron or its future momentum, but apparently incapable to shed more light on either the validity of the causality principle or the completeness of quantum mechanics. Only with the work of Grete Hermann would it become evident that the apparently failed thought experiment could be repurposed to deepen our understanding of quantum mechanics yet again.

6.5 Grete Hermann: Complementarity and Indeterminacy

Hermann clearly understood the dialectic nature of this new debate, demonstrating in the first sections of the *Grundlagen* that this dilemma could not be answered within the frame of quantum mechanics. The no-hidden-variables proofs so far advanced, Hermann argued, had all failed to recognise that the only unescapable objection to the existence of a hidden-variables theory would be a demonstration that quantum mechanics already causally explained all the events falling within its domain of application (see Sects. 3–8 of Chap. 15; see also Chap. 4). And there was the crux of the problem. On the one side, Heisenberg and his followers claimed that quantum mechanics was, although indeterministic, complete. On the other side, those believing, like Popper, in the existence of a causal theory, took quantum mechanics' limited predictions as evidence of its incompleteness. The notions of completeness, causality, and prediction were thus inextricably entwined. The solution to the debate could only come from a (dialectical) clarification of their relations. As Nelson had pointed out:

> In order to grasp [...] concepts clearly it is necessary, of course, to isolate them [...] to separate them from other ideas, to reduce them gradually to their elements, and through such analyses to advance to basic concepts. By holding fast to existing concepts, the philosopher guards himself against peopling his future system with the products of mere speculation and with fantastic brain children. (Nelson 1965, p. 31)

Hermann's *Grundlagen* is her attempt to separate from one another these three notions—completeness, causality, and prediction. Up to now, she argued, probabilistic theories had been interpreted as expressing an incomplete knowledge of nature, while the possibility to make precise and accurate predictions had been used as the criterion distinguishing theories revealing the true causes of natural phenomena from mere fantasies. But, Hermann urged her readers, it was essential to see that the principle of causality meant only to capture that 'nothing in nature occurs that is not in all physically determinable features, caused by (and this means: follows necessarily from) previous processes' (Chap. 15, p. 262). Causality did not analytically imply prediction. This in turn meant that a criterion other than prediction was necessary to determine if indeterministic theories, like quantum mechanics, could also be causally complete (i.e., if they could truly identify all the causes of the events falling within their domain of application, despite not being able to predict all of them) (see Chap. 15, pp. 241–242).

In other words, from Hermann's perspective, the only way to solve the dilemma at hand was to demonstrate that quantum mechanics already provided—albeit a posteriori—all the causes necessary to explain quantum phenomena. More specifically, this meant that an indeterministic theory such as quantum mechanics could be considered causally complete if it proved to be:

1. *Causally explanatory*, that is, the theory offers, at least a posteriori, causal narratives explaining the results of measurements (cf. Chap. 15, pp. 255–256).
2. *Indeterminist*: the theory cannot give causal narratives until observations are actually completed. (This is equivalent to saying that the 'observational context' plays a fundamental role in the measurement so that, in the construction of our 'retrodiction', we appeal to states of the object or measuring apparatus that were not (and could not be) present in the descriptions we had of them before the measurement (ibid., p. 256).
3. *Testable*. The causal narrative obtained *after* detecting a *first* system can be *indirectly* tested by using this cause to make an empirically verifiable prediction on a *second* system (in other words, it is possible to use a measurement as a state preparation procedure—more on this below) (ibid., Sect. 12).

Taken in the abstract, however, these requirements are meaningless. It is necessary to anchor the reasoning in experience which, even Nelson admitted 'is harder to do than an outsider might think' (Nelson 1965, p. 27). It is to Hermann's credit to have realised that the quantum electrodynamic description of the microscope experiment offered by Carl Friedrich von Weiszäcker (1931) could be used to explore the extent to which quantum mechanics met these proposed criteria. As Filk (Chap. 5, Sect. 5.3) explains, Hermann's description of the experiment added significantly to Weiszäcker's by expanding it to include the following three scenarios, differing from one another only with respect to the location of the photographic plate:

1. *The photographic plate is placed in the microscope's image plane*. In this case, the photon must be described as a spherical wave after its interaction with the electron. This enables physicists to measure the location where the interaction has taken place but, since the precise direction from which the photon enters the microscope remains uncertain, only an imprecise determination of the change of the electron's momentum is possible. In other words, the dual use of a corpuscular image (to describe the photon-electron interaction) and of the wave image (in the description of the photon's motion) gives the observer a better knowledge of the electron's position than the one available before the measurement, but a less precise knowledge of the electron's momentum (cf. Chap. 15, pp. 257–258).
2. *The photographic plate is located in the focal plane of the microscope*. In this situation, the photon is described by a plane wave, which enables us to use the microscope to measure the electron's momentum change at the time of the interaction, but not to ascertain the location of the collision (ibid., p. 258).
3. *No photographic plate is used and the photon is left free to pursue its course without any interference*. Quantum physics describes this situation through a combination of the photon's and the electron's wave functions, forcing us to relinquish the idea that in this case the electron and photon are two separate systems (ibid., p. 258).

The three scenarios show how quantum mechanics readily fulfills the first two criteria Hermann proposes to ascertain the causal completeness of apparently indeterministic theories. In the first two scenarios, a causal explanation of the measurement result can be given after the measurement, explaining how the wave travels to the photographic plate after interacting with an electron (criterion 1: causal explanation). It is, however, only possible to offer this causal explanation once the position of the plate is known and the result of the measurement obtained. As Filk so nicely puts it, the first two scenarios bring to the forefront the fact that

> The electron alone does not have a definite state immediately after the collision, but only 'relative states' with respect to the basis one chooses for the description of the photon. This state is conveniently chosen with reference to the type of measurement one intends to make (the 'observational context'), and therefore the 'relative states' of the electron also depend on this context. (Chap. 5, p. 76)

The third scenario reinforces this conclusion by showing how no causal description of the photon's or electron's motion after the interaction can be given until a measurement is made.[18] In other words, the three possible scenarios depend on the intended measuring set-up, a decision that can be made after the photon–electron interaction (criterion 2: indeterminism) (Chap. 15, Sect. 10).

The microscope experiment also shows, Hermann continues, how despite being an indeterministic theory, the results offered by quantum mechanics can be *indirectly tested* (criterion 3: testability). When the photon hits the photographic plate (the effect), we retrodictively assume that it followed from the photon's interaction with an electron (the cause). We can then test this hypothesis by using our results (e.g. the location of the interaction) to make a prediction as to the state of the electron *after* its interaction with the photon (for example, a prediction about the result of a future momentum measurement made on the particle). Hermann thus recognises and uses, although only implicitly, the fact that the microscope experiment is—as noted by Popper—not only a measurement procedure, but also a state preparation procedure—in order to confirm the causal narratives offered by quantum mechanics (Chap. 15, Sects. 10 and 12).[19]

As Hermann warns her readers, such 'backward deductions' or 'retrodictions' cannot, as Popper had hoped, be extended earlier than the photon–electron interaction. One cannot legitimately describe what happens between two consecutive position measurements on an electron or between a momentum measurement on the particle followed by a position measurement because such narratives fail to take into account both the duality of the quantum object and the fact that there are two different 'observational contexts at play'. Physically speaking, these descriptions of past paths are meaningless, 'empty' (see Chap. 15, Sect. 12).

Hermann's analysis significantly changed the parameters of the Heisenberg–Popper debate. Like Bohr, she refused to use the microscope experiment as a

---

[18] See Chap. 5 for a careful discussion of the ambiguities present in Hermann's discussion of this scenario.

[19] For a different interpretation of Hermann's notion of mediated testability, see the postface by Soler in Hermann (1935/1996, pp. 128–131).

justification of the indeterminacy relations, only using it to illustrate what she took to be direct consequences of the Complementarity Principle. But it was a much more complete illustration of the Principle than the one Bohr had offered years earlier (Bohr 1928). As Popper had shown, when presented as a mere position measurement—as Bohr did—the microscope experiment needed not be considered as an illustration of the 'blurriness' of quantum objects (suggested both by the Complementarity Principle and the indeterminacy relations obtained by Heisenberg from the quantum formalism). It simply suggested that a position measurement does not make an ideal momentum preparation procedure, a conclusion that made it easy to adopt an interpretation of quantum mechanics where 'indeterminations' were readily interpreted as limitations on our knowledge and where electrons do follow specific, continuous paths. It is only the three-scenario microscope experiment provided by Hermann that enables us to see why we cannot follow Popper in his conclusions since the 'observational context' of the experiment, specifically the location of the photographic plate, plays an unavoidable role in the explanation we ultimately give of our observations.

Hermann's description of the three possible outcomes of the microscope experiment furthermore transformed the debate surrounding the interpretation of the microscope experiment by suggesting—although somewhat implicitly—that the interpretation of the 'indeterminacy relations'—and hence of quantum mechanics in general—required two fundamentally different discussions: one about the limits imposed on the use of the measurement of an observable as a preparation procedure for its complex conjugate (as suggested by the microscope experiment) and a second discussion about the possibility to perform precise simultaneous measurements on conjugate quantities in order to determine whether Bohr was right to assume that it was impossible to simultaneously ascribe a precise position and a precise momentum to quantum objects. As we saw above, Popper had dismissed Heisenberg's analysis of the microscope thought experiment in terms of 'blurriness' of quantum objects, arguing that the microscope experiment associated an 'error' in measurement (the position indeterminacy) to the 'uncertainty' of a prediction (the momentum indeterminacy). Hermann's adoption of Bohr's tripartite notion of complementarity suggested instead that different—yet fundamentally related—kinds of indeterminacy relations might exist, each offering a mathematical representation of the various relations included under the banner of the complementarity principle (Sect. 13 of Chap. 15). Indeed, as Crull explains (Chap. 10), for Hermann the complementarity principle simultaneously pointed to:

1. *a need to describe quantum phenomena in terms of both the wave and the particle image*, illustrated in the microscope experiment by the initial description of the electron and photon as particles exchanging momentum and their subsequent representations as (spherical or parallel) waves after the interaction;
2. *an intrinsic restriction imposed on the application of either the wave or particle image*, limiting the precision with which conjugate quantities can be simultaneously ascribed to a quantum system. This was 'one of the most wonderful results of quantum mechanics' (Chap. 15, p. 267) for Hermann, who attributed it to the fact that all observations will disturb the quantum systems;

3. *the relation, embedded in the quantum mechanical formalism* that explained why, within the formalism, complex conjugate pairs—like position and momentum—are represented by Fourier transforms, a relation suggested by Heisenberg's derivation of the 'indeterminacy relations' from the quantum mechanical description of the electron.

While Hermann's essay suggested that different 'indeterminacy relations' relating different kinds of 'indeterminacies' or 'uncertainties' might exist, it would take over half a century before truly realising how complex the task of differentiating between these different expressions truly was. As Hilgevoord and Uffink later explained, quantum mechanics does contain at least two different notions of indeterminacy, one about what can be predicted of a quantum object and one about what can be inferred about its past. As they rightly pointed out, 'since the two uncertainties are conceptually quite different, the quantitative measures of these uncertainties must be defined quite differently' (Hilgevoord and Uffink 1990, p. 126). And to this day, the debate on how we should represent, evaluate, and interpret these indeterminacies continues, focusing on the question ultimately left unanswered by Heisenberg and Hermann as to the existence of an indeterminacy relation intrinsically linking the results of truly simultaneous position–momentum measurements that could ultimately confirm or infirm the blurry image Heisenberg had offered of quantum objects (see on this topic Busch et al. 2013; Ozawa 2003, 2004).

6.6 Welcoming Ignorance

The moral of the story, if there is one, is somewhat ironic. Hermann's own dialectical argument was not without weaknesses. As Filk (Chap. 5) rightly notices, Hermann did not seem to realise that measurement processes, such as the one described in the microscope experiment, may well 'create' the quantum states they reveal (the problem we refer to nowadays as wave function collapse). Overall, Hermann's essay was barely noticed in the literature (ibid.). It did not convince Heisenberg (see Crull 2010) and would not have satisfied Popper. By then, Heisenberg's original interpretation of the microscope experiment had long been discarded. As for Popper, he would soon come to reject his own deterministic interpretation of quantum physics (Popper 1967).

The later career of the microscope experiment—as a prime textbook example—was not more glorious. Popper was right to think that the liberal use made of the experiment as an illustration of the indeterminacy relations was problematic. Still today most accounts present the experiment as a straightforward joint measurement (something it is not), that illustrates a unique principle of indetermination (rather than a multiplicity of 'indeterminacy relations'), a problem hidden by a general inattention to how we evaluate 'measurement errors'. Few are those who know that there is no such thing as *the* microscope thought experiment and understand that behind this label lie a multitude of experimental set-ups, all playing different roles in the unfolding of our understanding of measurements, state preparation procedures, and entanglement.

We could be tempted to read the repeated failures of the microscope experiment as a cautionary tale demonstrating how imaginary experiments are nothing but misleading 'intuition pumps' that should not enter the scientific discourse. But such a conclusion rests on the long-entrenched assumption that knowledge is secured through successful thought experiments, not failed ones, an assumption the history of the microscope experiment presented here challenges. As Nelson reminded us: 'To Socrates the test of whether a man loves wisdom is whether he welcomes his ignorance in order to attain to better knowledge' (Nelson 1965, p. 25). Each failed instance of the microscope experiment did lead to important insights into quantum phenomena, from the uncovering of the indeterminacy relations, to the distinction between measurements and state preparation procedures, to entanglement.

It is true that the microscope experiment presents quite a challenge to those who claim that thought experiments are simple imaginary devices that enable us to readily grasp the laws of nature (e.g. Brown 1991). Presented in all of its intricacies, the microscope experiment would be pedagogically useless. Simplified, it is misleading. Unless, of course, a professor accepts the role of Socratic midwife, asking students what they mean by the position of the electron, how they would measure it, how they would evaluate the error of their measurements, what they mean by 'measurement error' and whether this 'error' is something that is a property of the object or a failure of the observer to know the world as it truly is. But then again, is this constant questioning not the original and constant role such thought experiments should play in a dialectic inquiry, namely to force us, again and again, into what Popper called the 'no-man's-land that lies between logic and physics' (Popper 1959, p. 215)?

References

Beller M (1999) Quantum dialogue: the making of a revolution. University of Chicago Press, Chicago

Bohr N (1928) The quantum postulate and the recent development of atomic theory. Nature 121(3050):580–590

Born M (1978) My life: recollections of a Nobel laureate. Routledge, New York

Brown JR (1991) The laboratory of the mind: thought experiments in the natural sciences. Routledge, New York

Busch P, Lahti P, Werner R (2013) Proof of Heisenberg's error-disturbance relation. Phys Rev Lett 111:160405

Cassidy D (1991) Uncertainty: the life and science of Werner Heisenberg. Freeman & Company, New York

Combourieu MC (1992) Karl R. Popper, 1992: about the EPR controversy. Found Phys 22(10):1303–1323

Crull E (2010) Grete Hermann and the gamma-ray microscope Gedankenexperiment, talk delivered at HQ-3: conference on the history of quantum physics, Berlin, 2 July 2010

Frappier M (2004) Heisenberg's notion of interpretation. Ph.D. thesis, University of Western Ontario. London, Ontario
Hacking I (1993) Do thought experiments have a life of their own? Comments on James Brown, Nancy Nersessian and David Gooding. In: Hull D, Forbes M, Okruhlik K (eds) PSA 1992, Philosophy of science association, East Lansing, MI, vol 2, pp 302–308
Hacohen MH (2000) Karl Popper: the formative years, 1902–1945: politics and philosophy in interwar Vienna. Cambridge University Press, Cambridge
Heisenberg W (1927/1983) Über den anschaulichen Inhalt der quantentheoretischen Kinematik und Mechanik. Zeitschrift für Physik 43(3–4):172–198. Translated as: The physical content of quantum kinematics and mechanics. In: Wheeler J, Zurek W (eds) Quantum theory and measurement. Princeton University Press, Princeton, NJ, pp 62–84
Heisenberg W (1931) Die Rolle der Unbestimmtheitsrelationen in der modernen Physik. Monatshefte für Mathematik und Physik 38:365–372
Heisenberg W (1949) The physical principles of the quantum theory. Dover, New York
Heisenberg W (1971) Physics and beyond: encounters and conversations. Harper & Row, New York
Hermann G (1935/1996) Les fondements philosophiques de la mécanique quantique. Vrin, Paris, translation of Hermann (1935b) by Schnell A. Edited with introduction and postface by Soler L
Hilgevoord J, Uffink J (1990) A new view on the uncertainty principle. In: Miller AI (ed) Sixty-two years of uncertainty: historical, philosophical, and physical inquiries into the foundations of quantum mechanics. Plenum Press, New York/London, pp 121–137
Hilgevoord J, Uffink J (2014) The uncertainty principle. In: Zalta EN (ed) The Stanford Encyclopedia of Philosophy, Spring 2014 edn. http://plato.stanford.edu/archives/spr2014/entries/qt-uncertainty/
Howard D (2012) Popper and Bohr on realism in quantum mechanics. Quanta 1:33–57
Kennard EH (1927) Zur Quantenmechanik einfacher Bewegungstypen. Zeitschrift für Physik 44:326–352
Leal F (2016) Introduction. In: Nelson (2016), pp 1–20
Milkov N (2012) Karl Popper's debt to Leonard Nelson. Grazer Philosophische Studien 86:137–156
Milkov N (2013) The Berlin Group and the Vienna Circle: affinities and divergences. In: Milkov N, Peckhaus V (eds) The Berlin Group and the philosophy of logical empiricism. Boston studies in the philosophy and history of science, vol 273, pp 3–32
Nelson L (1965) Socratic method and critical philosophy: selected essays by Leonard Nelson. Dover, New York
Nelson L (2016) A theory of philosophical fallacies. Springer, Cham
Norton J (1996) Thought experiments in Einstein's work. In: Horowitz T, Massey GJ (eds) Thought experiments in science and philosophy, Rowan & Littlefield, Savage, MD, pp 129–148
Ozawa M (2003) Physical content of Heisenberg's uncertainty relation: limitation and reformulation. Phys Lett A 318:21–29
Ozawa M (2004) Uncertainty relations for joint measurements of noncommuting observables. Phys Letts A 320:367–374
Popper KR (1940) What is dialectic? Mind 49(196):403–426
Popper KR (1959) The logic of scientific discovery. Hutchinson & Co., London
Popper KR (1967) Quantum mechanics without 'the observer'. In: Quantum theory and reality, Springer, pp 7–44
Popper KR (2008) After the open society: selected social and political writings. Routledge, London/New York
Robertson H (1929) The uncertainty principle. Phys Rev 34:573–574
Schrödinger E (1930) Zum Heisenbergschen Unschärfeprinzip. Sitzungsberichte der Preussischen Akademie der Wissenschaften, Physikalisch-Mathematische Klasse 19:296–303

Chapter 7
Challenging the Gospel: Grete Hermann on von Neumann's No-Hidden-Variables Proof

Michiel Seevinck

7.1 Introduction

In 1932 John von Neumann had published in his celebrated book, the *Mathematische Grundlagen der Quantenmechanik* (von Neumann 1932), a proof of the impossibility of theories that, by using so-called 'hidden variables', attempted to give a deterministic explanation of quantum mechanical behaviour. It is unclear how soon it was the case, but eventually von Neumann's proof became considered holy—it was received as Biblical wisdom that one should not challenge. As Belinfante has written: 'The truth, however, happens to be that for decades nobody spoke up against von Neumann's arguments, and that his conclusions were quoted by some as the gospel' (Belinfante 1973, p. 30).

However, in 1935 Grete Hermann did challenge this gospel by criticising the von Neumann proof on a fundamental point. This challenge was however not widely known at the time, and her criticism had no impact whatsoever. Thirty years later, Bell (1966) provided a critique of von Neumann's proof quite similar to Hermann's, but Bell's work did have great foundational impact.

In what follows, I shall go through the details of von Neumann's 1932 proof against the possibility of hidden variables and describe the reception of this proof—including Bell's 1966 criticism (Sects. 7.2 and 7.3). I shall then give Hermann's 1935 critique of von Neumann's argument, comparing it to Bell's critique (Sects. 7.4 and 7.5). Finally, I shall discuss the reception (or lack thereof) of Hermann's criticism, and speculate about why Hermann's anticipation of Bell's argument was not—and continues not to be—widely known (Sect. 7.6).

I wish to thank Guido Bacciagaluppi and especially Elise Crull, without whom this chapter would never have been written.

M. Seevinck (✉)
Nijmegen, The Netherlands
e-mail: m.p.seevinck@gmail.com

E. Crull and G. Bacciagaluppi (eds.), *Grete Hermann - Between Physics and Philosophy*, Studies in History and Philosophy of Science 42,
DOI 10.1007/978-94-024-0970-3_7

7.2 Von Neumann's 1932 No Hidden Variables Argument

In the fourth chapter of his 1932 treatise, von Neumann asked himself the following question: what reasons can be given for the dispersion found in some quantum ensembles? He presents two possible explanations for such statistical spreading (von Neumann 1955, pp. 301–302):

(Case I): The individual systems differ in additional parameters, not known to us, whose values determine precise outcomes of measurements (i.e., deterministic hidden variables). Given these additional parameters, the dispersion will disappear from the results.

(Case II): 'All individual systems [...] are in the same state, but the laws of nature are not causal.' Thus dispersion in measurement cannot be gotten rid of.

He continues to discuss Case I, noting first of all that no physical method exists by which one can divide a dispersive ensemble into dispersion-free ensembles. For example, assume you have an ensemble of radioactive atoms. They decay in different directions at different times. There is no physical method for separating the systems into sub-ensembles that are dispersion-free with respect to values for direction or time of decay. If we try to separate the atoms in such a way, there is unavoidable measurement disturbance which produces further dispersion.

However, it is possible to conceive of each such ensemble as composed out of two (or more) dispersion-free sub-ensembles. The heart of von Neumann's proof will be to demonstrate that even this latter scenario is impossible. Quantum mechanical ensembles cannot be thought of as dispersion-free or as a collection of dispersion-free sub-ensembles.

Informally speaking, note von Neumann's notion of a hidden variables theory. For him, such a theory is a causal theory that defines the state of the system 'absolutely' by supplying 'additional numerical data'; these additional data are the 'hidden parameters'. He writes (von Neumann 1955, p. 209): 'If we were to know all of these, then we could give the values of all physical quantities exactly and with certainty.' Accordingly, a hidden variables (or 'causal') theory is one which is 'in agreement with experiment, and which gives the statistical assertions of quantum mechanics when only ϕ is given (and an averaging is performed over the other coordinates).'

Von Neumann mathematically characterises these hidden variables as follows: every physically realisable state can be represented in principle as a mixture of homogeneous dispersion-free states. He gives the following two definitions (von Neumann 1955, pp. 306–307):

α An ensemble is *dispersion-free* if $Exp(\mathfrak{R}^2) = (Exp(\mathfrak{R}))^2$, for all $\mathfrak{R}$.

β An ensemble is *homogeneous* or *pure* if its statistics are the same as that of any of its sub-ensembles, i.e.,

$$Exp = a\,Exp_1 + bExp_2 \Longrightarrow Exp = Exp_1 = Exp_2.$$

Now suppose there are indeed homogeneous ensembles. Then if hidden variables exist (that means any dispersive ensemble can thus be split into two or more non-dispersive ones), then the homogeneous ensembles must be dispersion-free. In other words, no dispersive ensemble can be homogeneous. This is the unavoidable consequence of postulating the existence of hidden variables, according to von Neumann. And indeed, in classical Kolmogorov-type statistical ensembles, all and only dispersion-free ensembles are homogeneous.

But what about quantum mechanical ensembles? Is there a way to think of them in these classical terms? Von Neumann proves that this is not the case: all homogeneous ensembles are dispersive; there are no non-dispersive ensembles.

Let us consider his proof on this point in more detail. To show that there are no non-dispersive ensembles, he needs to consider a theory general enough to deal with statistical theories of the types described by both Case I and Case II above. (What we shall later see—and what Hermann focuses her criticism on—is that in making a certain assumption during his proof von Neumann loses the generality he begins with.)

Von Neumann implements this first by stipulating that every physical ensemble corresponds to an expectation functional, which is supposed to characterise that ensemble completely from a statistical point of view. There are several conditions this *Exp*-functional must satisfy, by von Neumann's lights (von Neumann 1955, pp. 313–314 for (0), (I), (II) and p. 311 for (A'), (B'); labels for all but 0 follow the original text).

(0) To each observable of a quantum mechanical system corresponds a unique hyper-maximal Hermitian operator in Hilbert space. This correspondence is one-to-one.[1]

(I) If the observable $\mathfrak{R}$ has operator R then the observable $f(\mathfrak{R})$ has the operator $f(R)$.

(II) If the observables $\mathfrak{R}, \mathfrak{S}, \ldots$ have the operators $R, S, \ldots$, then the observable $\mathfrak{R} + \mathfrak{S} + \cdots$ has the operator $R + S + \cdots$ (Note that the *simultaneous measurability* of $\mathfrak{R} + \mathfrak{S} + \cdots$ is *not* assumed.)

(A') If the observable $\mathfrak{R}$ is by nature a *nonnegative* quantity, then $Exp(\mathfrak{R}) \geq 0$.

(B') If $\mathfrak{R}, \mathfrak{S}, \ldots$ are arbitrary observables and $a, b, \ldots$ real numbers, then

$$Exp(a\mathfrak{R} + b\mathfrak{S} + \cdots) = a\,Exp(\mathfrak{R}) + b\,Exp(\mathfrak{S}) + \cdots$$

Von Neumann demonstrates on the basis of these assumptions that there exists a linear, positive semi-definite Hermitian matrix U_{mn} such that for any observable $\mathfrak{R}$,

$$Exp(\mathfrak{R}) = \sum U_{nm} R_{mn} = \mathrm{Tr}\,(\mathrm{UR})\,. \tag{7.1}$$

[1] Later on this assumption will be challenged in particular through modal interpretations of quantum mechanics, where the correspondence between observables and Hermitian operators is not one-to-one.

Thus every ensemble in quantum mechanics is characterised by a statistical operator known as the density operator (or density matrix).[2]

Von Neumann continues by considering whether there exist (i) dispersion-free or (ii) homogeneous ensembles among the density operators U (ibid., p. 320 ff.). For (i), the question becomes: what statistical operators U have $\mathrm{Tr}(UR^2) = [\mathrm{Tr}(UR)]^2$ for all R? It turns out that no U fulfill this requirement, and so no dispersion-free states can exist. In other words, for any quantum state one can always find an observable exhibiting dispersion. Regarding (ii), von Neumann proves that homogeneous ensembles do exist, and in fact they are the pure quantum states (in terms of density matrices, the one-dimensional projection operators). Thus, from (i) and (ii), all ensembles show dispersion, even homogeneous ones.

Now the question of hidden variables becomes whether the dispersion in the homogeneous ensembles may be explained by the fact that the states are mixtures of several states 'which together would determine everything causally, i.e., lead to dispersion free ensembles' (von Neumann 1955, p. 324). He goes on to answer his own question as follows:

> The statistics of the homogeneous [dispersive] ensemble [...] would then have resulted from the averaging over all the actual states of which it was composed [...] But this is impossible for two reasons: First, because then the homogeneous ensemble in question could be represented as a mixture of two different ensembles, contrary to its definition. Second, because the dispersion free ensembles [...] do not exist.

In conclusion, no homogeneous ensembles exist that are dispersion-free, therefore the assumption of the existence of hidden variables is refuted.

7.3 John Bell and the Standard View

John Bell paved the way for what now is considered the standard view on von Neumann's theorem. In 1964 (but published in 1966) Bell intended to show what the problem with von Neumann's argument was after he 'saw the impossible done' in Bohm's hidden variables theory (Bell 1982, p. 990). Bell traced the problem to von Neumann's assumption (B'),

$$Exp(a\,\mathfrak{R} + b\,\mathfrak{S}) = a\,Exp(\mathfrak{R}) + b\,Exp(\mathfrak{S})\,, \tag{7.2}$$

the assumption of linearity of expectation values for all possible observables. What is not often realised is that this linearity assumption holds true in quantum mechanics *irrespective* of whether the operators R and S commute. But as Bell well recognises, the problem is not that linearity fails for non-commuting operators. Rather, Bell reasons as follows: not only is this linearity relation generally true for quantum mechanical states, but it is also required by von Neumann for his hypothetical

---

[2]Gleason (1957) proves this also, but requires (B') only for commuting observables, with the extra assumption that the dimension of the Hilbert space be greater than or equal to three.

dispersion-free states. The expectation value for a dispersion-free state must equal one of the operator's eigenvalues, but eigenvalues *do not* generally combine linearly in quantum theory, unlike in classical theory. There are famous counterexamples to this end; for instance, Bohm and Aharonov (1957) had already presented the following case: the sum of two spin observables ($\sigma_x + \sigma_y$) will have eigenvalues $\pm\sqrt{2}$, whereas the spin observables independently each have eigenvalues ± 1. There is no way that you can add ± 1 to get $\pm\sqrt{2}$, and so the linearity assumption for *eigenvalues* fails in the quantum mechanical case, and this is where the trouble lies.

Bell writes:

> The essential assumption can be criticized as follows. [...] A measurement of a sum of noncommuting observables cannot be made by combining trivially the results of separable observations on the two terms—it requires a quite distinct experiment. [...] But this explanation of the nonadditivity of allowed values also establishes the nontriviality of the additivity of expectation values. The latter is quite a peculiar property of quantum mechanical states, not to be expected a priori. (Bell 1966, p. 449)

To wit,

> [These assumptions] are seen to be quite unreasonable when one remembers with Bohr 'the impossibility of any sharp distinction between the behavior of atomic objects and the interaction with the measuring instruments which serve to define the conditions under which phenomena appear.' (ibid., p. 447)

In a sense Bell's criticism becomes ironic: Bell is using Bohr to combat von Neumann arguing against hidden variables, in a judo-like manoeuvre. In his 1982 paper, Bell remarks that the linearity assumption

> [...] cannot possibly hold before averaging, for the individual results [...] are eigenvalues, and eigenvalues of linearly related operators are not linear related. [...] His 'very general and plausible' postulate is absurd. (Bell 1982, p. 994)

Here we see a subtle change in Bell's treatment of von Neumann's assumption: it is no longer just false, but it is 'absurd'. And now the standard view considers this allegedly superior proof 'silly':

> A few years later Grete Hermann, 1935, pointed out a glaring deficiency in the argument, but she seems to have been entirely ignored. Everybody continued to cite the von Neumann proof. A third of a century passed before John Bell, 1966, rediscovered the fact that von Neumann's no-hidden-variables proof was based on an assumption that can only be described as silly—so silly, in fact, that one is led to wonder whether the proof was ever studied by either the students or those who appealed to it to rescue them from speculative adventures. (Mermin 1993, pp. 805–806)

Perhaps it is too dismissive to call the argument 'silly', as the key assumption does hold in certain cases (e.g. Kolmogorov states). But we may conclude that von Neumann's proof is an unconvincing argument against hidden variables. Von Neumann himself does appear to have over-interpreted the significance of his result. He writes:

> In each state ϕ the expectation values behave additively: $(R\phi, \phi) + (S\phi, \phi) = ((R + S)\phi, \phi)$. The same holds for several summands. We now incorporate this fact into our general set-up (at this point not yet specialized to quantum mechanics). (von Neumann 1955, p. 309)

But the additivity rule for expectation values in the case of incompatible (non-commuting) observables cannot be justified in light of the Bohrian point that contexts of measurement play a role in defining the nature of quantum reality. As Bell puts it (Bell 1966, p. 449): 'There is no reason to demand it [linearity] individually of the hypothetical dispersion free states [...].' And thus there is no reason to demand that the dispersion-free states are of the form of a density operator U. Indeed, what von Neumann has shown is that if you do require this, you are committed to the trace rule, and thereby committed to representing statistical states using density operators.

I believe the most appropriate conclusion is captured in part by a recent point in Bub (2011) that von Neumann's proof does not *rule out* hidden variables, but rather holds for a *limited class* of hidden variables, namely, those that obey assumption (B'). Thus the proof does not fail on the whole, but works only for a restricted category of hidden variables. It is fair to say von Neumann appears to have missed this point.

To cast his proof in a more positive light, one might instead consider it a *completeness proof* in as much as quantum mechanics already includes all the states allowed by von Neumann's additivity postulate. This highlights that the additivity rule, especially for non-commuting observables, is not a trivial thing.

Von Neumann's argument might also be positively construed as a *consistency proof*, in that the additivity rule, although lacking justification, nevertheless *holds true* in quantum mechanics (Jammer 1974, p. 274). This is indeed surprising, for a priori no statistical relations are expected between $Exp(\mathfrak{R})$ and $Exp(\mathfrak{S})$. The additivity holds true in the quantum mechanical case because, as described by Belinfante (1973, p. 25): 'It so happens that the other axioms and postulates of quantum theory conspire to make $Exp(\mathfrak{R})$ [...] expressible as $\int \psi^* R \psi \, dx$'.

In conclusion, it does not count against von Neumann's *theorem* that the non existence of dispersion free states is so easy to prove in the case of the hidden variables he chooses to investigate (a lot easier than he thought, in fact).[3] The real issue is whether the premises in the proof plausibly capture an appropriate notion of hidden variables. Let us now turn to Grete Hermann's treatment of this proof, some thirty years prior to Bell's analysis.

7.4 Hermann's Critique of von Neumann's Argument

In Grete Hermann's 1935 essay, 'Die naturphilosophischen Grundlagen der Quantenmechanik', she includes a section entitled 'Der Zirkel in Neumanns Beweis', 'The Circle in Neumann's Proof' (Sect. 7 of Chap. 15). In this section she focuses on (B'), von Neumann's linearity of observables, writing that 'Neumann's proof stands or falls with this assumption' (Chap. 15, p. 252). This is thus the crucial assumption in her eyes also.

Hermann also comments on the problematic status of the additivity rule in light of the impossibility of simultaneous measurement of non-commuting observables—the above-mentioned 'Bohrian' point. She states:

[3]This point is addressed in Bacciagaluppi and Crull (2009) (*eds.*).

> The relation is, however, not self-evident for quantum mechanical quantities between which uncertainty relations hold, and in fact for the reason that the sum of two such quantities is not immediately defined at all [...] However, for the so-defined concept of the sum of two quantities that are not simultaneously measurable, the formula given above requires a proof [...] From this rule Neumann concludes that for ensembles of systems with equivalent wave functions, and therefore for all ensembles generally, the addition theorem for expectation values holds also for quantities that are not simultaneously measurable. (ibid., p. 252)

Von Neumann does indeed give such a proof. As Hermann explains:

> [S]ince each ensemble of physical systems can be decomposed into sub-ensembles whose elements agree with each other in terms of their wave functions, then it follows, first, that the theorem in question needs to be proved only for ensembles whose elements satisfy the condition of equal wave functions. (ibid.)

In other words: either we are already dealing with a case of pure states, or it can be decomposed into sub-ensembles corresponding to pure states and considered case by case. For such ensembles, von Neumann proves that they obey the following rule: $((R + S)\phi, \phi) = (R\phi, \phi) + (S\phi, \phi)$, where ϕ is the wave function of the observed system. It now follows that for all ensembles, the sum rule for expectation values holds—even for such quantities that cannot be measured simultaneously.

Thus, also Hermann recognises this is true for quantum mechanics. She states that the interpretation of the expression $(R\phi, \phi)$ will be crucial for the entire proof:

> [U]ntil the proof of the impossibility of new variables—which has yet to be given here—the expression $(R\varphi, \varphi)$ may denote the expectation value of $\mathfrak{R}$-measurements only for such ensembles of physical systems on which *this but only this condition* is imposed—of being in the state φ; to remain applicable, this [posit] must instead leave open whether this expectation value is also the same in all subsets of such ensembles that are selected from them on the basis of any new features. (ibid.)

What Hermann is saying here is that in evaluating this quantity in the state ϕ, which according to our assumption consists of an ensemble of identical wave functions, we must remain open to the possibility that this ensemble itself is not homogeneous with respect to new characteristics:

> But if one leaves this open, then one can no longer infer, from the asserted addition rule for $(R\varphi, \varphi)$, that also in these subsets the expectation value of the sum of physical quantities is the same as the sum of their expectation values. (ibid.)

'In this way, however,' she notes, 'an essential step in Neumann's proof is missing' (ibid.), and so Hermann is led to the following conclusion:

> If instead—like Neumann—one does not give up on this step, then one has implicitly absorbed into the interpretation the unproven assumption that there can be no distinguishing features, of the elements of an ensemble of physical systems characterised by φ, on which the result of the $\mathfrak{R}$-measurement depends. However, the impossibility of such features is precisely the claim to be proven. Thus the proof runs in a circle. (ibid.)

Hermann concluded, as would Bell, that von Neumann *ruled out* the existence of dispersion-free states by requiring without further justification the additivity rule also at the level of hidden variables.

7.5 Comparing Hermann and Bell

Three important points can be made when comparing Hermann's earlier treatment of von Neumann's proof to Bell's 1960s treatment of the same proof; each of these points highlights ways in which Bell's paper accomplishes more, and with greater strength, than Hermann in her 1935.

First, Bell and Hermann both use 'the Bohrian point' of incompatible observables to argue against the additivity rule: it cannot be reasonably assumed at the hidden variables level, but nonetheless von Neumann does so. Hermann concludes that the proof is therefore 'circular', which is a strong logical point. However, Bell's critique is much more powerful. He shows that if you apply this assumption (B'), this further entails the additivity of eigenvalues—and then Bell goes on to provide an explicit counterexample. Thus, although Bell and Hermann start at the same troublesome moment in von Neumann's argument, they give differing final verdicts; Bell's analysis points more clearly to the deeper assumption arising from (B') that gets the proof into trouble.

Secondly, Bell's 1966 paper received a great deal of attention perhaps owing to its providing a much more comprehensive analysis than was available to Hermann. For instance, Bell addresses other no-go theorems (e.g. that of Jauch and Piron), and includes discussion of the important theorem by Gleason (1957). Bell also provides, independently, a Kochen–Specker-type proof *avant la lettre* (Kochen and Specker 1967).

Thirdly, Bell's paper explicitly addresses non-locality issues by invoking Bohm's hidden variables theory (recall Bell's reaction in his 1982 paper to Bohm's theory: 'I saw the impossible done'), whereas Hermann's motivation for her Sect. 7 remains obscure. In fact, Hermann has up to this point in her 1935 essay listed the failure of various no-hidden-variables arguments, and includes von Neumann among these. Then, after demonstrating what she clearly views as a critical flaw in his proof, she writes that notwithstanding the strength of the mathematical formalism, it cannot be deduced that further undiscovered features with a different mathematical formulation are impossible. In other words, she seems to hold out hope for the possibility of hidden variables after cataloguing many failures to rule out just such a theory. But, remarkably, she then goes on to argue that quantum mechanics as it stands is in fact already complete.

Paparo (2012) has made an interesting case for why Hermann did not end up having this section on von Neumann play a significant role in her overall thesis—and it has to do with what was really at stake for Hermann. Paparo writes (p. 22), '[…] [Hermann] proceeds to show how mathematical and statistical arguments have failed to defend the causality principle and consequently, that *only philosophy* can answer the question of whether it is possible to overcome the limits in the predictability of quantum mechanics.' A little later she adds: 'The mathematical formalism alone is not able to answer the question of whether the limits in the predictability of quantum mechanics are insurmountable or only there due to our lack of knowledge.'

Thus it seems that Hermann did not want to settle the issue by developing yet another hidden variables theory, but rather wanted to show that a philosophical analysis is required for answering the question of causality in quantum mechanics. And I further suggest that Hermann was not trying to be revolutionary: it seems within her character to stress continuity over and above radical changes; she sought a continuous story from Kant on that could easily include the natural sciences.

7.6 The Reception of Hermann's Criticism

Hermann was clearly ahead of her time. Only after John Bell's 1966 paper does the limited applicability of von Neumann's no-go proof really become widely known. Yet Hermann's anticipation of Bell's criticism was surely known to certain key figures (see Chap. 8). For instance, Weizsäcker wrote a review of Hermann's 1935 essay and so would have encountered her section on the circularity of von Neumann's proof. And what of Heisenberg? He was surely aware of this argument in her essay, which was written at the end of Hermann's stay at his own institute in Leipzig. Thus the question becomes more mysterious as to why her poignant criticism was ignored at the time, much less now that we are aware of her work more broadly.

One might speculate on this point by recalling my earlier statements about the 'holy' status of von Neumann's proof, and Belinfante's quote. Similarly, Bridgman writes:

> Now the mere mention of concealed parameters is sufficient to automatically elicit from the elect the remark that John von Neumann gave absolute proof that this way out is not possible. To me it is a curious spectacle to see the unanimity with which the members of a certain circle accept the rigor of von Neumann's proof. (Bridgman 1960, p. 206)

Finally, Feyerabend adds the following by way of anecdote:

> [Bohr] came for a public lecture [...]. At the end of the lecture he left, and the discussion proceeded without him. Some speakers attacked his qualitative arguments—there seemed to be lots of loopholes. The Bohrians did not clarify the arguments; they mentioned an alleged proof by von Neumann, and that settled the matter [...] like magic, the mere name 'von Neumann' and the mere word 'proof' silenced the objectors. (Feyerabend 1996, pp. 77–78)

Despite this admittedly less-than-concrete evidence, it nevertheless seems plausible that given the status of von Neumann's work at the time, Hermann did not feel the need to challenge the hegemony. But perhaps more important is the mere fact that Hermann's treatise was published in an obscure series. The excerpts that appeared in 1935 in the important journal *Die Naturwissenschaften* (Hermann 1935a) did not include the argument against von Neumann, but focused on her Kantian ideas. Of course one might also speculate that Hermann's work on this topic was not made widely known because those with whom she interacted on this point—especially Bohr and Heisenberg—had some interest in preserving belief in the results of von Neumann's no-hidden-variables proof, since it supported their own ideas regarding the incompletability of quantum mechanics.

One might further posit various sociological factors contributing to Hermann's proof getting buried, or at least not receiving its due attention. For one, she was a woman in a time when women were not yet well-received in the scientific community. Furthermore, she was rather young, lacking in influential connections, and approaching quantum mechanics as an outsider (as primarily a philosopher and mathematician rather than physicist). She was also a political outsider and active dissident, and perhaps these reasons contributed in some small way to her work failing to achieve the level of notoriety it might have had.

But as Paparo (2012, pp. 67–68) has also argued, it may well have been that Hermann herself had a different agenda entirely:

> [...] I claim that one of the most relevant reasons has been overlooked in previous discussions on the matter. In my opinion, a primary reason is to be found in Hermann's personality and work, as she did not actively pursue the wider dissemination and understanding of her discovery. First of all, she published the critical analysis of von Neumann's proof written in small font, having earlier stated that anything in such a font could be readily skipped. Secondly, in later editions of the paper, the disproof is simply left out, stressing again the insignificance Hermann ascribed to it. Either way, the result of her critique to von Neumann's proof only served to show how it was impossible to answer the problem of the completeness of quantum theory on a physical level, and why a philosophical analysis was necessary. What Hermann particularly wanted to show is that quantum mechanics could still be seen as causal and complete, without having to assume some hidden causes.

I agree with this: Hermann herself was less invested in resolving the issue of hidden variables and causal indeterminacy from a physical perspective but rather from a philosophical point of view. And again, it seems within her personality that she did not desire the status of a revolutionary or radical, but wanted to present in her arguments a more conciliatory attitude.

Perhaps also Hermann's unfortunately limited role in the standard narrative—especially regarding the question of hidden variables—is due to the fact that Max Jammer, author of several seminal sources in the historical and philosophical foundations of quantum mechanics, and from whom many of us learned this material, discussed her criticism of von Neumann's proof, but in fact criticised her circularity claim (Jammer 1974, pp. 272–275).

It is hoped that with the translation of her 1935 paper into English and through the essays in the present volume, her substantial contributions to the early debates will become more widely known and studied.

Furthermore, it is hoped that a more thorough study of von Neumann himself will result from these investigations. As Redhead has said of von Neumann's 1932 work (Redhead 1987, p. 1), it is 'a book more frequently referred to than read by physicists because of its mathematical sophistication.' Or, as suggested by Ted Bastin,

> Well, I suppose that they regard von Neumann's book as a perfectly adequate formal treatment for pedants, people who like that sort of thing [formal mathematics]. They wouldn't read it themselves but they're glad somebody has done all that hard work! (Bastin 1977, p. 157)

Finally, James Albertson, who has provided an accessible and therefore well-studied Dirac formulation of the proof (Albertson 1961) is not at all critical of von

Neumann, and in fact relegates all assumptions—including the very problematic one discussed above—to an appendix.

The remedy is, then, not only to aid in the wider dissemination of Hermann's own work, but to study the primary sources, going back to von Neumann's *Mathematical Foundations* (in the original German as well as in English) and seeing what riches it yields.

References

Albertson J (1961) Von Neumann's hidden-parameters proof. Am J Phys 29:478–484

Bacciagaluppi G, Crull E (2009) Heisenberg (and Schrödinger, and Pauli) on hidden variables. Stud Hist Philos Mod Phys 40(4):374–382

Bastin T (1977) A clash of paradigms in physics. In: Duncan R, Weston-Smith M (eds) The encyclopaedia of ignorance. Pergamon Press, New York, pp 119–127

Belinfante FJ (1973) A survey of hidden-variables theories. Pergamon Press, Oxford

Bell JS (1966) On the problem of hidden variables in quantum mechanics. Rev Mod Phys 38:447–452. Reprinted as Chap. 1 of Bell (1987)

Bell JS (1982) On the impossible pilot wave. Found Phys 12(10):989–999. Reprinted as Chap. 17 of Bell (1987)

Bell JS (1987) Speakable and unspeakable in quantum mechanics. Cambridge University Press, Cambridge

Bohm D, Aharonov Y (1957) Discussion of the experimental proof for the paradox of Einstein, Rosen, and Podolsky. Phys Rev 108(4):1070–1076

Bridgman PW (1960) Review of de Broglie, Non-linear wave mechanics: a causal interpretation. Sci Am 203(4):206

Bub J (2011) Is von Neumann's 'no hidden variables' proof silly? In: Halvorson H (ed) Deep beauty: understanding the quantum world through mathematical innovation. Cambridge University Press, Cambridge, pp 393–408

Feyerabend P (1996) Killing time: the autobiography of Paul Feyerabend. University of Chicago Press, Chicago

Gleason AM (1957) Measures on the closed subspaces of a Hilbert space. J Math Mech 6:885–893

Hermann G (1935a) Die naturphilosophischen Grundlagen der Quantenmechanik. Die Naturwissenschaften 23(42):718–721

Hermann G (1935b) Die naturphilosophischen Grundlagen der Quantenmechanik. Abhandlugen der Fries'schen Schule 6(2):75–152

Jammer M (1974) The philosophy of quantum mechanics: the interpretations of quantum mechanics in historical perspective. John Wiley, Hoboken, NJ

Kochen S, Specker EP (1967) The problem of hidden variables in quantum mechanics. J Math Mech 17:59–87

Mermin ND (1993) Hidden variables and the two theorems of John Bell. Rev Mod Phys 65:803–815

von Neumann J (1932) Mathematische Grundlagen der Quantenmechanik. Springer, Berlin

von Neumann J (1955) Mathematical Foundations of Quantum Mechanics. Princeton University Press, Princeton, NJ. Translation of von Neumann (1932) by Beyer RT

Paparo G (2012) Grete Hermann: mathematician, philosopher, and physicist. Master's thesis, Utrecht University, The Netherlands

Redhead M (1987) Incompleteness, nonlocality, and realism: a prolegomenon to the philosophy of quantum mechanics. Clarendon Press, Oxford

Chapter 8
Grete Hermann's Lost Manuscript on Quantum Mechanics

Elise Crull and Guido Bacciagaluppi

8.1 Introduction

The story of how we became aware of Hermann's 1933 essay is as follows. In Weizsäcker's oral interview with Thomas Kuhn in AHQP (von Weizsäcker 1963), he mentions in connection with the discussion of Hermann's time in Leipzig and the 1935 essay that Hermann had sent a previous manuscript to Bohr and Heisenberg. However, no such manuscript is extant either in the Heisenberg archive or in the Bohr Archives, and could have been presumed lost.

However, at the May 2012 workshop on Hermann at the University of Aberdeen more details concerning the lost manuscript emerged in a wonderfully amusing letter written to Grete Hermann by Gustav Heckmann in December 1933 from Copenhagen.[1] Though the whole of Heckmann's letter is delightful in tone and content (see below, Chap. 11; the letter is translated in full as Chap. 13), of special relevance here are his comments about the reception of Hermann's 1933 essay by the Copenhagen

[1]Our special thanks to Dieter Krohn for bringing this letter to the attention of the workshop participants.

Work for this chapter was carried out in part while EC was a postdoctoral research fellow at the Hebrew University's Edelstein Center for the academic year 2013–2014. She wishes to thank the Edelstein Center for their generous support.

E. Crull (✉)
The City University of New York, New York, USA
e-mail: ecrull@ccny.cuny.edu

G. Bacciagaluppi
Universiteit Utrecht, Utrecht, The Netherlands;
IHPST and SPHERE, Paris, France
e-mail: g.bacciagaluppi@uu.nl

E. Crull and G. Bacciagaluppi (eds.), *Grete Hermann - Between Physics and Philosophy*, Studies in History and Philosophy of Science 42,
DOI 10.1007/978-94-024-0970-3_8

physicists. Heckmann writes that he had not had time to really study the essay yet, but had spent over an hour discussing it with Heisenberg.

Heckmann quotes Heisenberg's response to the essay as follows: '"In substance, she is certainly wrong", but "a fabulously clever woman"'. Heckmann also reports that Heisenberg, Bohr and Weizsäcker have all read her paper and are taking it 'absolutely and completely *seriously*', even wishing to compose a joint response. Though Heisenberg believes Hermann must still learn more physics, Heckmann writes that she will 'always find an open door and an open ear with Heisenberg'. Continuing from this remark Heckmann jokes: 'Use it yet, ere you knock at the harder door made of older wood: that of Bohr'. It is natural to assume that this open door and open ear extended by Heisenberg were what encouraged Hermann to travel to Leipzig the following year to study and discuss matters more intensively with both Heisenberg and Weizsäcker. Heckmann's comments also perhaps explain why Hermann did not visit Copenhagen for this purpose.

Thus in Heckmann's letter we learn not only that Bohr, Heisenberg and Weizsäcker had read her essay, considered it important and even wished to prepare a joint response to it, but furthermore we discover that a copy of Hermann's work had also been sent to Dirac. And indeed! It was in the Dirac archives at Churchill College, Cambridge that one of us [GB] found a letter to Dirac from Hermann dated 9 November 1933, to which was attached the lost manuscript, 'Determinism and Quantum Mechanics' (Hermann 1933).

In the cover letter to Dirac, Hermann writes enthusiastically about Dirac's *Principles of Quantum Mechanics*[2] and states that the stimulation she received through his work has 'kindled a desire in me to become acquainted with your opinion concerning several ways of thinking which in the study of quantum mechanics have imposed themselves on me more and more'. In particular, Hermann tells Dirac that she is keen to address the question of determinism in the new theory, but notes that while her interest is primarily philosophical, she focuses here on the entirely *physical* question, as yet unresolved, of whether indeterminism plays an essential role in the theory 'and thus requires a merely statistical treatment' of it. Since a discussion of Dirac's approach to quantum mechanics comprises a large portion of her essay, Hermann submits it to Dirac for his consideration and asks humbly for his opinion.

We do not know if and how Dirac responded to this 1933 analysis of indeterminism in quantum mechanics,[3] but even without considering Dirac's (and Heisenberg's, and Bohr's) response it is clear that this manuscript is a treasure. In the following, we shall give a description of the manuscript, comparing it to her more comprehensive essay of 1935. We shall also begin to investigate questions relevant to the historical and philosophical context within which Hermann's essay was composed. As we have argued elsewhere (and continue to emphasise), Hermann's role in the philosophical

---

[2]Dirac's book appeared in German translation in the same year as the first English edition (Dirac 1930); the second English edition followed only in 1935.

[3]We have checked the contents of Hermann's Nachlass with regard to her correspondence with Heisenberg, Weizsäcker and Van der Waerden in late 1933–early 1934, as well as the Dirac Archive, but found no evidence of his reply.

development of quantum mechanics can no longer in good conscience be ignored, as there is abundant evidence for her deserving significant notice in this context, among others.[4]

8.2 Summary of the 1933 Manuscript

The aim of Hermann's 1933 essay (as laid out in her Section 1) is to investigate, solely on grounds of the physics and its theoretical structure (and apart from purely speculative or philosophical assumptions) whether or not causality is in fact threatened by quantum mechanics; importantly, she will attempt to demonstrate this in a novel way as compared to prior attempts to salvage the law of causality, namely by showing that the aspects of the theory giving rise to indeterminism are in fact not logically necessary to the theory as a whole, and can therefore be excised without slightest injury to either the logical coherence or explanatory power of the theory. In this way, the theory can be fully and correctly understood as compatible with determinism.

Hermann rejects as spurious the claim that causality is preserved because the loss of predictability in quantum mechanics is due to the limitations imposed on our knowledge of the state of a system. If it is indeed impossible in principle ever to determine the causes of an event, then according to Hermann 'the claim that such a cause nevertheless exists is mysticism'. Hermann thus focuses on the question whether quantum mechanics prohibits in principle that results of measurements may be predicted from knowledge of initial conditions and laws (however specified), thus by-passing explicit discussion of (possibly controversial aspects of) the notion of causality.[5]

Hermann investigates one by one those particular moments in the quantum theory when indeterminism is typically invoked, and in each case demonstrates that the so-called indeterminism can be cut out of the theory without affecting what remains (explicitly, in terms of the theory's ability to account for observational data, or to use her Kantian phrase, in keeping with 'the data of experience').

Of course the first place one turns when discussing the supposed indeterminism of quantum mechanics is to Heisenberg's uncertainty relations (her Sect. 2). Hermann asks what it is, exactly, that is meant to be undetermined in these relations.[6] There are two potential answers to this question: upon measurement of one of a pair of

[4]Translations of both Hermann (1933) and Hermann (1935) are included in this volume as Chaps. 14 and 15, respectively. German editions of both and of a selection of Hermann's correspondence (among which the letter to Dirac, the letter from Heckmann and correspondence with both Heisenberg and Weizsäcker—including the latter's reply to Hermann's manuscript on behalf of Bohr) are included in a forthcoming volume edited by Herrmann (2017).

[5]In the 1935 essay Hermann takes care to disentangle causality and predictability, since determining the causes of an event need not mean determining them *in advance*.

[6]We use 'undetermined' here, because Hermann clearly takes the German 'unbestimmt' to (misleadingly) suggest the reading 'that which has not been determined but has a value' rather than 'that which does not have a determinate value in the first place'.

canonical variables (i) the exact value of the other remains undetermined, or (ii) the result of a measurement of the other cannot be predicted exactly.

Hermann regards (i) as incorrect because the very formulation of the answer implies the existence of an exact value, and this contradicts the necessity of the equal applicability of the wave and particle pictures to describe the system of interest. Answer (ii), though logically feasible, nevertheless allows for the possibility of discovering certain new properties that would enable the exact prediction of measurement outcomes (Hermann is here of course discussing what we now call hidden variables). Thus, all that is implied through this solution is a temporary inability to exactly predict outcomes—but this solution in no way excludes the possibility of hidden variables.

Hermann now turns to Dirac's formulation of quantum mechanics (Sect. 3), which undeniably (she writes) contains indeterminism. Again the aim will be to discover whether or not this indeterministic aspect is necessary for the formulation. In other words: can the locus of indeterminism be clearly identified and then cleanly excised from Dirac's presentation of the theory without in any way hindering either its logical coherence or its explanatory virtues? Hermann says yes, and yes again.

The moment of indeterminism in Dirac's formulation comes in with the interpretation of the coefficients in the decomposition of a wave function into eigenfunctions of a given observable, as the probabilities for finding upon measurement the corresponding eigenvalue of the observable; when multiple coefficients of the decomposition are non-zero, one cannot predict exactly the outcome of an eigenvalue measurement.

There is nothing as yet in Dirac's formulation that forbids the existence of hidden variables. This possibility, argues Hermann, is only shut out through Dirac's concept of maximal observation. However, she demonstrates that the concept of maximal observation alone is insufficient for applying the formalism to empirical data: one requires in addition a criterion ensuring the *maximality* of a given set of 'maximal observations'. Yet there is no such criterion available: we have not as yet discovered a method of guaranteeing that beyond any maximal observation there do not still exist unknown properties which could, either potentially or actually, more completely determine the state without modifying it in terms of other ('compatible') properties.

Yet Dirac's formalism succeeds in explaining the phenomena. Hermann concludes that the condition that states be defined via maximal observation must not be doing any of the explanatory heavy-lifting in the theory. Therefore this condition—which is, recall, responsible for the occurrence of indeterminism in this view—can be lifted out of the theory without affecting its coherence or diminishing its explanatory power.

Hermann admits we need enough of a concept of maximal observation to guarantee that the indeterminism arising in quantum mechanics has its origins in the Heisenberg uncertainty relations. She modifies Dirac's definition accordingly: a state has been defined through a maximal observation *with respect to* a given pair of non-commuting observables if further observation would result in the narrowing of either variable's eigenvalue spread to a degree violating the relevant Heisenberg relation.

In the case of classical mechanics, such a condition would restrict states to simultaneous eigenstates of all physical quantities. In the case of quantum mechanics, there can still exist states corresponding to superpositions of eigenstates; however, the spread of these superposed eigenstates (the range of eigenvalues with non-zero decomposition coefficients) can be uniquely determined using the sort of measurement described as maximal by Hermann. Maximal observations thus determine frequencies of measurement results, but this is compatible with determinism, since it does not rule out that finer subensembles determined additionally by as yet unknown physical traits might not display any spread of results at all.

After her Sect. 4 titled 'Neumann's Proof of Indeterminism', which arguably does a clearer job than the counterpart section in her 1935 essay of explaining the fault in von Neumann's proof ruling out hidden variables (we shall examine this more closely below, Sect. 8.3.3), Hermann concludes her 1933 essay with a positive thesis. In Sect. 5, somewhat misleadingly titled 'The statistical interpretation of quantum mechanics' (why misleadingly shall also be discussed below, Sect. 8.4), Hermann acknowledges that her deterministic—or rather, not-yet-proven-indeterministic—construal of Dirac's formulation leaves one with a problem: what is the intuitive meaning of the spread of, e.g., position values for an electron? (Recall from the section on the uncertainty relations that Hermann rejects the idea that electrons have determinate but generally undetermined values of position and momentum.)

To this she writes:

> The simplest intuitive picture that positively conveys this notion consists surely in imagining that the whole interval (q, dq) is filled with the smeared-out electron. Similarly one arrives at the notion that this electron has no single momentum, but rather is an object diffusing like a wave-peak, whose momentum values fill the interval (p, dp). Then the expressions $|c_i|^2$ of Dirac's theory give the density, as it were, with which a definite point of this position or momentum interval is occupied by the electron. (Chap. 14, p. 236)

Hermann then suggests an interpretation for this smearing-out that overcomes the problems (pointed out by Schrödinger himself) of an interpretation as mere spatial extension. For the latter, it is acceptable to further subdivide the interval over which the electron is smeared out to come to a more exact resolution of the system's variables; for the former, any such attempt to subdivide the interval violates the empirically established duality of the wave and particle pictures. 'This evidently means', she writes, 'that only the whole system, satisfying the Heisenberg relations with respect to its position and momentum intervals, can enter into interaction with other physical systems. In any observed physical effect of a system, thus, the system is always participating as a whole' (ibid.). And to finish this train of thought: 'The classical assumption whereby interaction between masses can be reduced to interaction between point masses, is thus supplanted by a more complicated assumption in quantum mechanics: interaction between physical systems presupposes that each of them exhibit an extension corresponding to the Heisenberg relations' (ibid.).

Her concluding paragraph follows on the heels of the above statements, and is worth repeating in its entirety (especially as it will be discussed in more detail below in Sect. 8.4):

> Since every observation of a system constitutes an effect of the system on the observer, it follows that the object of an observation can only be a system that is subject to these relations. Now if one performs an eigenvalue measurement on it—say of position (defined by an interval before the measurement)—then this measurement transforms the system into another that has one single position but is distributed over the entire eigenvalue space of momentum. The laws according to which this transformation takes place in detail, which

therefore determine which position is then observed, are at present unknown. However, it is not impossible that one of these days physics will get onto their trail. (Chap. 14, p. 237)

8.3 Comparison with the 1935 Essay

The first point of comparison to be made is that Hermann's 1933 manuscript makes sense of the strange logic of her 1935 essay, i.e., why she is at pains to list all attempts to causally complete quantum mechanics and demonstrate their insufficiency, but then continues to say that this search for a causal completion is moot because quantum mechanics is *already* causally complete. Especially Filk and Seevinck (this volume, Chaps. 5 and 7) have pointed out this odd about-face partway through the 1935 essay, and proffered various plausible explanations. Indeed, the 1935 essay's logic seems to cry out for explanation: one is led to expect Hermann to follow her intensive catalogue of hidden variables no-go arguments with an argument of her own *for* hidden variables, but she does not do this. Instead, she suggests that such endeavours are a waste of time and effort, searching for a causal completion where one already exists.

The 1933 paper allows us to make sense of the strange change of tack in 1935: while she has, it is clear, changed her thinking about the causal completeness of quantum mechanics, she still wishes to include in the latter essay some of the initial writing she had done regarding the general question of completability in 1933. And so she includes to a large extent the material from the beginning sections of her 1933 paper (namely, Sects. 1 and 2) in the first three sections of her 1935 essay. Also, Sect. 3 on Dirac in the 1933 paper significantly overlaps with Sects. 5 and 6 from 1935, and Sect. 4 on von Neumann with Sect. 7 in 1935.

As we have seen, in 1933 Hermann's goal is to argue that the indeterministic aspect of quantum mechanics can be cut out of (at least Dirac's formulation of) the theory without affecting explanatory power or logical cohesion. Her argument focuses importantly on the continuing possibility of hidden variables, and demonstrations to the effect that the impossibility proof of von Neumann has nontrivial flaws.

The question then arises: what caused her to change her mind in the intervening years? It is likely that it was largely due to her participation in Heisenberg's seminar at Leipzig and the many intense discussions she held during that time with both Weizsäcker and Heisenberg.[7] However, it may also have been due in part to a deepening of her understanding of Bohr's approach to quantum mechanics (perhaps through reading more of Bohr's papers, as we know was suggested to her by Heisenberg[8]), in particular a more nuanced consideration of Bohr's complementarity. This

[7]Indeed, on 16 June 1934 she wrote to her mother that Heisenberg had finally had the upper hand in their debates (Herrmann 2017, Part III, Letter 8).

[8]Heisenberg may have been referring to the collection of four of Bohr's essays, then (in December 1933) in preparation, to be published in 1934 under the title *Atomic Theory and the Description of Nature* (Bohr 1934).

idea and the relative character of observation play the principal roles in her later arguments for the causal completeness of quantum theory. Not only do such Bohrian considerations only arrive in her 1935 piece, but they are the key for her arguably radical departure from Nelson with her natural-philosophical conclusion regarding the splitting of truth.

The role of natural philosophy and, in particular, Nelsonian philosophy, really only comes to bear in a deep way in the arguments of the second essay. The work of 1933 is only nominally Friesian (cf. her opening paragraphs); Hermann begins from a Kantian desire to maintain causality in light of the new physics. She also offers a reading of Heisenberg's uncertainty relations as revealing the merely analogous status of classical concepts when applied to quantum systems, though this point about analogies is already largely contained in Heisenberg's own treatment of his relations (cf. Heisenberg 1927). The architecture of Hermann's 1933 essay—how she interprets, analyses, then answers the question at issue—also smacks of Nelsonian, natural-philosophic methodology (cf. discussion of this methodology and what it entails in Chap. 3.) Beyond this, however, one sees little in the way of Friesian transcendental idealism. The same is not so of the 1935 paper, which (arguably) considers the context-dependence of quantum mechanical 'truths' as impetus for developing a novel, anti-dogmatic natural-philosophical worldview extending beyond physics.

8.3.1 *Heisenberg's Uncertainty Relations*

As mentioned above, Hermann's treatment of the uncertainty relations changes little from 1933 to 1935, and in fact occupies the same place: these relations are discussed in Sect. 2 of both essays (in 1935 her treatment extends also into Sect. 3). While she has clearly added more detailed experimental considerations as well as more forcefully made her philosophical points in 1935 (including invocation of Bohr's correspondence principle), we wish to draw special attention to a portion of this discussion appearing in both of her essays (Sect. 3 of 1933; Sect. 5 of 1935) that forces us to revise a claim made in our 2009 paper (Bacciagaluppi and Crull 2009) regarding contextual versus non-contextual hidden variables. In that paper, we argued that in Heisenberg's 1935 response to EPR he was the first to recognise two distinct classes of hidden variables—those that would determine certain values of the observed system independently of observational context (in contemporary parlance: 'non-contextual hidden variables') versus those that would so determine aspects of the observed system but depend on the means of observation (today's 'contextual hidden variables'). We remarked that while Heisenberg's treatment of the latter class of hidden variables is rather 'hand-waving' and in fact incorrect, the historical point remains.

However, from Hermann's 1933 discussion of the uncertainty relations we find that she has preempted Heisenberg's division of hidden variables into two types (and indeed, one continues to speculate—as we do in the 2009 paper—the extent to which Heisenberg's thinking in his response to EPR was importantly influenced by

Hermann; we might now include this point about hidden variables among such). The proof of which: in the 1933 essay, Hermann points out that nothing within Heisenberg's relations precludes the possibility of 'discovering other, as yet unknown physical quantities, such that the knowledge of their values together with the position and momentum of the electron are sufficient to predict the result of the position measurement'. Fine thus far, but she continues from there with the following statements:

> These quantities need not necessarily be determinants just of the observed electron itself; they could also pertain to the measurement apparatus used for the position measurement, since this indeed exerts a demonstrable influence on the electron.
>
> The task of exactly predicting the result of the position measurement is thus vastly more complicated than classical mechanics assumed. (Chap. 14, p. 227)

A few paragraphs later she applies this consideration to a series of hypothetical position measurements made on an ensemble of electrons homogeneous with respect to q, dq, p, dp:

> But why should it be impossible now, in the cases where the position measurement gives the same value q' [...], to find a trait *that pertained to the electron or the measurement apparatus or both together already prior to the measurement*, and that was not present in the other cases which led to different measurement results—and then to use the presence of this trait as a basis from which to predict the occurrence of the result q' [...] in future measurements? (ibid.; emphasis added)

Thus, she explicitly considers a special case of what we now call contextual hidden variables: the idea that results of measurements on a system may depend not only on as yet undiscovered features of the system itself, but also on details of the measurement context (in this case on as yet undiscovered features of the apparatus).

8.3.2 Dirac's Formalism and Maximal Observation

Section 6 of Hermann's 1935 paper is simply entitled 'What are "maximal observations"?' and briefly conveys the essential point of its earlier incarnation: maximal observations certainly maximally determine—up to agreement with the Heisenberg uncertainty relations—the values of all *quantum mechanical* variables (in particular such variables as position and momentum). However, this does not preclude the existence of hidden variables, i.e. the existence of as-yet undiscovered features of a system (or apparatus) that would determine the results of those experiments that quantum mechanics classifies as 'measurements' of the quantum mechanical variables and so does not suffice as an argument either for or against the completeness of quantum mechanics.

Aspects of Hermann's clear treatment of the Dirac formalism in 1933 appear not in this section on maximal observation, but rather in the prior section dedicated to the interpretation of the wave function. However, there are a few aspects of the fuller discussion in 1933 worth mentioning.

The first thing to do is simply flag the part of her first paragraph in Sect. 3 of the 1933 paper, in which she mentions von Neumann's mathematical objections to Dirac's use of δ-functions as being irrelevant to her own discussion.

By way of comparison we may note that in Hermann's 1933, her understanding of the 'essential distinction' between classical mechanics and quantum mechanics—especially as stated in her Sect. 3 paragraph beginning with 'It seems reasonable to replace Dirac's requirement' (Chap. 14, p. 230)—undergoes significant alteration in the time between then and the writing of her 1935 essay. In the earlier work, Hermann seems to understand this difference between the old theory and the new primarily with regards to the distinct possibility of obtaining superpositions of eigenstates in the latter but only (pure) eigenstates in the former. In the 1935 essay, the 'essential difference' between the quantum theory and the classical theory involves observational context and other, more obviously Bohrian, considerations.

One notices also that Hermann's understanding of statistics has surely deepened in the intervening years between essays. Whereas statistical considerations form no explicit part of her 1933 considerations, her 1935 essay contains several paragraphs (Chap. 15, Sect. 4) describing the differences between symmetric and asymmetric wave functions, and the different statistical situations applicable to each. She considers how the statistics of symmetrical ensembles nullifies, in some sense, the notion of individuality, and this in turn affects distinguishability—and distinguishability is directly relevant to the question of the existence of hidden variables. In the end, however, she concludes that looking to statistics for a completion of quantum mechanics will succeed no more than any of the other attempts described in her first chapter.

8.3.3 Von Neumann's Proof

Hermann's treatment of von Neumann's impossibility proof against hidden variables in Sect. 4 of her 1933 manuscript is slightly different from her treatment in the 1935 essay. Before discussing it in detail, we wish to draw attention to the first of Hermann's epigraphs, which includes the statement that: 'it can be shown in a mathematically exact way that the established formalism of quantum mechanics allows for *no* such completion. If thus one wants to retain the hope that determinism will return someday, then one must consider the present theory to be *contentually false*' (Chap. 14, p. 223). While this statement is perhaps easy to overlook in the little-quoted paper by Born (1929) it comes from (where Born has just argued that Planck's constant represents a fundamental limitation to all measurements, and the statement might easily be interpreted along the same lines), Hermann's use of it here suggests that in fact it refers to von Neumann's impossibility proof, and thus indicates that it was already regarded as conclusive at least in Göttingen circles well before its publication in von Neumann's book.[9]

[9] Von Neumann's proof first appeared in fact in a paper presented by Max Born to the 11 November 1927 session of the Gesellschaft der Wissenschaften zu Göttingen (von Neumann 1927). Hermann

Hermann's 1935 treatment is discussed in detail by Michiel Seevinck in this volume (Chap. 7), so only the main points will be recalled here. Hermann begins by boldly announcing that von Neumann's proof begs the question of hidden variables by assuming that '*the expectation value of a sum of physical quantities is equal to the sum of the expectation values of the two quantities*' (Chap. 15, p. 252; emphasis in the original). As she explains, in classical physics this follows trivially because the sum of two quantities is defined in terms of the sum of their values (which are always simultaneously well-defined), but for quantities corresponding to non-commuting operators, the sum of the two quantities is given by the operator sum. Expectation values are nevertheless linear also in quantum mechanics, and, as Hermann puts it, von Neumann's proof of this relies on decomposing arbitrary ensembles into sub-ensembles characterised by the same wave function, then noticing that the usual formula for expectation values defined by a wave function is in fact already linear for arbitrary operators. But she points out that for hypothetical sub-ensembles characterised in addition by some yet unknown parameters, the assumption of additivity is unproven, and the non-existence of dispersion-free ensembles thus relies on an assumption proven only for quantum mechanical ensembles. Hence the circularity.

Note that Hermann adds a qualification to which we shall return below: these additional parameters would escape the axiomatic framework of von Neumann's theory, in which physical quantities correspond bijectively to Hermitian operators. To her own qualification she replies that it only underscores the fact that von Neumann's *mathematical* result cannot decide the *physical* question of the existence of such additional parameters.

As compared to Bell's criticism of three decades later, also discussed by Seevinck, Hermann's treatment appears to lack the further point that linearity is not only unproven, but is demonstrably false for any hypothetical dispersion-free ensemble, since the expectation values for physical quantities in these ensembles would equal eigenvalues of the corresponding operators, and eigenvalues of non-commuting operators do not generally behave additively.

Reading Hermann's 1933 version of her criticism, one gets a better overall feel for what von Neumann is doing, since Hermann's description is more detailed and explicit. And, perhaps strikingly, it attributes to von Neumann himself the insight that eigenvalues of non-commuting operators do not behave additively ('the eigenvalues of $R + S$ in no way need to be sums of those of R and S', Chap. 14, p. 234), so that if the sum of two physical quantities is defined in terms of the sum of the corresponding operators, additivity of expectation values has to be verified separately. She just refers to an 'instructive example' given by von Neumann, which must be that of his footnote 164, where he considers the energy of a bound electron as a sum of kinetic energy (a function of momentum) and potential energy (a function of position), and remarks that the procedure for measuring the energy (observing spectral lines) is quite distinct from measuring the momentum and the position of the electron and then computing the corresponding function of the resulting values (von Neumann 1955,

---

may also have known of it already, but her presentation in 1933 appears to follow that in Chapter IV of von Neumann's book (von Neumann 1932).

pp. 309–310).[10] Thus Hermann seems to be aware of the point Bell was going to bring home to the collective consciousness three decades later.[11]

As regards the claim of circularity, it hardly appears explicitly in the 1933 version, only in one passing remark, where Hermann writes: 'For these ensembles [the ones characterised by a given wave function]—*but only for these*—has Neumann proved the inevitability of dispersion' (Chap. 14, p. 234; emphasis added), and plays no further role in Hermann's considerations. Hermann's argument is simply that von Neumann has proved for ensembles characterised by quantum mechanical wave functions that expectation values are additive, but that it does not follow that they are additive for even finer sub-ensembles. In order to get more traction from von Neumann's proof, says Hermann, one would have to assume that the specification of any as yet unknown parameters in defining further sub-ensembles leaves the expectation values of all quantum mechanical quantities unchanged (i.e. that any hidden variables have no additional predictive power).

Admittedly, Hermann's passing remark is inaccurate: von Neumann's proof shows precisely that, even *allowing* for the possibility that apart from the usual quantum mechanical ones there be other ensembles defining additive expectation values, there are in fact no other such ensembles. And commentators (e.g. already Jammer 1974, p. 275) have complained that a claim of strict circularity is too strong. On the other hand, von Neumann's discussion is indeed limited to the framework of expectation values *for quantum mechanical quantities*, i.e. quantities that correspond to Hermitian operators, and his justification for considering this a mathematically adequate framework does rely on the 'proof of additivity' Hermann takes exception to.

Indeed, what von Neumann is doing in Chap. IV of his book ('Deductive Development of the Theory') and in the 1927 paper in which his impossibility theorem first appeared, is an axiomatic derivation of quantum mechanics as some appropriate generalisation of probability theory to the case in which quantities are not always simultaneously measurable. One of the axiomatic requirements in the construction of such a theory is that expectation values (appropriately defined) be linear for (appropriately defined) sums of physical quantities. If two quantities are simultaneously measurable, the sum of two quantities is trivially defined in terms of the sums of their values. But for quantities that are not simultaneously measurable it becomes *implicitly defined* by the requirement that expectation values be linear (a point stressed also by Bub (2010)). In order to construe quantum mechanics as such a generalised probability theory, von Neumann needs it to be true that the usual expectation values of quantum mechanics (defined in terms of wave functions) are in fact linear with respect to the usual operator sum, which can thus be taken as the explicit representation of the implicitly defined sum of physical quantities.

---

[10]Von Neumann's (1927) paper contains a similar example in footnote 9 (von Neumann 1927, p. 249).

[11]The question whether Bell is fairly criticising von Neumann for imposing an absurd condition on the hypothetical hidden variables has been fairly widely debated in the literature (already Jammer (1974, p. 273) points out the significance of von Neumann's footnote). On this, see the very convincing re-appraisal of von Neumann's proof by Bub (2010), as well as our further comments at the end of this section.

Von Neumann is thus taking an algebra of observables for which the usual quantum mechanical states define expectation functionals, and showing that the full set of expectation functionals for this algebra is simply the convex closure of the usual quantum mechanical states. Thus Hermann is certainly correct in claiming (as she explicitly does in 1935) that all von Neumann shows is that within his chosen mathematical framework there is no room for dispersion-free states, and that a proof of indeterminism based on von Neumann's theorem requires the additional assumption that the algebra of observables considered by von Neumann is the physically correct one.

Perhaps surprisingly, von Neumann seems to agree, because he considers this objection himself. He writes:

> Nor would it help if there existed other, as yet undiscovered, physical quantities, in addition to those represented by the operators in quantum mechanics, because the relations assumed by quantum mechanics (i.e., **I.**, **II.**) would have to fail already for the by now known quantities, those that we discussed above. It is therefore not, as is often assumed, a question of a re-interpretation of quantum mechanics, —the present system of quantum mechanics would have to be objectively false, in order that another description of the statistical processes than the statistical one be possible. (von Neumann 1955, pp. 324–325)

Von Neumann's relations **I.** and **II.** are the functional relations between observables, in Hermann's wording from her 1933 manuscript: 'If the operator r corresponds to the physical quantity R, the operator s to the physical quantity S, then for any function f the physical quantity $f(R)$ corresponds to the operator $f(r)$, and $R + S$ to the operator $r + s$' (Chap. 14, p. 232). That is, von Neumann is saying that if new physical quantities existed that allowed in principle for a deterministic prediction of measurement results, then e.g. the energy observable for the electron would *not* be the sum of the kinetic and potential energy observables, precisely because the value one would obtain in measurements on ensembles whose characterisation included these new parameters would in general not be the sum of the values obtained for the kinetic and potential energies.

Far from ignoring Bell's point, thus, von Neumann uses it in the opposite direction. For von Neumann, the existence of dispersion-free states would mean that the functional relations between operators do not capture the functional relations between observables, and in this sense quantum mechanics would be 'objectively false'. It is noteworthy that a stance similar to von Neumann's was taken also by Pauli and to a certain extent Schrödinger. In his letter of 9 July 1935 to Schrödinger (Pauli 1985, pp. 419–422) commenting on the claims in the EPR paper, Pauli points out that the existence of dispersion-free states would lead to contradictions with quantum mechanics, since (assuming the usual functional relations hold) it would lead to a continuous distribution of values for e.g. the energy of the harmonic oscillator or for angular momentum. Schrödinger took this point to heart and included it in his discussion of the EPR paradox in his celebrated first paper on entanglement in 1935 (Schrödinger 1935). Indeed, for him the EPR argument leads to a 'paradox' precisely in that while the assumption of locality establishes that the results of all possible experiments on either particle are predetermined, it is also the case that these results cannot be related functionally in the way the corresponding operators are related (cf. Fine (1994) and Bacciagaluppi and Crull (2009, Sect. 4.2)).

It is not clear whether one should follow von Neumann (or Born in Hermann's epigraph) in thinking that the addition of undiscovered quantities would make quantum mechanics false. For Hermann, quantum mechanics would still be correct in the sense that non-commuting operators correctly represent physical properties of quantum systems standing in specific relations of mutual ontological indeterminacy (not merely epistemic uncertainty), and in the sense that predictions based solely on physical data determining their wave functions are correctly captured by the statistical formulas of quantum mechanics. And the precise sense in which this is the case is the topic of the final section of her manuscript.

8.4 The 'Positive Thesis' in Hermann's 1933 Essay

Hermann's final section in the 1933 manuscript has no strict analogue in 1935, since it is a sketch of how a completed quantum mechanics might look like. In some ways, however, it is more about the interpretation of quantum mechanical wave functions, and may give insight into the way Hermann conceived of quantum mechanics even in 1935 when she settled on the view that it was already a causally complete theory.

First, however, a remark on terminology. Hermann's final section is titled 'The Statistical Interpretation of Quantum Mechanics'. This term, confusingly, has generally been used in two quite opposite ways, either in the sense of 'ontic-probabilistic interpretation' (quantum mechanics makes statistical predictions because it includes an irreducibly probabilistic element), or in that of 'merely statistical interpretation' (quantum mechanics makes statistical predictions because it fails to give an individual description of physical systems, much like classical statistical mechanics). The former use is traditionally associated with Max Born (whose Nobel prize in 1954 was awarded 'for his fundamental research in quantum mechanics, especially for his statistical interpretation of the wavefunction'), the latter with Einstein (whose favourite criticism of quantum mechanics in his later years was that for macroscopic systems quantum mechanics could correctly reproduce predictions about ensembles of systems but was silent about individual behaviours).[12] Perhaps even more confusingly, Born's use of 'statistical' was originally the same as Einstein's—indeed, perhaps tongue-in-cheek, Einstein always referred to his own statistical interpretation as 'Born's' statistical interpretation[13]—in the sense that Born first introduced it as an interpretation of the (asymptotic) wave function after a collision, as describing an ensemble of (bound or free) *stationary* states of the colliding particles, and *derived* from that the corresponding expression for the transition probabilities between the

[12] See e.g. Einstein (1953). The term has continued to be used in this latter sense even in recent years, notably by Ballentine (see e.g. Ballentine 1986).

[13] See e.g. again his Einstein (1953), written in fact for the festschrift in honour of Born's retirement from Edinburgh.

initial and final stationary states (for details, see Bacciagaluppi 2008). It is not entirely clear in which sense Hermann is using the term; however, since her intention in the final section is to ultimately discredit this interpretation, it is likely she has in mind the standard Bornian meaning for 'statistical' (which she also uses once in Sect. 2 to characterise Heisenberg's position) rather than Einstein's tongue-in-cheek use of the term.

The question Hermann addresses in her final section is precisely that of how to interpret the Schrödinger wave function, or equivalently the coefficients in a decomposition of the wave function. Since Hermann considers quantum mechanics to be correct (if incomplete), in particular the uncertainty relations, the position and momentum spreads of a wave function should correspond to something objectively real. But in a completion of quantum mechanics such as she envisages, an ensemble characterised by a given wave function need not be homogeneous, so that the position and momentum spreads would appear to be merely characterising the collective behaviour of the ensemble, and might not correspond to any objectively real property of the individual systems. As Hermann puts it, it is unsatisfactory

> that the interpretation of the position interval, which according to the entire framework of the theory is objectively determined by the state of the given electron, can only be displayed in a large ensemble of electrons, whose states need not even agree in every respect. (Chap. 14, p. 235)

The intuitive way of associating the spreads with individual systems, according to Hermann, is to think of individual systems as literally spread out in position and in momentum. In this way, indeed, 'the probabilistic statements drop out *of the interpretation* of the uncertainty intervals and decomposition coefficients, and appear—only insofar as they are justified by experience—as propositions of quantum mechanics' (Chap. 14, p. 236; emphasis in the original), i.e. probabilistic (in fact statistical) language enters only at a later stage—when we describe the behaviour of an ensemble of systems under measurement.

This, as Hermann notes, is of course the picture Schrödinger himself originally had of his wave functions, and a picture that he felt was ultimately inadequate.[14] Hermann suggests that one bite the bullet instead, and deny that a spread-out electron is made up of individual charge elements. This is a radical idea: that the novelty of the uncertainty relations lies in the fact that they define what counts as a physical system capable of entering into interaction with other systems. Hypothetical charge elements cannot interact on their own with other systems, and only systems as wholes can participate in interactions:

> The classical assumption whereby interaction between masses can be reduced to interaction between point masses, is thus supplanted by a more complicated assumption in quantum mechanics: interaction between physical systems presupposes that each of them exhibit an extension corresponding to the Heisenberg relations. (Chap. 14, p. 236)

[14] For instance, while treating the electron as a classically oscillating charge density leads to a useful semi-classical method of calculating the radiation emitted by an electron, one cannot describe the interaction of two charged particles as mediated by such semi-classical fields: the interaction of two charges is already fully described by the interaction term in the Schrödinger equation—see e.g. (Schrödinger 1928/2009, p. 411–414).

In her last paragraph, Hermann applies this insight to the case of a measurement—an interaction of a system with an observer. At all times during and after the measurement, the system must be subject to the uncertainty relations, thus at all times its position and momentum will be spread out accordingly. In this sense, Hermann is indeed committed to position and momentum not being both determinate at the same time, despite suggesting that *results of measurements* of position or momentum may always be pre-determined. Now, when, say, a position measurement is performed, the initial wave function of the system will change to one that is vanishingly narrow in position, and accordingly arbitrarily spread-out in momentum. The question of determinism and indeterminism refers to the laws that describe a measurement process in this sense. Hermann takes it she has shown that quantum mechanics itself is silent on this issue, and that future research may well establish that these laws are deterministic. To use modern terminology, what Hermann is envisaging is some form of collapse theory. The wave function is interpreted ontically (specifically in terms of position and momentum density). The collapse mechanism is empirically constrained to reproduce the statistical predictions of quantum mechanics if no further data are provided except the wave function of the system. But this mechanism may well turn out to be deterministic if we were to take into account any further 'hidden' variables of the system or the apparatus (or both).

Ending on a very speculative note, we might ask whether one should think of Hermann's 'positive' thesis in this final section as being superseded by her ideas on the causal completeness of quantum mechanics, or whether one might see here the seeds of certain concepts that would occupy centre stage in her 1935 essay that we shall discuss in Chap. 10: the quantum mechanical description becoming relative to the context of observation, and the 'splitting of truth' implied thereby.

References

Bacciagaluppi G (2008) The statistical interpretation according to Born and Heisenberg. In: Joas C, Lehner C, Renn J (eds) HQ-1: conference on the history of quantum physics, MPIWG, vol II, pp 269–288

Bacciagaluppi G, Crull E (2009) Heisenberg (and Schrödinger, and Pauli) on hidden variables. Stud Hist Philos Mod Phys 40(4):374–382

Ballentine LE (1986) Probability theory in quantum mechanics. Am J Phys 54:883–889

Bohr N (1934) Atomic theory and the description of nature. Cambridge University Press, Cambridge

Born M (1929) Über den Sinn der physikalischen Theorien. Die Naturwissenschaften 17(7): 109–118

Bub J (2010) Von Neumann's hidden variables proof: a re-appraisal. Found Phys 40(9–10): 1333–1340

Dirac P (1930) Die Prinzipien der Quantenmechanik. S. Hirzel, Leipzig

Einstein A (1953) Elementare Überlegungen zur Interpretation der Grundlagen der Quantenmechanik. In: Scientific papers presented to Max Born on his retirement from the Tait Chair of Natural Philosophy in the University of Edinburgh. Oliver and Boyd, Edinburgh, pp 33–40. Translation by Karanth D at http://arXiv.org/abs/1107.3701

Fine A (1994) Schrödinger's version of EPR, and its problems. In: Humphreys P (ed) Patrick Suppes: scientific philosopher, vol 2. Kluwer, Dordrecht, pp 29–38

Heisenberg W (1927) Über den anschaulichen Inhalt der quantentheoretischen Kinematik und Mechanik. Zeitschrift für Physik 43(3–4):172–198. Translated as: The physical content of quantum kinematics and mechanics. In: Wheeler J, Zurek W (eds) (1983) Quantum theory and measurement. Princeton University Press, Princeton, NJ, pp 62–84

Hermann G (1933) Determinismus und Quantenmechanik, manuscript. Dirac Archives DRAC 3/11

Hermann G (1935) Die naturphilosophischen Grundlagen der Quantenmechanik. Abhandlugen der Fries'schen Schule 6(2):75–152

Herrmann K (ed) (2017) Grete Henry-Hermann: Philosophie–Mathematik–Quantenmechanik. Springer, Berlin

Jammer M (1974) The philosophy of quantum mechanics: the interpretations of quantum mechanics in historical perspective. Wiley, Hoboken, NJ

Pauli W (1985) Scientific correspondence with Bohr, Einstein, Heisenberg a.o., vol. II, 1930–1939. Springer, Berlin. Edited by Meyenn K von

Schrödinger E (1928/2009) Wave mechanics. In: Bacciagaluppi G, Valentini A, Quantum theory at the crossroads: reconsidering the 1927 Solvay conference. Cambridge University Press, Cambridge, pp 406–431

Schrödinger E (1935) Discussion of probability relations between separated systems. Math Proc Camb Philos Soc 31(04):555–563

von Neumann J (1927) Wahrscheinlichkeitstheoretischer Aufbau der Quantenmechanik. Nachrichten von der Gesellschaft der Wissenschaften zu Göttingen, Mathematisch-Physikalische Klasse 1927:245–272

von Neumann J (1932) Mathematische Grundlagen der Quantenmechanik. Springer, Berlin

von Neumann J (1955) Mathematical foundations of quantum mechanics. Princeton University Press, Princeton, NJ. Translation of von Neumann (1932) by Beyer RT

von Weizsäcker CF (1963) Interview with T. S. Kuhn and J. L. Heilbron, Archive for the history of quantum physics. https://www.aip.org/history-programs/niels-bohr-library/oral-histories/4947

Chapter 9
Bohr's Slit and Hermann's Microscope

Guido Bacciagaluppi

9.1 Introduction

As discussed in detail elsewhere in this book, Grete Hermann's 1935 essay on quantum mechanics reflects to a large extent the intense discussions she had at the physics institute in Leipzig during the preceding year. Among the interesting aspects of Hermann's interaction with Heisenberg and Weizsäcker is the relation between her use of the γ-ray microscope, and the use and treatment of the microscope by Heisenberg and Weizsäcker, as discussed e.g. by Filk and by Frappier in this volume (Chaps. 5 and 6). This is also evident in Heisenberg's manuscript reply to EPR written in the summer of 1935, as discussed by Bacciagaluppi and Crull (2009, 2011) and references therein.

To briefly summarise some essential points of our previous discussion of Heisenberg's manuscript, the argument there combines considerations about interference (or transition probabilities) with considerations about the movability of the 'cut' between the quantum and classical description (about the freedom to apply the Born rule at different stages of the description of a measurement), as follows. Consider a system A and apparatus B that could be used to make two mutually exclusive measurements on A—of quantities q_A or p_A—and consider two possible placements of the cut, applying the Born rule directly to system A, or to the composite of systems A and B after they have interacted. In the case of a single measurement, say of q_A, the placement of the cut makes no difference: we get the same probabilities for the results. In this case, we can also conceive of hidden variables as determining the value of q_A already prior to measurement. But now consider using B to measure p_A instead, and consider placing the cut after the interaction between A and B, which

G. Bacciagaluppi (✉)
Universiteit Utrecht, Utrecht, The Netherlands;
IHPST and SPHERE, Paris, France
e-mail: g.bacciagaluppi@uu.nl

E. Crull and G. Bacciagaluppi (eds.), *Grete Hermann - Between Physics and Philosophy*, Studies in History and Philosophy of Science 42,
DOI 10.1007/978-94-024-0970-3_9

is thus treated quantum mechanically. We can think of the experiment as an interference experiment, in which the probabilities for the various possible values of q_A at the time of the interaction interfere to yield the probabilities for the values of p_A in the actual measurement. But, as Heisenberg had been arguing ever since 1927, if q_A actually had a definite value at the earlier time, then one would have to calculate the probabilities for the outcomes of the measurement of p_A at the later time by a straightforward application of the law of total probability, yielding no interference.[1]

Heisenberg's manuscript does not mention the γ-ray microscope by name, but it is clear from subsequent correspondence with Bohr that, when Heisenberg talks of an apparatus B that can be used to measure two conjugate quantities, he has the microscope in mind. In his final section, after summarising the argument against hidden variables he has just given, Heisenberg adds that it is essentially captured by the idea of the causal completeness of quantum mechanics as discussed in Grete Hermann's 1935 essay. Heisenberg does not elaborate in detail, but it is clear that his argument relies on the intuition that adding hidden variables to the quantum mechanical description would result in causal overdetermination.

In this chapter, I wish to draw a further comparison: between Hermann's use of the Heisenberg microscope and another famous use of a very similar thought experiment, namely Bohr's analysis of the suspended single slit in his own reply to EPR (Bohr 1935). The similarity of these two experiments was already pointed out by Jammer (1974, pp. 96–97): in both cases, immediately after the interaction we have two systems with a given total momentum and zero difference in position, and we manipulate the one in order to measure the other. (The only difference of note is that in the case of the microscope, we usually consider how the state evolves while the photon moves through the microscope, and the manipulation takes place only at a later stage.)

Other than in Heisenberg's case, there is no evidence that Hermann's discussion might have had any influence on Bohr.[2] Nevertheless, I shall argue that Hermann's use of different aspects of the classical pictures in the treatment of the Heisenberg microscope actually makes her treatment closer to Bohr's discussion of the suspended slit in his reply to EPR than to Heisenberg's treatment of the microscope in his own reply. In somewhat different terms: the causal stories as told by Hermann are in fact closer to the detailed physical analyses of the interaction between the particle and the screen as performed by Bohr than to the perhaps somewhat schematic picture of the cut as provided by Heisenberg.

---

[1]Cf. the classic textbook argument claiming to show that if 'particles' really went through the upper or lower slit, there would be no interference pattern in a two-slit experiment.

[2]In particular, it goes unmentioned in the correspondence between Bohr and Heisenberg in September 1935 on the latter's reply to EPR. Hermann did visit Bohr's institute in May 1935, however, and her work was discussed in a colloquium. In a letter of July 1936, Heisenberg replies to Hermann about an objection by Bohr relating to the movability of the cut, which Heisenberg, however, thinks should not affect Hermann's considerations. For Heisenberg's reply and his correspondence about it with Bohr, see Bacciagaluppi and Crull (2011), for his July 1936 letter to Hermann, see Herrmann (2017, Part III, Letter 16).

This in turn suggests that Hermann may be an especially acute interpreter of Bohr's views. I conclude this chapter by looking at Hermann's (and Bohr's) approach in the context of more general examples of measurement. The next chapter will then analyse in more detail Hermann's discussion of Bohr's complementarity and her attempt to incorporate it into a broadened construal of transcendental idealism.

9.2 Comparing Slit and Microscope

9.2.1 Hermann on the Microscope

Let us (yet again) briefly recall Hermann's treatment of the γ-ray microscope. As by now familiar, in the Heisenberg microscope thought experiment we initially have an electron confined to a given plane, i.e. its position within the plane is completely uncertain, while its momentum in the plane is known. We then illuminate it with a γ-ray photon of known momentum. Depending on where we place the photographic plate on which we record the scattered photon, we can use the microscope to measure (or to prepare: I return to this distinction in Sect. 9.3) the position or the momentum of the electron.

In Sect. 10 of her essay, Hermann considers three choices with regard to the placement of the photographic plate: (a) in the image plane of the microscope, (b) in the focal plane of the microscope, (c) nowhere at all.

In case (a), Hermann provides a causal analysis of the interaction based on a selective use of aspects of the wave and the corpuscular picture. This provides us both with a cause for the formation of the image on the photographic plate, and with the ability to predict the result of a subsequent measurement of position on the electron (which can be seen as an indirect check of the causal story given).

In case (b), she provides a causal analysis based on a *different* selective use of aspects of the wave and the corpuscular picture. This provides us both with a cause for the formation of the image on the photographic plate, and with the ability to predict the result of a subsequent measurement of momentum on the electron (which can again be seen as an indirect check of the causal story given).

The different selective uses of aspects of the classical pictures are specified by Hermann in her Sect. 12 as follows:

> How both conceptions [the wave picture and the particle picture] are made consistent with one another depends on the type of measurement: if the light is intercepted in the image plane of the observed object, then one is working in the wave picture with the conception of a spherical wave propagating from one point, and correspondingly ascribing a sharp position but a smeared exchange of momentum to the corpuscularly interpreted collision between electron and light quantum. If one carries out the observation in the focal plane of the microscope, then one deals with a parallel beam of rays, and accordingly in the corpuscle picture with a precisely determined exchange of momentum but an unsharp position. The single observational context that the physicist enters through observation of the photographic plate therefore determines which features of both pictures are used. (Chap. 15, p. 263)

Finally, in case (c) one obtains a linear combination of product wave functions (which we now call entanglement), and the photon and the electron each lack individual states. As Hermann points out, this process is not anschaulich.

9.2.2 Bohr on the Slit

Let us now turn to Bohr's treatment of his particle-and-slit experiment.[3] In his reply to Einstein, Podolsky and Rosen (Bohr 1935), before discussing the EPR example itself, Bohr discusses in detail the example of a particle passing through a suspended screen with a single slit. The initial momentum of the particle and of the screen (in the direction in which the screen can move) are known, thus their total momentum is known also after the particle has passed through the slit. And immediately after the passage, also the difference in position between the particle and the (narrow) slit in the screen is known (and is in fact zero). One can then measure either the position (immediately after passage) or the momentum of the screen, and this allows one to predict the position of the particle (in an immediately subsequent measurement) or the momentum of the particle, respectively.

As I understand the role the thought experiment plays in Bohr's reply to EPR, the suspended slit is further perfectly analogous to the EPR experiment, both in terms of the (near-)maximal entanglement between the particle and the slit,[4] and in the sense that the type of final state in which we prepare the particle (position or momentum eigenstate) depends on a free choice of manipulations performed on the screen *after* the particle has passed through the slit.[5] In my opinion this is the crucial analogy with the EPR case, which explains why Bohr spends so much time on this example.

Bohr's ultimate aim with it is to undermine EPR's criterion of reality, and he is arguing that although there is no 'mechanical disturbance' of the particle when one manipulates the screen, there is a non-mechanical 'influence' on the particle, to be precise, '*an influence on the very conditions which define the possible types of predictions regarding the future behavior of the system*' (Bohr 1935, p. 700, emphasis in the original). It is the latter that EPR would need to rule out in order to apply their criterion of reality. For the purpose of comparing how Hermann and Bohr treat their thought experiments, however, it is not essential how Bohr puts it to use further (and whether or not he is providing a successful reply to EPR!). What interests us here is the positive account, implied by Bohr's discussion, of how given the appropriate conditions certain 'predictions regarding the future behavior of the system' *can* indeed be made.

---

[3] I here make use of my previous analyses of Bohr's reply to EPR (mainly in collaboration with E. Crull) in Bacciagaluppi (2015, 2017) and Bacciagaluppi and Crull (2018).

[4] Only near-maximal, because the slit has a finite width.

[5] This point is emphasised also by Pauli in a letter to Schrödinger of 9 July 1935, in which he describes Bohr's (as yet unpublished) reply (Pauli 1985, pp. 419–422).

Bohr's phrase for these conditions is 'controlling the reaction of the object on the measuring instruments if these are to serve their purpose' (Bohr 1935, p. 697). What happens is that, by choosing to perform a position measurement on the screen, we can legitimately apply the 'idea of space location' also to the interaction between the particle and the screen. Specifically, we can say that immediately after the passage through the screen, the particle is at the same position as the slit, and we can thus predict with certainty also the position of the particle. Similarly, by choosing to perform a momentum measurement on the screen, we can legitimately apply the 'conservation theorem of momentum' also to the interaction between the particle and the screen, and since the initial total momentum was known, we can now predict with certainty also the momentum of the particle.

9.2.3 Bohr and Hermann Compared

A helpful context in which to read Bohr's discussion is Don Howard's now-classic analysis of Bohr's doctrine of classical concepts (Howard 1994). According to Howard, Bohr recognised that the 'uncontrollable exchange of quanta of action' destroyed the separability of system and apparatus (which—as stressed by Einstein—is necessary for the objectivity of a measurement), but he believed that this objectivity could be regained in each observational context.[6] Formally, this can be represented (Howard carefully chooses the word 'reconstructed', since this is not a formal representation Bohr himself would have used) as the system and apparatus getting entangled during the measurement interaction, with the entangled state being replaced with an appropriate mixture when we perform the corresponding manipulation of the apparatus.

Note that, indeed, when we perform a measurement of the position of the slit (whether or not we believe we are 'collapsing' the state of system and apparatus), the state of particle and slit becomes

$$\int_{-\infty}^{\infty} |x\rangle\langle x| \otimes |x\rangle\langle x|\, dx. \tag{9.1}$$

This corresponds to choosing to apply the idea of space location in describing the particle and slit, allowing us to selectively exploit the position correlations in the original entangled state. And when we perform a measurement of momentum on the slit, the state of particle and slit becomes

$$\int_{-\infty}^{\infty} |p_0 - p\rangle\langle p_0 - p| \otimes |p\rangle\langle p|\, dp. \tag{9.2}$$

[6] In the Como lecture, Bohr (1928, p. 580) writes: 'the quantum postulate implies that any observation of atomic phenomena will involve an interaction with the agency of observation not to be neglected. Accordingly, an independent reality in the ordinary physical sense can neither be ascribed to the phenomena nor to the agencies of observation'.

This corresponds to choosing to apply the conservation of momentum to the particle and slit, which allows us to selectively exploit the momentum correlations in the entangled state.

Howard's reading of Bohr now highlights aspects that one finds also in Hermann's treatment of the microscope. The element of Howard's analysis that is crucial for us here is that complementarity is not cashed out in terms of different choices of classical measuring apparatuses applied to a given quantum system, but in terms of *selectively* applying only certain aspects of the classical description to both system and apparatus. Hermann's treatment is explicitly in terms of selective application of different aspects of the classical description (which in this case would be the particle picture), and Howard's formal representation fits equally Hermann's treatment of the microscope. Furthermore, in Hermann the lack of Anschaulichkeit in the composite system after the interaction is explicitly formalised in terms of entanglement.[7]

By the same token, we can now discern a disanalogy between Hermann's and *Heisenberg*'s treatments of the γ-ray microscope. Although for Heisenberg the 'cut' is movable, given one placement of the cut one side is treated exclusively quantum mechanically and the other exclusively classically. Thus Bohr and Heisenberg have rather different views of the relation between classical and quantum descriptions, and from this point of view at least Hermann's account turns out to be closer to Bohr's account than to Heisenberg's.

From a different point of view, however, Hermann and Bohr are disagreeing, namely as regards the implications of their analyses for the notion of causality. It is well-known that Bohr associated causality with the conservation theorems, and thus expressed the view that causality and the space-time picture were complementary (Bohr 1928, p. 581). For Hermann, instead, *both* the application of the conservation theorems and that of the idea of space location allow one to tell causal stories. This is explicitly carried out in her analysis of the Heisenberg microscope, but can be equally carried out in the case of the particle and slit. In the case of the momentum measurement, Hermann's story is the same as Bohr's: momentum is exchanged between the particle and the slit in accordance with conservation of total momentum. In the case of the position measurement, the causal story is arguably (and quite simply) that the particle goes through the screen at the particular location of the slit, and this then causally determines the results of our separate measurements of position on both the screen and (if we wish) on the particle.

The upshot is thus that we can read Hermann as endorsing the framework of complementarity, while at the same time arguing that the doctrines of complementarity and causality can be reconciled in a natural way.

---

[7]Of course, insofar as we can see Hermann as explicating the concept of complementarity, the highlighting of analogies between Hermann and Bohr lends indirect support to Howard's analysis in the first place (cf. also Chap. 10).

9.3 Beyond Slit and Microscope

The examples discussed by Hermann and Bohr are not the most general examples of quantum measurements, because in both cases the final state of the two systems considered is (approximately) maximally entangled. I shall now look at more general cases and ask whether the approach(es) of Hermann and Bohr are still applicable.

9.3.1 Measurements of the First Kind

The first generalisation to consider is to measurements of the 'first kind' with arbitrary initial states. This terminology is from Pauli's 1933 handbook article (Pauli 1933), which I shall follow also in the formal treatment of measurements (with slight adaptation for ease of exposition). Measurements of the first kind are the usual ones of the form

$$|u_n\rangle|U_0\rangle \mapsto |u_n\rangle|U_n\rangle, \tag{9.3}$$

where the $|u_n\rangle$ are the eigenstates of the measured observable, $|U_0\rangle$ is the initial state of the 'apparatus' (which could be another microscopic degree of freedom, but at some point of a measurement chain will eventually be macroscopic), and the $|U_n\rangle$ correspond to the different read-outs.[8] Thus in particular,

$$\sum_n c_n|u_n\rangle|U_0\rangle \mapsto \sum_n c_n|u_n\rangle|U_n\rangle, \tag{9.4}$$

and upon reading off the value n, the system is left in the eigenstate $|u_n\rangle$ (yielding a repeatable measurement). A measurement of the first kind is a preparation, but it is also a 'measurement' in the sense that it makes possible 'an unambiguous conclusion back from the measured value to the quantity of the system under consideration before the measurement' (Pauli 1933, p. 97).

Note there are two ways of reading this last statement. The unremarkable one is that if one assumes that the system is initially in an eigenstate $|u_n\rangle$ as in (9.3), the measurement can establish with certainty which one. But one can read the statement also in terms of Hermann's thesis that even if the system is in a superposition of eigenstates $|u_n\rangle$ as in (9.4), one can conclude that the measurement result was caused by one particular value of the measured quantity. Thus we see that, indeed, Hermann's discussion of the Heisenberg microscope can be extended straightforwardly to the case of general measurements of the first kind with arbitrary coefficients c_n in the state of system and apparatus.

---

[8]Note that Pauli (unrigorously) treats also continuous observables as if they were discrete, but the treatment is meant to cover both.

It is not obvious whether Pauli himself intends the weaker or the stronger reading, but it is clear that he distinguishes between the preparatory aspect of the measurement (which is the aspect of paramount importance in the context of EPR) and the aspect of the measurement qua measurement. And perhaps also Bohr has this in mind when he talks about needing to control the (relevant aspects of the) interaction between the system and the measuring apparatus in order for the latter to 'serve its purpose'.

What about even more general cases? Modern-day measurement theory provides a very wide-ranging extension of the notion of measurement, including in particular so-called 'unsharp measurements' of observables and 'joint unsharp measurements' of incompatible observables.[9] An easy example of the former is a Stern–Gerlach measurement, if we take into account the fact that the particle being deflected is in general not described by a wave function with compact support, but (say) by a Gaussian with tails that are infinitely extended (if small in amplitude). This means in particular that the different output beams always have some non-zero overlap, so that even a particle in an eigenstate of spin in fact has a non-zero chance of being detected in the 'wrong' beam. The measurement procedure thus includes an irreducible and *irreducibly probabilistic* error, with in general an associated disturbance of the spin state.

I shall return to these cases below, but in order not to proceed anachronistically, I shall continue to follow Pauli's treatment, and discuss first the further cases that Pauli himself describes.

9.3.2 Measurements of the Second Kind

Besides measurements of the first kind, Pauli discusses also what he calls measurements of the 'second kind', as follows. He starts off by considering a completely general unitary interaction between system and apparatus, which we can write as

$$|u_n\rangle|U_0\rangle \mapsto \sum_{m,k} c_{mk}^{(n)}|u_m\rangle|U_k\rangle =: \sum_k |v_k^{(n)}\rangle|U_k\rangle, \tag{9.5}$$

[9]Since traditional quantum observables are self-adjoint operators, one can associate them one-to-one with their spectral measures, which are projection-valued measures (PVM) over the real line. Nowadays, observables are identified with more general positive-operator-valued measures (POVMs). 'Unsharp' measurements of a traditional observable are represented by commutative POVMs (all of whose elements share the same spectral measure), while non-commutative POVMs can be thought of as a form of joint realisation of unsharp measurements of more than one traditional observable. For general and thorough treatments of modern measurement theory, see Busch et al. (1991) and Busch et al. (1997); for the last point about interpreting non-commutative POVMs, see Cattaneo et al. (1997).

so that

$$\sum_n c_n |u_n\rangle|U_0\rangle \mapsto \sum_n c_n \sum_k |v_k^{(n)}\rangle|U_k\rangle =$$

$$= \sum_{n,k} c_n |v_k^{(n)}\rangle|U_k\rangle =: \sum_k |\psi_k\rangle|U_k\rangle. \quad (9.6)$$

He then imposes the condition that one can unambiguously retrodict the value of the measured quantity from the reading of the apparatus, in the form that for each k there is at most one n such that $|v_k^{(n)}\rangle \neq 0$. In other words, the set K over which the index k ranges decomposes into disjoint sets K_n such that

$$|u_n\rangle|U_0\rangle \mapsto \sum_{k\in K_n} |v_k^{(n)}\rangle|U_k\rangle. \quad (9.7)$$

Let us first specialise this assumption further and assume that each K_n contains exactly one element, which we can relabel n. Thus,

$$|u_n\rangle|U_0\rangle \mapsto |v_n^{(n)}\rangle|U_n\rangle, \quad (9.8)$$

and $|\psi_n\rangle = c_n|v_n^{(n)}\rangle$. In this case, on the strong reading of retrodiction, for each read-off n we can tell a causal story of how it was determined by the state $|u_n\rangle$ of the system. Unlike a measurement of the first kind, however, such a measurement disturbs the state of the system, causing it to change to $|\psi_n\rangle$.

We see thus that Hermann's approach can be extended also to measurements of the second kind (at least under the further restriction we have introduced). And again—echoing Bohr—the relevant aspect of the interaction can be reconstructed, and the interaction followed by the read-off counts as a measurement of the given quantity (as well as a preparation of one of the states $|\psi_k\rangle$—which incidentally need not be mutually orthogonal).

One might worry about our further requirement (9.8). Is it not the case that if we impose Pauli's original, less stringent condition, we generally have a one-to-many correspondence between the states $|u_n\rangle$ of the system and the states $|U_k\rangle$ of the apparatus? In that case we can no longer tell a causal story of how the former determine the latter. This is a delicate point. It is in fact the case, but that need not mean that the readings of the apparatus are uncaused!

Indeed, the fact that one value of the measured quantity may correspond to several values of the pointer's position can be understood along the lines that one is measuring, say, a discrete quantity such as the energy of the harmonic oscillator with a continuous pointer. The energy determines the position of the pointer to within a certain interval, and that is all we expect it to do. If there is a cause for the exact value of the position of the pointer within that interval, we shall expect it to be extraneous to the measured system, and lie rather with the apparatus itself. If we imagine the pointer as macroscopic (which at some point along the measurement chain it arguably is), then the manipulation of the apparatus leading to our reading off the value is presumably a measurement of the first kind (say, of the pointer's position), and thus the read-off *always has a cause*, which is simply the position of the pointer.

A similar conclusion is obtained if we subdivide the (actual) manipulation of the apparatus (in the case of a general measurement of the second kind as defined by Pauli) into two (actual or conceptual) stages: we first perform a non-maximal first-kind measurement on the pointer, with eigenprojections $P_n := \sum_{k \in K_n} |U_k\rangle\langle U_k|$. This is an implementation of a measurement of the second kind (on the system) that satisfies our further constraint (9.8), and in which each result finds its cause in the value n of the measured quantity. We then perform a further, finer measurement on the pointer, with eigenprojections $|U_k\rangle\langle U_k|$. It is now clear that the original system of interest, while it has determined the new initial state of the pointer, plays no further role in the analysis of this finer measurement. This measurement's results find their cause purely in the value $k \in K_n$ of the pointer observable.[10]

We have thus arguably extended Hermann's analysis (and similarly Bohr's) to cover arbitrary measurements of the second kind.

9.3.3 General Preparations

The preceding discussion now suggests also the way ahead in the completely general case in which we take Pauli's ansatz (9.5), but Pauli's condition of unambiguous retrodiction is *not* satisfied.[11] Pauli clearly (at least implicitly) allows for this possibility in (9.6): a measurement of the pointer with result k will prepare the system in the state $|\psi_k\rangle$, irrespective of whether the condition on retrodiction is met or not.[12]

Here, Hermann can simply bite the bullet. If in such a case we ask what determines the reading of the pointer, the answer is again just that it is the position of the pointer itself. But now there is *no causal story* to tell about how the system even partially determines the reading of the apparatus. Causality is safe, because the reading has a cause, but one rejects the interpretation of the interaction as a measurement of any quantity on the system. It is true that the interaction between system and apparatus has changed the state of the apparatus, but that can be thought of as a process (described causally by the Schrödinger equation) that is already concluded by the time we perform the measurement on the pointer. Only then do we enter a specific context for an observation, and are confronted with the (solvable) task of finding a cause for the result of that observation.

[10]In modern measurement theory, the first stage corresponds to the most general measurements described using projection-valued measures, while the combination of the two stages corresponds already to a very special case of a POVM.

[11]More precisely, the case in which the condition is not satisfied with respect to *any* basis $|u_n\rangle$ (otherwise we are trivially back to the preceding case).

[12]Note that precisely such a general case of preparation is actually described in the EPR paper, in their Eqs. (7) and (8), just before the special example of the EPR state (Einstein et al. 1935, p. 779). (This is somewhat ironic, given the conclusions we shall eventually arrive at below.).

Incidentally, in Pauli's treatment such a procedure is not even described as a measurement. Indeed, the only 'measurements' Pauli explicitly discusses are measurements of the first and of the second kind, which by definition do satisfy his retrodiction condition. It is plausible to suggest that Pauli did not think such more general interactions could be meaningfully described as 'measurements'. And if we translate our extension of Hermann's treatment into Bohr's language, we again see why one might resist interpreting them as measurements. Indeed, in these very general cases there is arguably also *no classical picture* (not even a partial one) that one could employ as a reconstruction of the interaction between system and apparatus. At a pinch, as in our example of an unsharp spin measurement above, one could stretch Bohr's approach and modify the classical pictures to include also irreducibly stochastic components. To our modern sensibilities, this is merely an addition of classical noise, so the resulting picture is still 'classical'. Indeed, it even allows us to tell a causal story about how the spin determines the results, but now in terms of probabilistic causality. Hermann would presumably have rejected all talk of probabilistic causality, but even Bohr might have resisted a modification of his viewpoint of complementarity requiring adding stochasticity to the classical physical pictures he lay so much emphasis on. Indeed, in this case the 'reaction of the object on the measuring instruments' would not be *fully controllable*, and the measuring instruments would no longer 'serve their purpose'!

There is one wrinkle in this picture, however. Recall that in Hermann's discussion, in the case of a measurement of the first (or indeed second) kind, if we then go and perform a further measurement on the system and find the state $|\psi_k\rangle$, we can take this as an indirect confirmation of the causal story about the original interaction between the system and the apparatus. Indeed, that same causal story provides an explanation of how come the reading of our apparatus and the state in which the system is left are in fact *correlated*. In the general case, as I have argued above, we can always find a causal explanation for the read-off of the pointer. Furthermore, if upon having found the pointer in the state $|U_k\rangle$ we perform a measurement on the system of either the first or second kind of the projection onto $|\psi_k\rangle$, we can give a causal analysis also of this measurement. Indeed, the state $|\psi_k\rangle$ will be the cause for the system testing positively for its presence. But now, *precisely* because there is no longer a detailed causal story to tell about the interaction between system and apparatus, there is no longer a causal explanation for the *correlation* between $|U_k\rangle$ and $|\psi_k\rangle$. We have causes whenever we need them, but not always common causes when we would like them.

The analogous point, we now recognise, must be made about Bohr: while in the EPR case, the choice of different manipulations on the apparatus always corresponds to a measurement of the first kind of some quantity or other, in general a different choice of manipulation by the experimenter will correspond to a preparation procedure that has no interpretation as a measurement of the first or second kind, and there is no classical picture of how our choice influences the 'conditions which define the possible types of predictions regarding the future behavior of the system'. This somewhat spoils Bohr's reply to EPR, since the failure to provide such a classical picture (or at any rate one without stochastic noise!) leaves unexplained why we should be able to make some predictions with *certainty* on the system…

9.4 Conclusion

The Heisenberg microscope and Bohr's suspended slit are two justly famous thought experiments. In Hermann's hands, the former takes a shape that arguably makes the direct comparison between them especially fruitful. Such a comparison suggests that Hermann is a shrewd advocate of Bohr's complementarity (in spite of differences of opinion on the issue of causality), and our further analysis has yielded some new insights into the scope and possible limitations of Hermann's and Bohr's approaches.

The next chapter will examine how Hermann herself explicitly discussed Bohr's complementarity and tried to incorporate it more fully into the framework of transcendental idealism.

References

Bacciagaluppi G (2015) Did Bohr understand EPR? In: Aaserud F, Kragh H (eds) One hundred years of the Bohr atom. Series M—Mathematica et physica, Scientia Danica, vol 1. Royal Danish Academy of Sciences and Letters, Copenhagen, pp 375–394

Bacciagaluppi G (2017) An Everett perspective on Bohr and EPR. In: Faye J, Folse H (eds) Niels Bohr in the 21st century. Bloomsbury Academic, London, pp 289–302

Bacciagaluppi G, Crull E (2009) Heisenberg (and Schrödinger, and Pauli) on hidden variables. Stud Hist Philos Mod Phys 40(4):374–382

Bacciagaluppi G, Crull E (2011) Translation, with introduction, of W. Heisenberg: 'Ist eine deterministische Ergänzung der Quantenmechanik möglich?'. Preprint posted on the PhilSci Archive. http://philsci-archive.pitt.edu/8509/

Bacciagaluppi G, Crull E (expected 2018) 'The Einstein Paradox': the debate on nonlocality and incompleteness in 1935. Cambridge University Press, Cambridge

Bohr N (1928) The quantum postulate and the recent development of atomic theory. Nature 121(3050):580–590

Bohr N (1935) Can quantum-mechanical description of physical reality be considered complete? Phys Rev 48:696–702

Busch P, Lahti PJ, Mittelstaedt P (1991) The quantum theory of measurement. Springer, Berlin

Busch P, Grabowski M, Lahti PJ (1997) Operational quantum physics. Lecture notes in physics, vol 31. Springer, Berlin

Cattaneo G, Marsico T, Nisticò G, Bacciagaluppi G (1997) A concrete procedure for obtaining sharp reconstructions of unsharp observables in finite-dimensional quantum mechanics. Found Phys 27(10):1323–1343

Einstein A, Podolsky B, Rosen N (1935) Can quantum-mechanical description of physical reality be considered complete? Phys Rev 47:777–780

Herrmann K (ed) (2017) Grete Henry-Hermann: Philosophie–Mathematik–Quantenmechanik. Springer, Berlin

Howard D (1994) What makes a classical concept classical? Toward a reconstruction of Niels Bohr's philosophy of physics. In: Faye J, Folse H (eds) Niels Bohr and contemporary philosophy. Kluwer Academic, Dordrecht, pp 201–230

Jammer M (1974) The philosophy of quantum mechanics: the interpretations of quantum mechanics in historical perspective. Wiley, Hoboken, NJ

Pauli W (1933) Die allgemeinen Prinzipien der Wellenmechanik. In: Handbuch der Physik, vol XXIV/1, 2nd edn. Springer, Berlin, pp 83–272. New edition with historical notes by N. Straumann (Springer, Berlin, 1990)

Pauli W (1985) Scientific correspondence with Bohr, Einstein, Heisenberg a.o., vol II, 1930–1939. Springer, Berlin. Edited by Meyenn K von

Chapter 10
Hermann and the Relative Context of Observation

Elise Crull

10.1 Introduction

Those who are acquainted with Grete Hermann's 1935 essay on the natural-philosophical foundations of quantum mechanics have, rightly, understood one of her main aims to be in line with that of many neo-Kantians: to preserve the law of causality in light of quantum mechanics' apparent indeterminism. Because Hermann's solution to the question of quantum mechanical indeterminism—and, relatedly, the question of quantum mechanical completeness—is uniquely posed and answered by appeal to retrodictive causality, what little philosophical scholarship has been done on Hermann's 1935 essay has focused on this aspect of her work.

Historians, on the other hand, have tended to regard more closely the natural-philosophical tradition in their analysis of Hermann. The locus of their investigations has been to understand the extent to which this tradition's specific interpretation of Kantian categories as *analogies* influences Hermann's discussion of the physics, as a student in Nelson's Friesian school.

It is clear Hermann's 1935 paper yields riches for each discipline her analysis touches upon. What I aim to do in the following is suggest that while the above-mentioned investigations of her paper have gone a long way to showcase these riches, perhaps one of her most novel—and I argue, central—claims gets lost between

Work for this chapter was carried out in part while I was a postdoctoral research fellow at the Hebrew University's Edelstein Center for the academic year 2013–2014. I wish to thank the Edelstein Center for their generous support.

E. Crull (✉)
The City University of New York, New York, USA
e-mail: ecrull@ccny.cuny.edu

E. Crull and G. Bacciagaluppi (eds.), *Grete Hermann - Between Physics and Philosophy*, Studies in History and Philosophy of Science 42,
DOI 10.1007/978-94-024-0970-3_10

the historical perspective (specifically neo-Kantian or natural-philosophical) and the philosophical one (specifically philosophy-of-physics). More plainly, one might be tempted to read Hermann's investigation of quantum mechanics as a means to a premeditated end, namely the salvaging of causality. But one can discern in this work a far deeper claim: that what she calls the 'most important lesson from quantum mechanics'—the relative context of observation—applies not just metaphorically and only to quantum mechanics, but to the full range of natural knowledge.

Seen in this light, her claims about causality in quantum mechanics are not the primary end of her investigations, but rather a means to a much more significant end, which is her appreciation of what few physicists at that time understood of quantum mechanics: the thorough-going dependence of quantum phenomena on the specific context of their observation. Additionally, her final chapter on the natural-philosophical significance of these physical considerations is not merely an addendum—not just a hasty attempt to invoke her Nelsonian training—but is rather a thesis with multiple significant facets. One facet is indeed the attempt to reconcile this modern physics with a Friesian interpretation of Kant's antinomies. A second facet of the relative context of observation is Hermann's nuanced appreciation of the way in which quantum mechanics is intrinsically unlike classical mechanics; this aspect of her work reveals a unique and particularly insightful understanding of Bohr's doctrine of complementarity and the correspondence principle.

Yet another novel facet is exposed when we read Hermann as genuinely applying her insights from quantum mechanics to the natural-philosophical thesis of 'die Spaltung der Wahrheit'—the splitting, or fracturing, of truth. Not only does exploration of this aspect reveal in what sense Hermann's project is distinct from other neo-Kantian attempts to reconcile quantum mechanics with causality, but it releases us from the burden of seeing the (problematic) thesis of retrodictive causality as her most significant contribution: the importance of her notion of retrodictive causality is diminished when we explore alternate valences of her project. Furthermore, such an understanding of Hermann's 1935 paper brings to the fore the insightfulness of her interpretation of Bohr, and of the sense in which one might read Bohr's twin principles (correspondence and complementarity) as Kantian—a long-standing debate in the literature. Perhaps most interestingly for philosophers of physics, these notions ultimately undergird the deep ways in which Hermann appreciated before Heisenberg, Schrödinger and other fathers of quantum theory the central significance of entanglement and context dependence for modern physics.

In order to establish my claim about these broader consequences of Hermann's 1935 work, I will first highlight the centrality of the role that the relative context occupies in precisely those portions of her essay one might initially assume are dedicated to the pressing concern of preserving causality (Sect. 10.2). Then I will visit in detail her second chapter, in which she discusses Bohr's two principles and explores the quantum-classical relationship, which forms the foundation of her third, natural-philosophical chapter. A corollary to this exploration of Chapter II will be a brief historiographical detour regarding Bohr's 'degree' of Kantian-ness, specifically as displayed by complementarity, and to what extent Hermann qua neo-Kantian is in a unique position to capture Bohr's (likely) intended philosophical nuances (Sect. 10.3). In Sect. 10.4 I explore the further 'facets' I have claimed come to light when we place the relative context of observation at centre stage throughout the whole of Hermann's 1935 work, including in Sect. 10.4.1 the quantum versus classical modes of description, in Sect. 10.4.2 a Friesian/Nelsonian interpretation of quantum mechanics, and most importantly, in Sect. 10.4.3, how quantum mechanics demonstrates the failure of any attempt at a unified, absolutist worldview—in the sciences and beyond. I conclude (Sect. 10.5) with a call for future work on Hermann's (1935) paper to explore in particular the novelty of her approach to modern physics in comparison to those of her contemporaries in different neo-Kantian schools.

10.2 The Relative Character of Quantum Mechanics

Hermann is explicit in the introduction to her 1935 paper that the *starting place* of her investigation will be to address the 'challenge' to the a priori concept of causality arising from quantum mechanics. However, she ultimately expresses her course of investigation in more general terms: 'to be discussed, starting from the lessons of experience, are the natural-philosophical implications of the physical achievements' (Chap. 15, p. 240). In the next three subsections—corresponding to the three chapters of the 1935 paper—I rely on textual support to ground my claim that the true heart of Hermann's work has less to do with causality and a great deal more to do with a radical new feature of the natural world brought to light specifically by quantum mechanics, to wit—the relative context of observation.

10.2.1 Chapter I, 'The Limits of Predictability'

The first chapter of three is primarily concerned, as advertised, with causality and the limits of predictability. The details of this chapter are certainly important for the purpose of reconciliation with Kant, and also here Hermann lays the groundwork for later chapters with a thorough consideration of duality experiments and the nature of the uncertainty relations. I wish to draw attention to a few crucial comments that touch upon her ultimate end of using the relative context of observation required by quantum mechanical explanation to argue for the fracturing of truth, at all scales.

In Sect. 2, following the introduction of Heisenberg's uncertainty relations, Hermann writes the following:

> The independence of the measurability of the two quantities that holds in the classical theory is thus upset. Whereas classically the state of a system can be expressed through a mere enumeration of the values of all occurring physical quantities, the quantum mechanical formalism employs novel symbols in the description of the state that express the mutual dependence of the determination of different quantities.

> These symbols, the wave functions of physical systems, and the mathematical formalism that prescribes the correct rules for their combination, follow the classical theory closely because of Bohr's *correspondence principle*. The classical description is compatible with the quantum mechanical one insofar as its quantities remain undetermined to such a degree that the indeterminacy relations are fulfilled. (Chap. 15, p. 244)

Bracket for the time being the interesting Kantian terminology of 'symbols' introduced here; more shall be said on this in Sect. 10.3.1 below. In this excerpt Hermann already points to a key difference between the quantum mechanical picture and the classical one: the independence of measurability of various quantities. In quantum mechanics one has 'mutual dependence of the determination of different quantities'. Yet this difference between the theories does not render them incompatible, thanks to the uncertainty relations.

The next few sections criticise several attempts to show the incompletability of quantum mechanics via hidden variables theories; these sections and their role in Hermann's causal completeness argument have been discussed elsewhere in detail (see, for example, Chap. 4). What is relevant here is that what she calls (in the heading to her Sect. 9) the *solution* to the causal completeness question is not retrodictive causality, but 'the relative character of quantum mechanics'.

In her tenth section, Hermann reviews the well-known thought experiment of the gamma-ray microscope. This is described in detail elsewhere in this volume (cf. Filk in Chap. 5 and Frappier in Chap. 6), but I wish briefly to touch upon the most interesting aspect of Hermann's description of the experiment. As noted by both authors cited above, Hermann's account of the microscope is novel in introducing in addition to the 'first case' (wherein the measurement event occurs in the image plane) and 'second case' (wherein the measurement event occurs in the focal plane), a fascinating 'third case' in which there is no measurement event at all—the photographic plate is removed to infinity. As I highlighted in early work on Hermann's 1935 essay (e.g. Crull 2010), in this third case Hermann clearly recognises the *inseparability* of two systems that had previously interacted—cf. the paragraph in small font at the heart of Sect. 10 (Chap. 15, p. 258).

Though this particular insight of Hermann's is worthy of its own detailed consideration (and indeed Filk and Frappier have begun this work), what I wish to highlight here is that it is not obviously *causal* considerations which led her to this extraordinary insight, but rather her nuanced understanding of the relative character of quantum mechanical observations (which will receive more detailed attention from Hermann later, especially in her Sects. 13–15). In particular, it is her considerations in sections leading up to the tenth specifically regarding the 'symbol' of the wave function and its limited ability to determine at once all the relevant quantities of interacting systems that ground this concluding statement of her paragraph about the 'third case': 'Through this linear combination the light quantum and the electron are thus not described each by itself, but only in their relation to each other. Each state of the one is associated with one of the other' (Chap. 15, p. 258). It is clear that Hermann is considering precisely the same notion of separability (or rather, lack thereof) true of two previously interacting systems that famously troubled Einstein.[1]

The final section of Chapter I discusses von Laue's and Schrödinger's hopes (as articulated in their individual 1934 *Die Naturwissenschaften* papers: von Laue 1934; Schrödinger 1934) that the limitations of predictability stipulated by Heisenberg's uncertainty relations were merely provisional epistemic limitations, not fundamental physical ones. Though she will later deny quantum mechanics' incompleteness on grounds of its already being causally complete (albeit retrodictively), this is not in fact the main reason she cites here. Instead, the argument Hermann deploys against these eminent physicists for the completeness of the quantum mechanical formalism—as perhaps the reader has already anticipated—is premised on the relative nature of quantum mechanics. She writes:

> Control of the arising disturbance does not fail because the formalism is still defective with respect to the explanation of this disturbance and thus in need of completion, but because the explanations it provides—which are complete and therefore not liable to emendation—are valid, as are all quantum mechanical statements, only relative to a certain observational context. In fact, to the one in which the disturbance in question is considered, thus the one that comes about in the first place through the observation. The explanations for the disturbance provide a foothold for predictions only to one who has performed this observation, and thus finds himself in this observational context; the outcome of the observation itself consequently cannot be predicted with their help. (Chap. 15, p. 260)

10.2.2 Chapter II, 'The Natural-Philosophical Situation'

Section 12 is titled 'Causality and Quantum Mechanics', and in it Hermann continues the line of thought begun in the previous section, describing why causality *qua predictability* is impossible:

> Since every physical description and explanation of processes is valid only relative to its respective observational context, so the calculation of, say, a corpuscle trajectory combining the variables of different observational contexts into one representation remains physically vacuous precisely to the extent that it exceeds the uncertainty relations: it is uncheckable and provides no grounds for future predictions. (Chap. 15, p. 263)

Importantly, Hermann follows this argument with a note relating in what way one must 'amend' Heisenberg's 'controversial' treatment of time-of-flight measurements:

> The controversial representation Heisenberg has given of this situation—that it is purely a question of taste whether one should assign physical reality to such a calculation of trajectories—is therefore to be amended in the sense that this calculation disregards the quantum mechanically crucial relation each specification of a physical variable bears to the

[1]Consider Einstein's own description of separability in his famous letter to Schrödinger, written after Hermann's paper (and also the EPR paper) was published—on 19 June 1935: 'the second [system], together with everything that pertains to its contents, is independent of what happens with respect to the first [system] (separate subsystems)' (Einstein 1935).

> respective observational context, and in this sense becomes physically meaningless. Conversely, the possibility of checking even only indirectly the causal claims that arise in the interpretation of a measurement process depends on the fact that this interpretation does justice to duality, and thus implicitly to the relative character of quantum mechanics. (Chap. 15, pp. 263–264)

When Hermann writes that the calculation of the sort mentioned above, i.e., 'a corpuscle trajectory combining the variables of different observational contexts into one representation' is not a viable option for completing the quantum mechanical formalism, she is saying so quite plainly on grounds that a context-independent (perhaps one might say, *non-contextual*) description of quantum processes is impossible because the calculation 'disregrads the quantum mechanically crucial relation each [...] physical variable bears to the respective observational context, and in this sense becomes physically meaningless'.

Here one can already discern a whiff of *contextuality* in the specific sense that anticipates Bell's 1966 work on hidden variables (Bell 1966). Indeed, while I and others (e.g., Crull 2010, and Seevinck in Chap. 7) had previously claimed that Hermann anticipated Bell's analysis of von Neumann's proof against hidden variables, it now seems clear she anticipates Bell in this more subtle sense as well. This is particularly interesting in light of Heisenberg's draft of a response to the EPR paper. In Bacciagaluppi and Crull (2009), we argued that in the version of the cut argument given in this 1935 draft response by Heisenberg he introduces for the first time (in anticipation of Bell by some thirty-odd years) the contextual/non-contextual hidden variables distinction. However, this draft was written by Heisenberg in July, and we know that he not only read but wrote the preface for the published version of Hermann's essay *before* the summer of 1935—indeed, Hermann's essay appeared two months prior to the publication of EPR itself. Thus it would seem that the credit for first noticing the possibility of contextual hidden variables goes not to Heisenberg in July of 1935, but to Hermann in the winter of 1934/1935 (or even that of 1933/1934; cf. Chap. 8). And this profound insight is once again a result of her thorough appreciation of the relative context of observation in quantum mechanics as opposed to classical mechanics, not to the thesis of retroactive causal relations.

Having already discussed the fuller implications of the relational context for the thinking of eminent physicists like von Laue, Schrödinger and Heisenberg, Hermann moves on to a critique of Popper's probabilistic ensemble interpretation. In a lengthy footnote in Sect. 12, she argues as follows:

> [...] [Popper] misunderstands that because of the duality experiments the applicability of the classical conceptions is limited according to the uncertainty relations already for every single elementary process, and that accordingly wave functions can in fact be used for describing the state of individual systems. That this use of wave functions is consistent with their probabilistic interpretation is based once again solely on the relative character of the quantum mechanical way of description: on the one hand, the wave function is completely determined by the values of those physical quantities that have a sharp value within the momentary observational context for the system. In this respect it characterises the system quantum mechanically relative to the observational context present. On the other hand, the probability interpretation of the wave functions yields those variables that remain of the classical-intuitive description according to the correspondence principle and that fix

> which statements can be made for the passage from one observational context into another. (Chap. 15, p. 263)

In sum, the section in which Hermann is most concerned with defending Kantian causality (Sect. 12) refers almost ad nauseam to the relative nature of observational context as an explanation, and not only for the conflation of causality qua predictability with causality qua cause-and-effect, and to the mistaken (but heretofore unrecognised as such) assumption in classical physics that one can predict the evolution of systems with certainty (cf. Chap. 15, pp. 262–264). She concludes the section with these words:

> For the result of the previous investigations is just this: the opinion prevailing throughout, and even plausible in, classical physics—that gapless causality and the possibility of in-principle unlimited future predictions are inseparably linked with one another—has proven to be false. It has been refuted by the demonstration of the merely relative character of the description of nature, which confirms afresh the assumption of thoroughgoing causality, but has broken once and for all with the hope of arbitrarily sharp predictions. (Chap. 15, p. 265)

The central role of relative context here is hard to miss. In fact, it is the ground upon which rests her entire argument regarding the causal completeness of quantum mechanics.

10.3 Complementarity, Correspondence and Kant

As I demonstrated in the previous section, Hermann's discussion is highly focused on the relative context of observation—so much so that the purported Kantian-ness of Hermann's approach has assumed a negligible role. The Kantian nature of Hermann's paper only truly comes into consideration *after* her discussion of causality is largely finished, when she digs further into the implications of quantum physics' context dependence. In this section I discuss these implications. But first, a primer on Kant's use of two crucial terms: 'Symbol' (symbol) and 'Anschauung' (intuition).[2]

10.3.1 'Symbol' and 'Anschauung'

How do the data of experience become transformed into objects of scientific knowledge? For Kant it is a matter of applying the concepts of pure intuition (space, time) to the data in order to form empirical intuition (pure intuition coupled with sensation). In other words: the sensory data comprising our experiences are perceived and rendered intelligible within the a priori framework of space and time, and also through the relational category of causation; when these data are interpreted with the aid of intuition, we are able to claim we have understood the experience (Kant 1781).

---

[2]I have relied on Chevalley (1994) for the following subsection; if there are misunderstandings of Kant, the fault is solely mine.

What happens when there are no sensory data available to us, as with the in-principle unobservable objects of quantum mechanics? In this case we have no empirical intuition, only pure intuition with which to produce understanding of the sort that gives rise to scientific knowledge. In the Third Critique (Kant 1790), Kant explains this by delineating two modes of Darstellung (representation). The first is *schematic*, and applies to the direct, demonstrative perception of sensible objects that leads to understanding. The understanding applies intuition, and this results in the objectification of the sensory data, and objectification is what enables (scientific) knowledge of the data of experience.

It is the second sense of representation that applies in the case where we lack sensory data. This type of representation is called by Kant *symbolic*.[3] In this mode, we apply the concept for which no sensible intuition is available (such as the unanschaulich—unintuitive—concepts of quantum mechanics that caused Schrödinger great disquiet) to some sensible intuition that serves as the symbol. The relation between the symbol and its associated unintuitive concept is formed by *analogy*. Analogy, for Kant, is the structural identity of relations standing between dissimilar relata. He writes in the Third Critique (as quoted in Chevalley 1994, p. 44):

> Symbolic presentation uses an analogy (for which we use empirical intuitions as well) in which judgment performs a double function: it applies the concept to a sensible intuition, and then it applies the mere rule by which it reflects on that intuition to an entirely different object, of which the former object is only the symbol.

For example, we lack sensible intuition regarding the concept of position in quantum mechanics, so we relate it to the classical concept of position (using, e.g. in the Dirac formulation, the position operator as symbol) but understand that our application of the position operator to the unintuitive concept of a quantum system's 'position' *can only be* analogous to the intuitive, immediately perceivable (ergo understandable) relationship between a position 3-vector and the *actual* position of a classical system.

With this background in place, a closer reading of in particular Hermann's second chapter will demonstrate how her Kantianism, coupled with her focus on the relative context of observation, leads her to a nuanced understanding of Bohr's two principles: complementarity and correspondence.

10.3.2 Hermann on Bohr's Two Principles

In Hermann's understanding, the correspondence principle is thoroughly Kantian: she describes 'intuitive classical concepts' as forming a bridge between the data of sensation and the formal statements of the theory. This is precisely what is required by Kant to understand unintuitive concepts—symbols (i.e. the quantum formalism in

---

[3]Chevalley cautions that one must not understand Kant's 'symbol' to be on a par with what mathematicians or logicians usually mean by the term: the latter typically refer to 'the conventional designation of concepts by signs or words' (Chevalley 1994, p. 44).

the guise of Dirac's observable calculus or Schrödinger's wave mechanics—or even Heisenberg et al.'s matrix mechanics, though she only explicitly mentions the two former schemata) are used to functionally relate, via analogy, the intuitive sensible objects of classical physics to the unintuitive, mediately sensible (or insensible) objects of quantum physics. It is in this way that classical concepts play a well-defined, necessary, but not strictly speaking *correct*, role in interpreting the quantum data of experience. Hermann writes:

> According to the *correspondence principle*, the intuitive classical concepts still form the bridge between the data of sensation and the formulas of the theory. Without them one cannot obtain from a measurement a viable starting point for theoretical inferences; they provide the key to the interpretation of unintuitive quantum mechanical formulas and thereby make possible their application to experience. (Chap. 15, p. 266; emphasis original)

So the correspondence principle becomes, in Hermann's Kantian interpretation, a necessary means by which to bridge the gap between intuitive classical concepts and unintuitive quantum mechanical concepts. But knowing that these symbols are indispensable is only the first step—Hermann must now demonstrate the precise scope within which these symbols (and associated concepts) are applicable. Enter complementarity, a doctrine which Hermann understands as 'essentially three mutually distinct but content-wise related relationships' (ibid.): (a) wave versus particle complementarity, (b) uncertainty relation/non-commuting variable complementarity, and what she considers the *essentially* complementary relationship, (c) that of intuitive-classical modes of description and the (unintuitive) quantum-mechanical mode of description.

This last relationship is explicitly Kantian, but the first two also depend on Hermann's Kantian version of complementarity: wave and particle pictures are mutually-constraining, complementary pictures precisely *in virtue* of their both being 'classical-intuitive constructions of processes in space and time' (ibid.). These constructions are understandably constrained, therefore, when applied to quantum phenomena; the duality experiments provide evidence of this.

Likewise, the second relationship's complementary nature—codified in Heisenberg's uncertainty relations, and so existing both within the wave picture and within the particle picture—is rooted in Kantianism, in the following way: this sort of complementarity only arises once the experimental context has been decided, and a choice has been made about within which picture the system will be observed. This choice already cuts off the experimenter from univocally determining all of the system's variables (cf. her Sect. 13 for more on the connection between causality, determinism and the context of observation).

However, it is the very last sense of complementarity Hermann clearly deems the most interesting and crucial aspect of Bohr's doctrine. Hopefully given the above, one is already prepared to understand just why this particular relationship is emphasised by Hermann: by her lights it is thoroughly Kantian, and undeniably arises from the special characteristic of quantum mechanics she had been at pains to highlight in previous sections: the relative context of observation.

Classical concepts, or the classical-intuitive mode of description, are fully deterministic, entirely intuitive and can be validly applied—*within the limits* circumscribed by senses (a) and (b) of complementarity. What occurs beyond this range? This is when the quantum mechanical formalism, expressed using the symbols of the wave function and Dirac's observable calculus, enters. While they cannot, in virtue of containing the previous senses of complementarity, be fully determinate nor 'directly intuitive', these symbols bridge the gap and, as Hermann puts it:

> Through this departure from intuition, the quantum mechanical formalism succeeds in the step, where classical physics fails, of combining into a single state description the apparently mutually contradictory intuitive variables of the system, and of thereby capturing *the strictly causal course of natural-law relationships* for the states so characterised. (Chap. 15, p. 266; emphasis original)

Of course, this causal completeness comes at the price of intuitiveness. But that is a price Hermann is happy to pay—unlike Schrödinger. And one might speculate that her ability to turn from intuitive concepts in this regard is due to the nuanced Kantian framework that allows her not only to resolve apparent conflict but to highlight 'one of the most wonderful results of quantum mechanics [...] that every observation is associated with a *disturbance of the observed system* due to the interaction of the system with the measuring apparatus and, given the only limited applicability of the classical concepts, is *uncontrollable* to a certain degree' (Chap. 15, p. 267; emphasis original).

One can already appreciate that Hermann's employment of the phrase 'classical concepts' is a great deal more nuanced than a claim about mere semantics, as some have often claimed (if not regarding Hermann, then Bohr's habit of utilising the same terminology). Indeed, she argues in her Sect. 14 that complementarity is not only sufficient from a Kantian perspective for explaining the unintuitive yet causally complete picture provided by quantum mechanics—she argues that complementarity is also *necessary*, precisely in its application of classical concepts to unintuitive quantum processes. But it *cannot* be classical concepts simpliciter that are necessary, and the reason for this is as follows:

> [T]he quantum mechanical formalism is on the one hand *physically closed* in the sense that it completely states the natural-law relationships in what happens [...] On the other hand, it characterises physical systems only *relative to the respective context of observation* in which the physicist stands towards his object; so it is excluded from intuitive interpretation, which is possible only where physical processes can be unambiguously construed as motions in space and time. The hope of being able to preserve from the classical theory certain firm pillars upon which the construction of a new intuitive picture of nature could rest, is therefore moot. (Chap. 15, p. 268; emphases original)

And yet Schrödinger's complaints about the possibility of a completely quantum-mechanically (and therefore unintuitively) described system are unjustified: immediately after the above quotation, Hermann emphasises that 'the quantum mechanical formalism ultimately signifies *no detachment from intuition*'. How she can claim this is in virtue of her subtle Kantian understanding of 'classical concepts' and their necessary involvement in any quantum mechanical description, via the correspondence principle. Thus, clinging to classical concepts in the blunt sense of concepts directly and intuitively referring to sensible objects is 'moot' (and so we must understand the role of such concepts in the *correspondence principle* to be nuanced in just the

Kantian way she suggests). But neither will the other extreme of doing away entirely with classical concepts succeed, for then 'the meaningful association between the data of sensation and the posits of a physical theory' is eliminated (ibid.)—and so we must understand the role of classical concepts within *complementarity* to be nuanced in precisely the way Hermann suggests. Hence the thorough-going dependence of Bohr's two principles, on Hermann's reading, upon Kant's 'symbol' and 'intuition'.

10.3.3 Historiographical Corollary

Bohr exegesis is controversial territory, to put it mildly. However, one of the more historically fascinating aspects of Hermann's 1935 paper, at least by my lights, is that it adds grist to the mill of Bohr scholars like Chevalley who see embedded in his writings a nontrivial Kantian influence. If Hermann, a brilliant neo-Kantian mathematician and contemporary to Bohr (and frequent interlocutor with members of his inner circle, e.g., Heisenberg and Weizsäcker) interpreted Bohr's principle of correspondence and complementarity to be so obviously—nay, necessarily—Kantian, then that is something worth pondering. I begin this pondering, albeit cursorily, in this subsection.

According to Chevalley, after 1928 Bohr consistently used the language of 'symbol' and 'intuition' when speaking of how one obtains objectivity in quantum versus classical mechanics (Chevalley 1994, p. 35). Ultimately we rely on objectivity in science, for objectivity is that property of knowledge which allows for intersubjective invariance of content; without achieving objectivity in our scientific theories, we have no way to apply the contents of that theory to a broader class of phenomena and obtain scientific explanations which are invariant across observers. As Bohr himself would argue using the doctrine of complementarity, intuitive classical concepts only apply (and thus the analogies only have strength) within certain bounds, namely, within particular contexts of observation.

Likewise relativity: in general relativity in particular, velocity and position and other such classical terms are no longer directly applicable to the data of experience, and instead we rely on symbols to relate unintuitive concepts into intuitive language using classical concepts once again. This explains why, in many of Bohr's writings on quantum mechanics, he often invokes general relativity alongside quantum mechanics—to highlight the similarity of these two nonclassical theories as regards the limits of space-time continuity (cf., for example, Bohr's response to EPR in Bohr 1935).

It should be noted, however, that for Bohr and for Hermann it is clear that the loss of intuitiveness does not imply the loss of objectivity. Instead, in theories reliant upon unintuitive symbols (like relativity and quantum mechanics) the *conditions* for objectivity are altered, in that an objective description necessarily requires inclusion of the relevant observational context—a thing not needed in classical, intuitive theories. Not only are the conditions of objectivity altered, but the interpretation of this property in new physical theories strongly discourages the reification of concepts that is often implicit in classical theories. In other words, because the process of obtaining objectivity in relativity and quantum mechanics is mediate, indirect and unintuitive, the old innocuous assumption about reading one's ontology off the concepts of the theory becomes greatly complicated. Our remove from direct perception of the phenomena in these latter theories is even greater than it was even for classical objects in the schematisation of Kant.

Complementarity, in Chevalley's analysis, is the name Bohr gives to the relationship between intuitive, space-time visualisable processes (rendered in the language of classical concepts) and unintuitive, non-visualisable, context-dependent processes found in quantum mechanics and relativity. Hermann's analysis of complementarity (especially in the third sense) sits nicely with this. Chevalley asserts that already after the refutation of the BKS theory, Bohr turns to this Kantian language of symbol/intuition in order to explain quantum phenomena and also to evaluate the quantum-classical divide (more on which below).[4]

One should notice, as this will be an important point of departure for Hermann both from Bohr and from her neo-Kantian school, that Bohr seemed to believe an accumulation of complementary perspectives would somehow bring one nearer to the truth. As Chevalley writes, 'Bohr's contention seems to have been that the loss of Anschaulichkeit ought to be compensated for by a strategy of multiplying different languages and perspectives' (Chevalley 1994, p. 42). Heisenberg also seemed to understand Bohr's interpretation in this way—that one could compound truth, so to speak, through myriad perspectives:

> [...] only by using a whole variety of concepts when discussing the strange relationship between the formal laws of quantum theory and the observed phenomena, by lighting this relationship up from all sides and bringing out its apparent contradictions, can we hope to effect that change in our thought processes which is a sine qua non of any true understanding of quantum theory. (quoted in Chevalley 1994, pp. 42–43)

Hermann's disagreement with this view should already be evident from portions of her 1935 text quoted above in Sect. 10.2; the analysis of Hermann's complementarity also hints at her conviction regarding the complete absence of any absolute, unified truth. Why she considers this 'composite' stance to truth not only wrong-headed but in-principle impossible will be made even clearer when we turn to discussion of her final chapter.

With respect to Hermann's division of complementarity into three parts, she anticipates later historians who attempt the same—e.g., Bitbol and Osnaghi (2013). Indeed, not only does she recognise these different aspects of the doctrine, but explains how each of them must be interpreted in order to work, and does not suggest (rightly, I would argue) that the concepts need reduce to one another. Thus, she avoids the problem pointed out by Bitbol and Osnaghi that the first 'sense' of complementarity entails a serious difficulty regarding simultaneous possession of values for non-commuting variables. They suggest that a satisfactory alternative interpretation of this notion may come about if we

---

[4]Note, however—as Chevalley does—that in Bohr (and so too in Hermann, as we shall see) this divide is *not* intended to coincide with the object–instrument divide. On Bohr's position, see Howard (1994).

> focus on the observations that are *possible* when only the experimental preparations, but not the measurement to be performed, has been fixed [...] Along this line of thought, one may reconcile mutual exclusions and completion of conjugate variables without logical inconsistency: mutual exclusion pertains to *actual* experimental arrangements, whereas completion (or exhaustiveness) refers to *possible* measurements. (Bitbol and Osnaghi 2013, p. 157; emphases original)

Again, Hermann not only anticipates this problematic reading of complementarity but resolves it with appeal to the relative context of observation.

Hermann's transcendental-idealist narrative about the acquisition of scientific knowledge in quantum mechanics significantly alters certain other claims made by Bitbol and Osnaghi in their recent paper on Bohr and Kantian epistemology (Bitbol and Osnaghi 2013). For example, they write this on the question of gaining scientific understanding: 'When it comes to quantum mechanics [...] things are not as simple. Indeed, Bohr's point can be formulated precisely by saying that the *transcendental* approach of knowledge becomes unavoidable when one is concerned with quantum phenomena' (p. 154). They go on to explain (in some detail) the primary arguments Bohr deployed to this end, which amount to the following: (i) one must assume that at least some aspect of the measurement apparatus is free of Heisenberg uncertainty, and (ii) no account of experience is possible without assuming a particular conceptual framework (linguistic or otherwise). While it may or may not in fact be possible, or even correct, to interpret Bohr's 'point' in all this to involve Kantian transcendental idealism, we have already in Hermann a lovely, clear argument for the inescapability of the uncertainty relations at any level of description (disagreement with (i)), yet the necessity of involving the context of observation in one's theory of measurement (agreement with (ii)).

Though Hermann's position on the issues Bitbol and Osnaghi have labeled (i) and (ii) above has already been seen to some degree in the preceding discussion of her 1935 paper, Hermann's commitment to the necessity of a fully quantum mechanical description will come out all the more clearly with her invocation of Heisenberg's cut (more on which in Sect. 10.4.1). In this argument Hermann is often (and oftentimes explicitly) following Bohr and Heisenberg's lead; inasmuch as she claims to report from a position proximate to these figures their own views on the uncertainty relations, the cut, and the context of observation as a necessary component in scientific descriptions of quantum phenomena, historians of Bohr and Heisenberg would benefit from perusal of her language and style of argumentation.

In closing, I think it safe to claim that a close reading of Hermann's 1935 paper adds important, new fodder for the on-going debate regarding Bohr's philosophy. In this section I have hastily sketched but two examples of Bohr scholarship that would be affected if Hermann's 1935 paper were incorporated in the historical narrative: one example is of Hermann's interpretation of Bohr roundly supporting Chevalley's analysis, while the other is of Hermann's interpretation of Bohr somewhat modifying the analysis of Bitbol and Osnaghi. All this aside, however, I note that regardless of the degree to which one believes Kant influenced Bohr's thinking, Hermann has at least done this: she has shown us, from the perspective of an insider with philosophical expertise, the contours of a viable Kantian interpretation of quantum mechanics.

10.4 Exploring New Facets of the 1935 Paper

As I claimed in the introduction, zeroing in on the importance of the relative context of observation in Hermann's 1935 paper exposes new facets of this work. Among them, I here focus on her approach to the quantum–classical divide (including her version of Heisenberg's cut argument) and her radical claim that this novel feature of the natural world demonstrated in quantum mechanics—the splitting of truth—carries through to other arenas of natural philosophy. Between these two topics, I briefly touch on the novel Friesian analysis Hermann provides in her third chapter, which signifies a departure for her teacher and mentor, Leonard Nelson. Though this aspect of her paper has indeed been discussed by others in this volume (cf. Chaps. 2, 3 and 4), I will argue specifically that Hermann's motivation for this new version of Friesianism stems at least in part from her thesis of the relative context in quantum mechanics.

10.4.1 The Quantum–Classical Divide

Hints about Hermann's clear understanding of the quantum–classical divide appear already in the first chapter, in Sect. 9. There, in the midst of describing her thesis of (retrodictive) causal completeness, she makes the following fascinating claim:

> For no natural process is it completely excluded that in some context it may be considered and accordingly interpreted only as part of a measurement process. The classical causal reasoning that, in the interpretation of the observed measurement outcomes, leads from the measuring instrument to the observed system thus need not be broken off there, since it is always possible that this system in turn may serve as measuring instrument for some other system with which it has interacted. (Chap. 15, p. 255)

Though it seems clear from historical investigation of primary sources that such an understanding of the arbitrary choice of system as thing-measured and system as measuring-thing was in the air prior to Hermann's writing down the above statement, one finds strikingly similar lines of thought in Heisenberg's own reaction to the question of completeness in his response to EPR (cf. Bacciagaluppi and Crull 2009, 2011, 2018). Importantly, we notice that Hermann's statement about the arbitrary cutting-off of the measurement process is made in support of her argument that hidden variables (in her words, 'new as-yet undiscovered features') are at best redundant, at worst pointless. Continuing from the above quoted passage:

> If, e.g., an elastic collision has taken place between two bodies, it is sufficient to measure the change in momentum that one has experienced, in order to determine also that of the other. If one of them is measured, it can thus be considered on the one hand as the object of this measurement, on the other hand as a measuring instrument for the determination of the other, in which case one then calls upon the explanations provided by the classical theory of elastic collisions for the occurrence of this particular change of momentum, and precisely in so doing [one] has proven the attempt to explain this process through new as-yet undiscovered features to be pointless. (ibid.)

One recognises here the beginnings of the 'cut' argument often employed by Heisenberg, but altered importantly: Heisenberg uses the movability of the cut to demonstrate the impossibility of hidden variables, ergo the *incompletability* of quantum mechanics, whereas Hermann uses the movability of the cut to argue for the *completeness* of quantum mechanics, ergo the impossibility of hidden variables. Hermann's argument seems to be that since the choice of measuring device is arbitrary, the causal chain (which can only be traced after the fact) nevertheless cannot be predicted beforehand.

> Thus, if after the reading of the measuring instrument one explains backwards the position of its needle through a theory of the measurement process, this [explanation] traces the process of measurement back to states of the measured system and the measuring instrument that were not, and could not, have been included in the preceding description of these systems. The description of these systems is thus not univocal in quantum mechanics. (Chap. 15, p. 256)

This failure of univocality is precisely what bothered Einstein about the theory, and what led Heisenberg (among others) to declare the theory 'incompletable'. But Hermann opts for a different route, and her motivation is both Kantian (as we have seen) and pivots on the relative context of observation uniquely required by this new physics. Her conclusion following from this failure of univocality? That through the measurement process itself, because it involves interaction between various systems arbitrarily labelled 'object of interest' and 'measuring device', a new context for physical observation is created, 'in which both systems are presented to the observer in a new way that cannot be uniquely predicted from the previous one' (ibid.).

And now it is understood why in an important sense Hermann's solution to the apparent conflict with causality implied by quantum mechanics is not retrodictive causality, but rather the unique, entirely nonclassical feature of quantum mechanics that the description allowed by this formalism is relative to the context of observation, and this imposes limits on our ability to trace causal chains as in classical physics. Thus far Hermann's approach to the cut argument.

But can Hermann really be said to endorse a fundamentally quantum view of the world due to the relative context? Establishing this aspect of her thinking will be important for understanding the scope of her claim about the splitting of truth—i.e., whether she intends the splitting to occur throughout natural philosophy or only within certain disciplines, and whether she intends the splitting to be metaphorical or literal.

In Sect. 15, Hermann notes that scientific explanations (what she calls 'explanations of nature') are grounded in—and therefore stand or fall according to—observations. But observation by itself does not generate knowledge on behalf of the perceiver. A further step is required, and it is on the whole an interpretive one within science, for merely everyday objects like chairs and tables can be properly judged on grounds of observation sans interpretation.

Explanations in classical physics are much like our experiences of such everyday objects, and become organised in our perception in terms of synthetic a priori categories, formating such experiences statically in terms of spatial and temporal relationships, and dynamically in terms of causal relationships. Thus we believe such classically-derived explanations reflect phenomenal structure with integrity. Explanations in quantum mechanics, as discussed above, require more work for the Kantian—but are still entirely possible:

> Despite all the failures of the classical theories, quantum mechanics also retains the method for proceeding from observation to the explanation of nature through the construction of such models and of the causal regularities valid within them. Forced by contrary experiences, it has only broken with the single assumption that these models describe the course of nature *objectively*, that is, independently of the observer and the manner in which he observes. The intuitive physical models, and indeed those of classical physics, appear to be still indispensable for the explanation of nature, even if they merely serve as *analogies* for it. (Chap. 15, p. 269)

Objectivity is thus only retained in as far as intuitive, classical models are relied upon for providing natural explanations, even if these models are considered analogies (but analogies in a precise, Kantian sense). Furthermore, a single intuitive model can no longer provide the full natural explanation of an observed event: quantum mechanics teaches us that one can only obtain a partial, nonobjective model from a given observational context. Because the only sorts of models we can obtain in quantum mechanical modes of explanation are relative to the context of observation, so too the natural knowledge derived from such models can at best be relative. And this is quite a radical point in Hermann.

As an aside, it is somewhat startling (but not surprising) to see that when Hermann examines the quantum mechanical side of the cut, she very clearly recognises that the measurement process must lead to an unfactorisable state of affairs—but again, this proto-description of entanglement is couched in the Kantian language of symbols:

> The result [of the interaction between object and measuring apparatus] will generally be a linear combination whose individual terms each connect one state of the object to one of the measuring apparatus. These various possibilities, among which observation decides, are combined on an equal footing by a symbolic addition into a combined wave function that develops in a phase space determined by the degrees of freedom of both systems. But this means that one must forgo *within this observational context* the intuitive tracking of each of the two individual systems. (Chap. 15, p. 270; emphasis original)

10.4.2 Friesian Quantum Mechanics

Hermann concludes her second chapter (also her Sect. 15 on the quantum vs classical modes of description) by emphasising once again that intuitive classical concepts necessarily play a role in quantum mechanics—the very specific, indispensable role of analogies. This allows her to segue to her third and final chapter, titled 'Transcendental Idealism', in which she defends the Friesian/Nelsonian understanding of Kant's categories as analogies. It is only with this nuanced, limited application of the a priori categories of space, time and causality that Hermann's arguments up to this point can run through.

In particular, Hermann argues (Sect. 16) that a Nelsonian resolution to the antinomies is the only way to render not just modern physics (like relativity and quantum mechanics) consilient with Kant, but even *classical* physics. For instance, if one were to take the relational category of causality strictly literally (as opposed to analogously, as Fries insisted), one would be forced to locate a cause for every process at increasingly infinitesimal time intervals in order to carry out ‘a strict application of the concept pair “cause–effect”’ (Chap. 15, p. 272); because this is impossible, such a literal application of Kant’s law of causality also is impossible.

The way one circumvents the problem of infinitely small causal chains is by applying differential equations to descriptions of natural law-like relationships. However, the nature of these equations is such that they, too, cannot substantiate a literal reading of the pure intuitions of space and time. Her argument here comes down to recognising the inevitable dependence of differential equations on only approximately local (but imprecisely definable) regions of space and time, but also these equations’ dependence on the values of those variables *not* differentiated over, which are also changing over the imprecisely-defined local space-time regions.

She then provides a rather fascinating proof relying on the failure of any precise physical characterisation of either the parts or the whole of a physical system (cf. Chap. 15, p. 273), that the space-time point to which one assigns various determinable variables of a physical system ‘does not harbour the bearer of these properties; otherwise one would be able to determine them from it [the space-time point] alone’. Her striking conclusion from these failures of literal application of the Kantian categories is not just to undergird a Friesian reading of the categories, but an even deeper ontological claim:

> One can thus say nothing as to the specification of the properties of physical objects or events that determine these as they are constituted in themselves; rather, the alleged properties of physical systems in truth only specify certain relations between the parts of the system, without these parts being themselves unambiguously specifiable. (Chap. 15, p. 273)

Notice that this espousal of relationalism comes directly out of *classical* considerations. It will certainly, then, be compounded when applied to quantum mechanics and the necessity of complementary modes of description. As we learn from Hermann’s Sect. 15, the necessity of the cut does not entail the failure of an entirely quantum mechanical description. The partitioning implied by the cut of an experimental set-up—*regardless of whether that cut happens to align with the division into classical versus quantum modes of description*—will render ‘unambiguously specifiable’ those properties of *subsystems* created by the cut, but *not* within a specific context of observation. Indeed, the preservation of this, but only this, relation among experimental contexts is what Hermann had stressed in the prior chapters as the solution to the apparent failure of causal continuity.[5]

The task of explicitly linking Nelsonian transcendental idealism with the discoveries of quantum mechanics is taken up by Hermann in her Sect. 17. I shall not discuss

[5]Although I will not discuss it here, the final section of Heisenberg’s draft response to EPR is dedicated to a discussion of the relative context of observation, and Heisenberg cites Hermann’s 1935 paper as his primary source for these considerations (Bacciagaluppi and Crull 2011). It would be most interesting to compare in greater detail, side-by-side, Hermann and Heisenberg on this point of overlap/interaction.

this section, as it is treated already in Chap. 4. I merely remark that the arguments she makes therein only lend support to claims made earlier in this chapter, specifically regarding Hermann's Kantian-ness, her argument for the indispensability of classical concepts yet the maintaining of objectivity within quantum mechanics (but in importantly nuanced ways that perhaps more correctly grasp Bohr's own thinking), and of course the centrality of relational context to her thinking.

10.4.3 The Splitting of Truth

As we have just seen, Hermann clearly argues from physical grounds for Kant's a priori categories to be read as mere analogies, in keeping with the Nelsonian school. However, her precise and intentional deployment of the term 'analogy' in these sections must not mistakenly lead one to assume that her paper in the final section—on the splitting of truth writ large—*also* should be considered a mere analogy to the (literal) splitting of truth in quantum mechanics which necessarily follows from the relative nature of observation.

Indeed, Hermann's language in Sect. 18 is strong enough to speak unambiguously against a reading of the link between quantum mechanical lessons and 'truth' as analogous or metaphorical. Consider her claim in the very first sentence that the splitting of truth is an 'even deeper connection between the results of quantum mechanics and the reflections of the critical philosophy' (Chap. 15, p. 276), which is followed with a declaration of independence from Nelson by denying the existence of his long-sought universal (and universally axiomatised) science:

> The proof in transcendental idealism that natural knowledge is inadequate for capturing reality but rather only picks out, in an incomplete way, relational networks whose foundations remain indeterminate within the scope of this knowledge—opens the way for the possibility of different mutually independent yet mutually compatible modes of confronting reality through perception. Only with insight into this possibility is the understanding of the actual structure of human perception disclosed, which—no matter how one might force it—is irreconcilable with the postulate of a universal science comprising all areas of perception.

On what grounds does she make these claims? On grounds of quantum mechanics. She is explicit on this point in the next paragraph (Chap. 15, pp. 276–277), where she explains that the fracturing of worldviews (language she borrows from Apelt) is not only also to be found in quantum mechanics, but is in fact demonstrably *extended* by it. The nature of this extension not only serves to further Hermann's departure from Nelson's absolutism, but is entirely due to—as the reader will by now have guessed—a familiar motif:

> The relative character of the quantum mechanical description of nature leads to this, that already in the purely physical treatment of natural systems various representations appear side by side, none of which claims absolute validity, rather which are all valid only relative to the respective context of observation, and precisely because of that can exist in harmony with one another despite their differences. From this point of view, the natural-philosophical novelty of quantum mechanics is describable thus: the splitting of truth goes deeper than philosophy

> and natural science had previously assumed. It penetrates into the physical knowledge of nature itself; instead of merely delimiting its scope against other possibilities for grasping reality, it separates various equally legitimate representations within the physical description that cannot be unified into a single picture of nature. (Chap. 15, pp. 276–277)

Quod erat demonstrandum.

10.5 Conclusion

In this chapter I have argued that when one reads Hermann's 1935 essay as primarily an investigation into the meaning and implications of the relative nature of quantum mechanics—not only for quantum mechanics, but all the way into natural-philosophical considerations that include ethics—certain new dimensions of her work appear with greater clarity. Among these are (i) her particular Kantian interpretation of Bohr's complementarity and correspondence principles, (ii) her unique understanding of the quantum–classical divide, (iii) the failure of Kant's a priori categories of space, time and causality to apply literally even for obtaining classical natural knowledge, and (iv) the splitting of truth. Reading Hermann's paper in this way also relieves the pressure of those who see all too clearly the significant problems with the concept of retrodictive causality yet felt compelled to considered this conversial concept her main result.

All this is merely the beginnings of exploration, however. Avenues of further work that touch upon points raised in this paper include, but are not limited to, these:

- I suggest (with malice aforethought) that Hermann's newly discovered 1933 paper (cf. Chaps. 8 and 14) will add strength to certain issues raised herein, in particular by illuminating her changing opinions about the question of completeness and determinism in quantum mechanics.
- Along the same lines, it would be fruitful to investigate her later works (e.g., Hermann 1937; Henry-Hermann 1953, 1985) with an eye towards the facets discussed here, especially the degree to which her relationalism and belief in the fracturing of truth or worldviews applies to, say, ethics.
- There is much work to be done comparing Hermann's particular Friesian interpretation of Kant, as it plays out in the context of her discussions of philosophy of physics, to her contemporaries belonging to other neo-Kantian schools who also wrote on such matters, among them Schlick and Carnap, and perhaps even Reichenbach and Cassirer.
- In Ryckman (2005, p. 41) the author suggests that a lecture by Planck in 1908, titled 'Die Einheit des physikalischen Weltbildes' ('The unity of a physical worldview'), greatly influenced various neo-Kantian thinking about relativity. It would be interesting to investigate whether the same lecture caused ripples for neo-Kantians thinking about quantum mechanics; I note specifically the obvious disagreement with Planck's paper suggested by Hermann's final section of the 1935 paper.

- I have suggested, but not carried out, a thorough analysis of the ways in which Hermann's work—which was written during a time of intense collaboration and discussion and tutelage with and under Heisenberg—influenced Heisenberg's thinking on questions like incompleteness, hidden variables, and ontology.[6]
- Related to the prior point is the suggestion that a more thorough study be made of Weizsäcker's work (specifically his post-war propositions regarding the deterministic nature of the wave equation and measurement theory) in relation to Hermann's. We know from multiple sources that these two figures were in contact at various points of their lives, and specifically discussed philosophy of quantum mechanics.

Hopefully these cursory suggestions are yet sufficient to demonstrate—in their number and potential interest to philosophers and historians alike—the importance of integrating Hermann into the foundations of quantum mechanics narrative.

References

Bacciagaluppi G (2008) The statistical interpretation according to Born and Heisenberg. In: Joas C, Lehner C, Renn J (eds) HQ-1: conference on the history of quantum physics, MPIWG, vol II, pp 269–288

Bacciagaluppi G, Crull E (2009) Heisenberg (and Schrödinger, and Pauli) on hidden variables. Stud Hist Philos Modern Phys 40(4):374–382

Bacciagaluppi G, Crull E (2011) Translation, with introduction, of W. Heisenberg: 'Ist eine deterministische Ergänzung der Quantenmechanik möglich?' Preprint posted on the PhilSci Archive http://philsci-archive.pitt.edu/8509/

Bacciagaluppi G, Crull E (expected 2018) 'The Einstein Paradox': the debate on nonlocality and incompleteness in 1935. Cambridge University Press, Cambridge

Bell JS (1966) On the problem of hidden variables in quantum mechanics. Rev Mod Phys 38:447–452

Bitbol M, Osnaghi S (2013) Bohr's complementarity and Kant's epistemology. Bohr, 1913–2013, Séminaire Poincaré XVII:145–166

Bohr N (1935) Can quantum-mechanical description of physical reality be considered complete? Phys Rev 48:696–702

Chevalley C (1994) Niels Bohr's words and the Atlantis of Kantianism. In: Faye and Folse (1994), pp 33–55

Crull E (2010) Grete Hermann and the gamma-ray microscope Gedankenexperiment, talk delivered at HQ-3: conference on the history of quantum physics, Berlin, 2 July 2010

Einstein A (1935) Letter to E. Schrödinger, 19 June 1935. In: Archive for the History of Quantum Physics, M/f No. 92, Sect. 2–107

Einstein A, Podolsky B, Rosen N (1935) Can quantum-mechanical description of physical reality be considered complete? Phys Rev 47:777–780

Faye J, Folse H (eds) (1994) Niels Bohr and contemporary philosophy. Kluwer Academic, Dordrecht

Henry-Hermann G (1953) Die Überwindung des Zufalls: Kritische Betrachtungen zu Leonard Nelsons Begründung der Ethik als Wissenschaft. In: Specht M, Eichler W (eds) Leonard Nelson zum Gedächtnis. Verlag Öffentliches Leben, Frankfurt am Main/Göttingen, pp 25–111

Henry-Hermann G (1985) Die Überwindung des Zufalls. Meiner, Hamburg

[6]For more on Heisenberg's thoughts regarding these issues, see Bacciagaluppi and Crull (2009) and Bacciagaluppi (2008).

Hermann G (1935) Die naturphilosophischen Grundlagen der Quantenmechanik. Abhandlugen der Fries'schen Schule 6(2):75–152
Hermann G (1937) Über die Grundlagen physikalischer Aussagen in den älteren und den modernen Theorien. Abhandlugen der Fries'schen Schule (Neue Folge) 6(3):309–398
Howard D (1994) What makes a classical concept classical? Toward a reconstruction of Niels Bohr's philosophy of physics. In: Faye and Folse (1994), pp 201–230
Kant I (1781) Kritik der reinen Vernunft. In: Kants gesammelte Schriften, vols 3–4. De Gruyter, Berlin
Kant I (1790) Kritik der Urteilskraft. In: Kants gesammelte Schriften, vol 5. De Gruyter, Berlin, pp 165–485
von Laue M (1934) Über Heisenbergs Ungenauigkeitsbeziehungen und ihre erkenntnistheoretische Bedeutung. Die Naturwissenschaften 22(26):439–441
Ryckman T (2005) The reign of relativity: philosophy of physics 1915–1925. Oxford University Press, New York
Schrödinger E (1934) Über die Unanwendbarkeit der Geometrie im Kleinen. Die Naturwissenschaften 22(31):518–521

Part III
Discussions at the Hermann Workshop

Chapter 11
Panel Discussion

Dieter Krohn, Rene Saran and Fernando Leal

Dieter Krohn: We thought we should first introduce ourselves and tell you how we were connected to Grete Hermann. I met her in the early '70s. In those years she was still prepared and willing to do some seminars with those people in Germany who were interested in becoming facilitators of Socratic dialogues. She did that together with Gustav Heckmann. And I was one of those participants, and met her. Later on I became a member of the Political-Philosophical Academy, it was in the late '70s, and she was the chairperson of that (PPA, as we call it). I experienced her in many discussions then.

One specific incident I might mention, although it is very private; but it sheds light on her as a person. My father had excellent knowledge and practical experience as a craftsman, but when he lost his job he started a business and he was a bad businessman. So he went bankrupt. And I was a guarantor to him, so I had to pay a lot of money. When Grete got to know about it, through accident, she immediately offered me the money and said 'You can pay it back whenever you like'. And so I was

The following is a transcription of a panel discussion held in the morning of Sunday, 6 May 2012, as part of the Grete Hermann Workshop, University of Aberdeen, Scotland. The initial transcription was made by Thomas P. Scott and edited by Guido Bacciagaluppi and Elise Crull. All footnotes in this chapter are editorial.

D. Krohn
Philosophisch-Politische Akademie, Bonn, Germany
e-mail: dieter.krohn@gmx.de

R. Saran
Society for the Furtherance of Critical Philosophy, London, U.K.
e-mail: rene.saran@btinternet.com

F. Leal
Universidad de Guadalajara, Guadalajara, Mexico
e-mail: ferlec@hotmail.com

E. Crull and G. Bacciagaluppi (eds.), *Grete Hermann - Between Physics and Philosophy*, Studies in History and Philosophy of Science 42,
DOI 10.1007/978-94-024-0970-3_11

out of all the difficulties. She helped me a lot. That was, I think, a very, very prominent feature of her character that is seldom recognised. So that is my relationship to Grete Hermann.

Rene Saran: I wanted to make just three very small points to start with, which in a way relate me to Grete. One is that the year when I was born, in 1921, that happened to be the same year I now learn she first heard Nelson give a lecture. So that is point number one!

The second one is that during the war when she was a refugee in Britain—and by the way I was myself a refugee from Nazi Germany as a child, while she was already an adult. She was about 20 years or so older than I was. During the war when I was in my early twenties, she lived for some time in the same community house as I did. We used to live in community houses; I am a child of communities, really. I lived in communities all my life until about the early 1960s, when I lived more like a family. She lived in the same house outside London in a suburb, a small town which one could reach easily from London. We had all fled from London to avoid the bombs. Grete lived in the same house, so I really got to know her also at the personal level.

But in addition to that, as Dieter has mentioned her link with Socratic dialogue, I suppose she is one of the people who fed into my interest in Socratic dialogue, about which—right towards the later part of my life (it was published less than ten years ago)—I edited a book with a German colleague, which draws on the German literature which we had translated into English (Saran and Neisser 2004). In terms of the Nelson/Heckmann method in Socratic dialogue, the modern twentieth century method—Grete would have been one of the stepping stones or influences in terms of my interest in that, because during the war I participated in a dialogue that she led. I don't remember what it was about or what we talked about because it is so long ago, but I do remember one thing. As quite a lot of you know German, you will see the point. Grete and I had a dispute because—the dialogue was in German by the way—I kept on hearing the word 'Prinzipchen' and I kept on saying, 'But it isn't "Prinzipchen", it is "Prinzipe", and 'It can't be "Prinzipchen", because it is not little, it is so big'. Remember I was in my early twenties myself. (The ending 'chen' in German makes something small.) I thought they kept on saying 'Prinzipchen'.[1] They weren't saying that, I just misheard it. It took a long time for me to realise what my mistake was. That was the second memory I have of Grete, this little thing about 'Prinzipchen'.

The third thing, which I think long-term is much more significant, is that later in life I have been one of the mainstays, or one of the continuing forces, within the British charity called Society for the Furtherance of the Critical Philosophy. I am not a philosopher by the way, I am a political scientist, but never mind, I know quite a little bit from hearing and talking to people like Dieter and Fernando, who know philosophy much better than I do.

The original foundation document for SFCP, this charity—the signatories were Minna Specht and Grete Hermann, and the reason for that was that the British charity

[1] The German word 'Prinzip' has two possible plurals: 'Prinzipien' and 'Prinzipe', the first of which sounds very much like the diminutive 'Prinzipchen'.

Fig. 11.1 Caption reads: 'The residents of Østrupgård, spring 1935. The adults, from left to right: Liselotte Wettig, Grete Hermann, Gustav Heckmann, Minna Specht'. Photo courtesy of Rene Saran

SFCP was given in its trustee document a much broader brief, but the initial push to set it up was the Nelson/Specht School for children which had emulated the original Walkemühle. I think we saw a picture from the Walkemühle in one of the presentations. And up there (indicating Fig. 11.1) is the picture from this book about the period in Denmark, which has Grete right in the centre at the top there of the children and Minna and Gustav.

That school was in Denmark, where they had taken flight with the group of children from Walkemühle who had difficulties in going back to their homes. Many of their parents were active illegally against the Nazis or were living abroad, and it was all difficult. So Minna took those children who needed it to the emigration school in Denmark. Then, when it came about in 1938 that it looked highly likely that Denmark was also going to be occupied by Nazi forces, the question was: what should Minna and the school—Grete was working at the school at that time—what should the school do? And it was decided—Minna had met a very interesting man from Wales who had a charity in Wales for unemployed miners. He said, 'I have got a house which I am not using any more in Wales and you can have that for the school'. So that is why the school went to Wales.

But the school needed a financial solidity behind it. SFCP, this charity which Grete helped to found, was actually initially founded, in terms of practical purposes, to finance the school. But right from the start the trustees allowed us to have a much wider brief than the school, and eventually in the '90s—Patricia Shipley, who is sitting over there, was a trustee at the time and she is now, together with Fernando, one of our honorary fellows for academic advice—in the '90s Pat and I went to the Charity Commissioners and said—look, the school had to close. It was closed during the war when there was all this fear in Britain about people of German nationality.

One after another (of the teachers) came to be interned, and what was to happen to the children? So the school was eventually closed, but initially the SFCP was concerned with school finance and then at a later point it developed into a much wider brief.[2] That is my third connection because I have spent many years—I am just about withdrawing from it now, although I am a trustee of the SFCP—I was for many years the chair and the secretary, so in a kind of lineage of connections that connects me to Grete also. So that is sort of the personal side.

Fernando Leal: I met Grete Hermann three times in my life. I will tell you a little bit of each. I came to Germany as a student of philosophy in 1975. I was 21 at the time. My first shocking experience was when I was trying to read all those books in the wonderful library of the philosophy department in Heidelberg. One day I came to a room which contained books on philosophers of the nineteenth and early twentieth century. Then suddenly I saw nine yellow volumes, very thick ones, and the whole set was *Leonard Nelson's Collected Works* (Nelson 1970–1977). I said 'Who is this man?' I asked this question because Nelson's name had all but disappeared from the histories of philosophy that were then current.

I was very intrigued, especially so because Volume IV was called *Critique of Practical Reason*, and I thought 'What a cheeky man! He calls his volume the same as Kant. This is shameless!' So I took it out and opened it and it says: 'Dedicated to Hilbert'. At the time I was very interested in mathematical logic, and I said: 'What is this? A Critique of Practical Reason dedicated to Hilbert?' And it said, 'It is dedicated to Hilbert because I am trying to make ethics into a science'. Dear me, that was too much! As you can imagine I closed the book and put it away on the shelf, not knowing what to think.

But of course I came back. At some point—at the time I had a scholarship from the Social Democratic Party—I was invited to a political seminar and I met someone in one of the informal meetings in the pub. One of us (we were a bit tipsy by that time) said 'Nelson', and the other said 'What do you mean, Nelson? You know Nelson? You have read Nelson?' We marvelled at the fact that we could find someone in Germany who had actually read Nelson. Well, to make the story short, he organised a seminar and I was invited as a student to give a lecture on Nelson.

I did that and one of the people in the audience was Susie Miller, who was the wife of Willi Eichler, the political leader of the movement after Nelson died. Somehow she liked what I said. So she took out from her pocket a ticket to a very special conference in Bonn. This was a kind of closed event, for initiates I suppose—for people who belonged to the Social Democratic Party. It was a conference in honour of Kant. It was 1981, and you will remember that the *Critique of Pure Reason* was published in 1781, so it was in honour of the publication of the *Critique of Pure Reason*. Susie told me, 'I think you deserve to go instead of me'. So I was dispatched to Bonn to go to this conference. There were only three speakers. The first speaker was Helmut Schmidt—I don't know if the name rings a bell, he was the Chancellor at the time—and of course he was a wonderful speaker. I don't imagine that there

[2]From the Charity Commission Pat and Dieter learnt that the closure of the school did not constitute a problem for the SFCP.

are many heads of state who can actually give a lecture on Kant, but he was one of them. Then came Paul Lorenzen, some of you may perhaps have heard his name. He was the most famous professor of philosophy at the time, I would say. He also gave an impressive performance. And the last speaker was Grete Hermann, and she outshone both of them. It was incredible. It was so impressive that I was speechless. That was my first experience of her.

I did not know anything about her except that she was one of the editors of the collected works (of Nelson). But there she was, in all her glory, giving this wonderful speech about Kant and about Nelson and about the foundations of liberal socialism. You may not know, but Grete Hermann was one of the architects of the programme of the Social Democratic Party after the war, so she was basically saying what the Social Democratic Party was about at the time—things have changed, but at the time. I am talking about Helmut Schmidt's Germany. It is the only one I actually lived through as a student. It was a wonderful time. So that was the first time.

A few months later I was invited to a kind of annual conference they had in Germany, 'Thought and Action' it was called—'Geist und Tat'. It was a kind of internal affair of the Philosophisch-Politische Akademie. Of course I was thrilled by the fact that I was going to meet Grete Hermann, this wonderful speaker. Although I was quite a bit shy, I wanted to talk to the woman. By that time I had already started to read some of her stuff. Everybody told me she *is* the philosopher in the Nelson circle. I didn't know anything about the 1935 paper, nobody talked about that. She was known among that crowd as basically a practical philosopher. Anyway, I had been reading Nelson for quite some time then, and I had a lot of doubts, as every student has. When I tried to express my doubts to some of my professors in Germany, they either had not known anything about Nelson, had never read him; or they had read Nelson, but kept the matter quiet. There is a wonderful expression in German: to kill someone by silence or through silence—'totschweigen'. One of my professors actually told me that Nelson was killed by silence. So, some of the professors knew him but they did not talk about him; somehow it was not done at the time.

Anyway, I wanted to talk to her because I thought this is my one chance to talk to someone who has actually read Nelson and could perhaps clarify some points. So I mustered all the courage I could and went up to her and asked her if she could grant me ten or fifteen minutes of her precious time. That is when I experienced her from a different perspective. As I said, the impressive lecturer was first, and here was this very serious woman, very upright-standing: she looked at me and said, 'O.K., after lunch we can walk'. And that's what happened—we took a walk in the meadows after lunch for about fifteen, twenty minutes. At first of course I was very eager and I was talking very fast and expressing my doubts. She listened, gravely, looking alternatively at the ground and at me. When I stopped she said: 'Young man, you have to study more'. So, basically she said that I had got Nelson wrong, and that all my questions would be answered by themselves if I continued studying him. So that is the second aspect of Grete. She was not very forthcoming.

The third time happened about a year later, when I was invited to a more informal conference in Bonn which had more of a political character. At the time I was reading a book that is perhaps all pop biology—I didn't know at the time—anyway, it was all about how plants might have feelings. Now, the thing that makes it important is that Nelson has in his ethical books an argument for vegetarianism. And basically the argument is about how if animals can feel pain, then we are not allowed to kill them and eat them. Of course the argument is more complicated—philosophical arguments are always very complicated—but basically that is the point. And now there was this book claiming that there were some experiments that proved that plants could have feelings and could feel pain. I didn't know whether that was true, but it was intriguing—and it connected directly with vegetarianism, the argument for it, and the practice following from the argument. Then I thought: well, I have to ask Grete—what does she think about this? I saw on one occasion she was having breakfast by herself. She was often by herself—I think she was a bit intimidating—anyway, I sat down next to her. I explained about this book. I said: 'Perhaps the science is bogus and all this is wrong, but it is a possibility, perhaps plants feel pain. In that case Nelson's argument, as I understand it, could be extended to plants and then what would you do?' And she looked at me very seriously and said, 'We will eat stones!' That's consistency for you!

Dieter Krohn: This picture (indicating Fig. 11.1) gives me the chance to insult the English and flatter the Scots. You see, it is Grete here, Minna Specht here, and Gustav Heckmann here. I found a letter by Max Born, 7th of February 1937, a letter to Gustav Heckmann. You need to know that Gustav Heckmann did his doctoral dissertation under the supervision of Max Born, and since then they were very, very close friends, even after the Second World War when Max Born lived in Hannover, and they regularly met.[3] So here in '37 Max Born writes to Heckmann (in German of course, so I have to read it in German first and try a translation that will, of course, not be a literal translation): 'Wir haben es hier bei den Schotten unerhört gut' (because he was in Edinburgh at the time)—'We enjoy life here with the Scots'. 'Stadt und Land gefallen uns'—'We like the city and countryside'. 'Und die Menschen sind menschlicher als in England'—'The people are more humane than in England. Could you not transfer your school to Scotland? But', (this is the sentence for you:) 'Aber vielleicht ist Wales auch gut'—'But maybe Wales is good as well'. This was 1937, when Gustav Heckmann and his school had already looked for a building here (in Britain) and thought about coming.

This is a kind of a chance to lead over to a short history of the organisations that have been mentioned so far—just to give you a chance to understand everything, to see where it belongs. Some of the organisations connected with Nelson have been mentioned already. 'Geist und Tat': that was always the motto—'thought and action'. Nelson not only thought that ethical socialism was the right political way, but he also founded at least two organisations. First, the International Youth League (Internationaler Jugendbund). Then, when all of the members of this International Youth League were expelled from the Social Democratic Party, he founded a party of his own, the Internationaler Sozialistischer Kampfbund. You see that the international component is in both of those organisations; that was very important.

[3]More precisely, Born lived in Bad Pyrmont, not far from Hannover where Heckmann was professor at the Pädagogische Hochschule.

I have an anecdote in connection with that. When Grete Hermann started to work together with Nelson on the publication of his *System der philosophischen Ethik und Pädagogik* (Nelson 1932), she used to write in the specifically German handwriting called Sütterlin, and he was strictly against it, because that was no international handwriting; so she changed her writing to a type that could be read everywhere. That was the Internationaler Sozialistischer Kampfbund (ISK). There is an English translation which sounds really…

Rene Saran: Funny?

Dieter Krohn: Yes, funny now, if not even worse: Militant Socialist International (MSI). The interesting thing is: my impression from the documents and from what I can read there is that they, especially Nelson, first wanted to educate people so that they would become ethical politicians. We all know how difficult that is, to combine ethics with politics. His idea was: we need people who are really ethically perfect, let's say, and then we can start to change society and maybe politics. If the historical situation had not changed in Germany, those members would have been in the first place trained, educated to become independent thinkers who were able to make their own decisions and act in accordance with the ethical principles.

International Youth League, Militant Socialist International—they had a kind of boarding school in which adults were educated or trained to become responsibly active in politics. And they had a school for children. When the situation became dangerous in Germany—the Nelsonians had very early seen the danger in the development of German politics—the teachers at that boarding school decided to close the adult department and go to Berlin and publish a daily paper there, to influence politics. Their intention, their aim was that the parties on the left, especially the Social Democrats and the Communists, should unite against Hitler, because that was the only way to prevent him from gaining power. That they did—published a paper every day—a small group, and the members of the ISK in Germany distributed, sold that paper everywhere. And Grete was one of those who worked for that paper. She lived in Berlin in the Inselstrasse.

In 1932 they published the paper for a whole year. And then of course in 1933, when Hitler came to power, it was absolutely banned, forbidden. The 'journalists', those responsible for that paper, had to leave Germany as quickly as possible. That was the ISK. The ISK then went to France first. The school with the children, with Minna Specht, Gustav Heckmann, then later on Grete, went to Denmark. The ISK had a branch in Britain already. They then, in 1940, founded the Society for the Furtherance of Critical Philosophy. First it was called—and you can see the difference—Society for the Furtherance of *the* Critical Philosophy: there is just one critical philosophy. And this is the English branch, so to say.

The most influential and important members of the ISK came to Britain as refugees. And after the war most of them went back to Germany and founded the Philosophisch-Politische Akademie again. Because you need to know that Nelson not only had those two political organisations, the International Youth League and Militant Socialist International, but as a kind of background he had founded the Philosophical-Political Academy. Minna Specht was the head of this Academy after

Nelson's death in 1927. The Academy was the owner of that boarding school. After the war when everybody got back what the Nazis had stolen, they had to re-found the organisation. Let me come back to Grete Hermann: she, then, after Minna Specht, who was the first head of the PPA, became the second head after Minna Specht's death.

Let me combine in a few sentences Grete's biography with the institutions I have just mentioned. In '21 she first met Nelson. In '25 she did her dissertation, as we know, under the supervision of Emmy Noether. I believe this dissertation, or some of its findings, are still seen as a foundational paper for computer algebra. Then she was asked by Nelson in '25 to help him edit his book. She was not asked by Nelson to join the party. She had written what we call her Staatsexamensarbeit—that is another thesis you have to write if you want to become a teacher in a German school—she had written her Staatsexamensarbeit under Nelson's supervision, and it was on transcendental idealism. So this was the step towards philosophy and the first publication: transcendental idealism. After that, Nelson found she was the right one to help him with editing his text; they worked together.

He never demanded what he demanded of those who wanted to become members of the political party. Because the requirements were very strict: vegetarianism has been mentioned, no alcohol…

Rene Saran: No smoking?

Dieter Krohn: No smoking? It was not in their statutes: it was expected, no smoking. High membership fees: you were allowed to keep what you really needed to lead a decent life. Everything else above that you had to pay into the party's coffers. So the party has never been more than say, about 300 people—it really meant something if you joined this party. This is one of the reasons why nearly all of them worked against Hitler, and when Hitler had come to power still worked: they either had to leave the country because they were too well known there, or they went into the resistance movement and did a lot of very successful actions in Germany. Of course, the Gestapo found some of them, and there were trials against them, and some were sent into concentration camps. But on the whole, compared to other resistance groups, they were very successful because of the discipline they had and the training they had. For me, someone born in the last year of the war, it is very impressive to know that these were Germans as well.

Grete worked together with Nelson, and the demands he did not ask of her she obeyed secretly. So, without telling Nelson, she became a vegetarian and she left the church. And only after Nelson's death in '27 she joined the party. Before that she worked on a philosophical level only. After Nelson's death she went to the Walkemühle where Minna Specht was, and worked together with Minna Specht on the issue of more works by Nelson.

So, it is rather a very, very limited time of her life in which she worked on physics. She felt the demand, Nelson's demand, to do something about the causality problem. He couldn't really stand it that someone doubted this principle. And this she must have felt, because she started to work on that problem after Nelson's death. But then, at the beginning of '32, she worked for the newspaper, *Der Funke—The*

Spark. And after that she went to Denmark, and worked in the kitchen and in the garden, and only sometimes went to conferences; but not in Germany, she couldn't go there, but to other conferences. If you look at the list of her publications you will find some publications that might interest you. For instance, in 1936 she wrote a paper on das Kausalproblem, 'Zum Vortrag Schlicks', so she dealt with Schlick's ideas on the causality problem (Hermann 1936). Some more about this at a Congrès Descartes, a publication in 1937 as well (Hermann 1937). Then this went out of her scope, interestingly. Only after the war, from time to time—this is from 1950 for a current political discussion—another paper by her which is so clear that I could understand part of it! It was meant for the educated public maybe. It is in this journal *Geist und Tat*, that is usually translated *Thought and Action*: a short paper on 'Ethik und Naturwissenschaft' ('Ethics and Science'; Henry-Hermann 1950), in which she explains, in simple words and very clearly, what you can read in her 1935 paper. Wonderful!

Some more pieces of information: when Grete lived in Britain after having come with the school, she first worked in a vegetarian restaurant. Rene, you can tell us more about that I am sure. We have to mention the connection with the Society for the Furtherance of Critical Philosophy; she also worked on ethical questions and got a fee from the Society—yes, she got a fee for that; they all didn't have any money and were unable to really earn money, so for this work on ethical questions, they paid her a fee, and we are to be very thankful for that. She got the money in 1950; she had finished her *Politics and Ethics* that was published in '45 (Hermann 1945). It is a very interesting paper, or rather a book of 85 pages. And I read one of the passages as a kind of judgement about some of the physicists in Germany—I will quote that later on.

Rene Saran: Listening to Dieter brought other things into my mind, one of which, for example, I now find quite staggering. All the time I knew Grete—until she went back to Germany after the war, when she had this fantastic position to influence young teachers who were coming after the Nazi era (some of them I also had known whom she influenced at that time). All that time, up until then, I knew her really as a member of the group, like I knew all the members of the group, certainly in England; it was a much smaller group in Britain, who knew each other. And many of the German members in Germany also knew each other. Because it was small, in the personal sense one got to know each other quite well. Many of us, not all of us, lived in these community houses. I had no idea until the later period of my life that Grete was such a gifted scientist; it wasn't talked about. Now, post hoc, I find that amazing. I don't really know what the explanation for it is. That is one thing that I remembered.

The other thing I thought about and I talked to Fernando before this workshop about—he suggested I should tell you this story because it is really quite a nice story. Grete told me herself that when she was working in the school in Denmark she had a conversation with one of the children, called Peter Nemenyi—I knew a number of the children. Peter Nemenyi became, and his father (Paul) had also been, a well-known mathematician. (Paul) Nemenyi's son was Bob Fischer, who became a world-famous chess player.

Grete had this little conversation with Peter Nemenyi when Peter was four years old. He came running to her in the garden—Dieter mentioned that she worked in the garden—and said, 'Grete, I have to know; what is the highest number?' Grete probably looked at him in her usual way, and said: 'I don't know'. This little boy ran away, and for five or six weeks Peter refused to talk to Grete. Then he came running again and he said: 'Now I know why you said you didn't know, because there isn't one!' I found that such a revealing story because it demonstrates indirectly actually the Socratic Method, that people have to find out for themselves. She was quite confident that this kid, who was mathematically obviously very gifted, would find it out by himself, and he did.

I knew Peter, and I had been in touch with Peter before he died. His years were 1927–2002. Not so long before he died he took up contact with me by e-mail. He was living in America—he had been a professor in America. He took up contact with me through another contact in our group, Nora Walter, who is in that picture of the children at the school. Peter was there also. He said: 'Rene, I want some material about Socratic dialogue. I have been talking to my friends here where I live in America',—one of the cities, I forget which one—'and I have talked to them a lot about the school because it was such a memorable time for me, and they all want to know more about it. So can you send me some material? I don't have anything to give to them, to show to them'. So I sent him some stuff and we had perhaps three or four e-mail exchanges, this is years and years after he had met me, because in 1933 my mother had left the country immediately because she undoubtedly would have been arrested. By the way, my father, who was not in the group anymore then, although he originally had introduced my mother to the Nelsonians—he didn't do that and he was arrested the morning after the Reichstag fire and was imprisoned and luckily came out after six months.

To come back to the school: there is Peter (pointing to Fig. 11.1) whom I had met there as a child because at twelve I had gone to Denmark and been with the children, because Minna Specht had asked my mother to come to Denmark to help her learn Danish, because she also spoke Danish, which she learned when she was expecting me—she had gone to Denmark to eat decently because in Germany there was this terrible crisis after the (First World) War. So Peter contacted me and I sent him this literature. Then shortly after his death the people who were having to see to his papers found my address in his computer. They did not know where I lived because the address does not show which country you are from. They wrote to me, so I actually wrote a little obituary of the little I knew about Peter and his time in the school, and I told this story about Grete and Peter.

I want perhaps to add, because it might throw light indirectly on Grete's personality: we were often accused in our political group—which I joined, I think, when I was about 21 or 22—but the children of our community were not indoctrinated. That is important to remember. The adults were trained to be militant socialist fighters, but the children were not. Of course the education in the school was profoundly influenced by the Nelsonian philosophy. It was set up in order to practice in education what Nelson thought philosophically was the right thing in education. I was there as a very small child. I went to the Walkemühle as a child in the school at the age of about two-and-a-half or three and I was there for three years. I don't remember meeting Grete at that time. But I have very nice memories, for example, of Leonard Nelson's father, who was a lawyer and was living in the Walkemühle as a retired man. He was also very musical and used to play the piano beautifully. So Nelson obviously came from—I'm sure you all know this—a very cultured family.

But to come back to the ways of the group—after all, as we have heard from Dieter, in 1927 or 1928 after Nelson's death she joined this group, the German part of it—and the British part was an offshoot of the German part. There were these demands, and Grete perhaps expressed her conformity to these demands in a way that gave, or helped to give, this impression of her being stiff and distant. That is how it seems to have affected her, as far as I can judge, but she was actually, in some ways, at the personal level a very gentle person. I think most people probably didn't experience that, and Fernando has told us how he perceived her as a student. You, Dieter, experienced her in the Academy, where she was the chairperson of the Academy, where she probably will not have shown these sorts of features.

My experience of her is as a young woman in the early twentieth century, looking for the answers to life's questions. I remember one or two conversations, which I won't relate because they are too personal, where I went to her with these baffling conceptions in my head as to what I was to do about something—remember, it is a characteristic of the Nelsonian movement that theory and action are just two sides of the same thing: if you believe something you also have to do it. That's how I grew up in that, and I think it impregnated itself on my personality. I used to be able to go to Grete as a young person among other mentors of the older generation and ask her advice on really pressing things, and she would be very understanding; she wouldn't say I had to read more books, like she did to Fernando, but she talked to me and tried to explain to me what aspect, what point, or what reason in particular I hadn't thought about, and it helped me a lot. So there was that kind of personal relationship that I had, which I had with quite a few people in the group, both the senior ones and the contemporaries in my age group, because we actually all had pretty close relations in terms of communicating with each other; that was one of the features of the work of the group, because we all believed in the same ideals. It was a terrific commitment, a commitment to social justice.

Patricia Shipley: You mentioned Nelson's father. What about Grete's mother? In answer to my probing, Dieter—yesterday, when I was working informally behind the scenes, I asked him about Grete's relations with her family. She had a large family. Six or seven siblings perhaps?—Dieter, you responded they were very close. What about her relationship with her father?—Well, so and so. What about her relationship with her mother? —Very close relationship. There was an interesting anecdote that you gave me: that Grete was in Hamburg on some kind of Nelsonian duty. She was discouraged from going to her mother's bedside, something like that—you can correct the anecdote—and she was very upset about that, I guess she was very upset, and she said that this is an example of the inhumane side of Nelson and his philosophy.

I raise this because we haven't come to the elephant in the room, which is values. Whether there are interesting value differences between the sexes. I don't mean to imply that these are fixed and immutable. I was thinking yesterday of what you were saying briefly about philosophy as a way of life. I have been struggling a bit with what Grete understood by science and how it compared with Nelson's view of what science was. I understand that he wanted to establish some kind of scientific basis to his ethics. I think this is a very different conception of science than Grete had; his more abstract, hers more concrete. It is more grounded, I think, in the concrete realities of everyday life, and I believe she brought a particular female perspective to bear on that. That's all I want to say at the moment.

Dieter Krohn: Just to tell you the story: I haven't mentioned that Grete, as long as she could travel in Germany—that means up to 1936—visited groups of the party and had philosophical discussions with them. So she had a philosophical discussion with those in the resistance group in Hamburg as Pat mentioned. Afterwards she wanted to visit her mother. She was very close to her mother and she wanted to visit her, but then the group in Hamburg asked her to stay on for another night or two nights, to carry on with the discussion. It was clear for her that she had to do it because private matters are not as important as political matters; that's the Nelsonian idea. So she did not go to Bremen, which was near Hamburg, to see her mother. This would have been the last chance to see her mother because she died during the war. Afterwards, after the war, when she worked on Nelson's ethics and revised it, changed it, she contemplated this incident and then it was a clear example to her for the rigidity of Nelson's ethics and how wrong it was. But I suppose we will hear more about Nelson's ethics this afternoon, so we should maybe not elaborate on that now.[4]

Fernando Leal: Do you want to say something about it?

Dieter Krohn: I think we need to have a look at her physics properly as well. I have an interesting letter, because it is interesting to you as physicists. I cannot *not* quote it.

Rene Saran: Given this is a workshop—we are friends the three of us—it doesn't really matter in what sequence things come.

Patricia Shipley: Can I just briefly follow on from that? There was quite a lot that was said yesterday about why, given that Grete made what seems to have been a significant contribution to the debate about quantum physics—this concept of hidden variables and so on—why did she seem to be lost from the history of science for so long? A list of possible reasons for this was given and the weight of evidence I think seems to be in favour of the political angle, which may well be the case. I think, having thought quite carefully about this, there may well have been a sexist background. If that had something to do with it, I don't think it was the primary reason, it could have

[4]Fernando Leal's Chap. 2 in this volume was originally split into two presentations, the second of which, on 'Complementarity in Ethics?', was given in the afternoon after the Panel Discussion and introducing the General Discussion.

been a secondary reason. I don't think she was an honourary male; I think she was fully identified as a woman. I think this is a particular resource and perspective that is brought to philosophy which is very valuable. I think you can read a lot of it in this particular essay of hers on 'Conquering Chance' (Henry-Hermann 1991) when she contrasts causes with reasons, for example. It is clearer in another publication of hers I have been reading recently, the one in *Ratio* on the significance of the study of behaviour for the critique of reason (Henry 1973a, 1973b). I do recommend those particular publications for a wider perspective on our subject this weekend.

Roberto Angeloni: I would like to return to the relations between Grete Hermann and politics; her ideas about politics. She wanted to find a unity between the socialist party and the communist movement. But about her doctrine: what does she think about the Marxist doctrine and in particular about historical materialism? Which are the relations according to her—if we have some quotations—between quantum mechanics and historical materialism?

Fernando Leal: Well, you have to remember that at the time, in the '20s and '30s, the Social Democratic Party in Germany was Marxist. That was one of the reasons why the original political organisation founded by Nelson was expelled from the party; it was because they were not Marxists. The reason they were not Marxists is because they had this ethical foundation. As you know, for Marx and the Marxists ethics is just part of the superstructure: it is ideology. So there was disagreement and even a struggle on this question. There is a very remarkable small book by Nelson in which he actually attempts to refute Marxism. It is called *Die bessere Sicherheit* (Nelson 1927); it was translated into English as *The Better Security* (Nelson 1928).[5] It is in this booklet he tackles all the questions about historical materialism and tries to refute them and says the only foundation for political action has to be ethical. That was also Grete's position.

Roberto Angeloni: And the relationship with quantum mechanics?

Fernando Leal: I don't know anything about that! Perhaps there is one connection, but not between quantum mechanics and historical materialism. There is one connection between quantum mechanics and ethics.

Roberto Angeloni: She was very critical about materialism, and Marxism too. That was the way in which her scientific approach influenced her political ideas.

Fernando Leal: Indeed.

Roberto Angeloni: How can we measure our ethical values according to her; those values that are the basis of our education, thanks to which we can build the

[5]This booklet by Nelson is a refutation of the Marxist demand for socialism for being non-ethical, arguing that socialism can only be warranted for ethical reasons.

perfect society? Which is the way we measure these values? If the values are not the product of economic mechanism in history, those values must come from somewhere.

Elise Crull: What grounds the ethics if not a political agenda, if not a religious agenda? Something like that: whence the values?

Roberto Angeloni: Where do values come from?

Fernando Leal: To go into that we would need to have a seminar—if you want to understand Grete's position.

Gregor Schiemann: The values are not materialistic at all: they are rational. So you have arguments for them, and arguments are not historically dependent. There is a way of deciding and producing values which is not reducible to the material world.

Roberto Angeloni: How do you teach those values?

Gregor Schiemann: You teach the Socratic method. Everybody has their own values because they are rational, and we are rational beings, so ask the people and you will hear what they have to say, and the outcome is these values.

Rene Saran: These values in which Grete was also convinced—and I was also convinced of them, in a very different way, because I am not a philosopher—they were very deeply entrenched in my being because I belonged to this group and this group was founded on these values. *But* we claimed at least—one does not always live up to all one's claims—we claimed at least to be non-dogmatic about the values. And particularly through the use of and practice of the Socratic method we, in seminars, probed the meaning of those values to us, each of us, through group work in the dialogue. For me that is a very important part of the Socratic method and this fostering through the method of what our book has called 'enquiring minds' (Saran and Neisser 2004).

For example, in the trust deed of the SFCP the concept of dogmatism is rejected; the charity is not permitted to support any system of religious belief which is dogmatic. So I don't know whether that helps with regard to the values and the ethics. I never felt, as a person who was a member of these groups and worked with other people in them, that the values were imposed on me, although some of the rules that stemmed from these values were pretty strict. I didn't actually feel them as strict because they were part of my way of life. If I wanted to I could question any time—and of course the group became more relaxed, even after Grete's death—but I would think to some extent even during Grete's later years—the attitude within the group, the way of thinking, was evolving. It became less—I have called it 'rigid', but the rigidity wasn't dogmatic: it could always be questioned, and that was very important.

Patricia Shipley: You mentioned that these values were deeply ingrained in your way of being. Is this really what Susie Miller was referring to when she used the word

'Einsicht'? And it is a word that was used by Susie as well in relation to the resistance work that they did (cf. Shipley and Mason 2004, p. 22). It is not insight—that is a literal translation from English, isn't it? It is more than that, isn't it?

Rene Saran: Conviction, being convinced, belongs to it: having rehearsed in your own mind, or in a group with others, the reasons that favour this system of values.

Patricia Shipley: Was it self-evident to you?

Rene Saran: Not self-evident to the extent that I never questioned them. In a way, it was part of our tradition.

Gregor Schiemann: It is a special way to see a thing. It is not only conviction: you can be convinced of everything you have arguments for, but this Socratic dialogue aims at values which everybody has. So it is Platonic in origin. It helps people to come to insights which everybody potentially has.

Dieter Krohn: It is a difficult matter to explain Socratic dialogue and Grete Hermann's ethics. Just briefly, to come back to Roberto's questions: with ethics it is the same as with epistemology. You start off from your concrete judgements and through regressive abstraction you arrive at some result. What Grete thought is that what you arrive at is that you feel you have a duty. Duty is the result of the first analysis.

But what is the content of that duty? That depends—there we have the difference between Nelson and Grete Hermann—that is not clear in every case and unchangeable, but it depends on your 'practical experience' as she calls it; practical in the sense of ethical experience. This is, of course, dependent on historical matters, on your personal experiences. But there is one thing, according to Nelson and according to Grete Hermann as well, that you have to take into account as well, and this is the 'principle of fair adjudication'. The principle of fair adjudication is a kind of weighing up of interests of those who are concerned or who have got some interest in an action which you have to take. How you see these interests which you have to balance against each other depends again, of course, on your practical experience and so on. But this is an ethical principle.

Fernando Leal: There is an essential difference between Socratic methods, because here we don't have an ethical intellectualism as with Socrates.

Dieter Krohn: That is the difference with the Platonic ideas; those ideas do not change, but here our ethical insights can change according to the situation in which we have to take those decisions.

Elise Crull: And that's Nelson?

Dieter Krohn: Nelson wouldn't say so. For him everything is clear. Instead, Grete used her idea of practical experience to modify Nelson's system of ethics in a way that the core of truth in the Nelsonian philosophy could be kept—could be freed from its misleading absolutist claims and its true meaning vindicated (cf. Heckmann 1985, p. xi). This idea that everything is clear if we only think correctly and methodically is what has to be put away.

Martin Jähnert: How did the school situate itself—after the war, within Germany—especially with the 1968 student revolts?

Dieter Krohn: That is interesting. We only had the Philosophical-Political Academy after the war, just perhaps twenty people, mostly at that time Nelsonians who had studied under Nelson. They reacted very differently; there was no clear position of the PPA. Susie Miller was rather critical, but Gustav Heckmann was very sympathetic with the students, he discussed things with the students. It was interesting because—among the members of the PPA, we had one Minister of the Interior for Saxony, who of course as a Minster of the Interior could not be very sympathetic with the student revolt. But we had another one who left the Social Democratic Party because the SDP expelled the SDS—the students' body. We had one who was one of the fathers of the German constitution after the Second World War, who protested against Willi Brandt when he instituted the Berufsverbote (professional bans): there was a time in Germany when nobody who was suspected to be a communist was allowed to get a job in the state system, which included teachers. One left the party, the other one defended it, so that was the situation within the PPA—it was no longer a political organisation. The charter explicitly stated that no political matters should fall within the aims of the PPA, but only scientific matters: books, publications, conferences, seminars, and similar activities. So things had changed a lot.

There is an interesting point to be related regarding Grete Hermann and also Weizsäcker and Heisenberg and Heckmann.

Patricia Shipley: On Heisenberg, Niels Bohr, Copenhagen and all that stuff, you told me yesterday about Grete's modesty and gave concrete examples of that. I have the impression that she would have gone there, she would have seen all that going on, all the competitive spirit, competing for resources, the political argument. I think she was beyond that, she was on another mission; this was not meaningful to her. There were much more meaningful things she could do with her life. So she starts looking at philosophy, especially practical philosophy: philosophy as a way of life.

Dieter Krohn: I wouldn't agree with that interpretation of her life, because when she came back to Germany she became director of the (Bremen) College of Education in April 1947. She was a director of that institution for quite a while, until she retired. Then in 1953 a very important pedagogical committee was founded in Germany, the so-called Deutscher Ausschuss für das Erziehungs- und Bildungswesen, which was the committee that discussed the structure of Germany's school system, university system. They wrote a lot of important and influential papers. She was a member of that committee, which only had twenty members. So she was very active in politics. She was for the Union of Teachers in Germany the one who led the committee on the principles of education. She was active in the union therefore. So she was still very, very active.

Patricia Shipley: I didn't mean to convey the impression of some kind of angel disappearing up into the ether. I think she was very grounded. That's not what I meant at all.

Giulia Paparo: I have a question and a remark. I will leave the remark to later because I guess maybe your (Dieter's) letter will be connected to my remark. The question is more biographical: what about her husband? She got married in England.

Rene Saran: Oh, that's very easy to answer. It was not a real marriage, like with many people at that time, including people in our political group. They married in order to get British nationality, because it was very difficult to live in Britain without the possibility to take paid work. So they married for their very existence. It happened to my mother; it happened to Susie Miller who married my boyfriend of that time… It is not to say that Susie Miller and I were in any way estranged, because I knew exactly what the reasons were!

Patricia Shipley: Could she have been incarcerated if she didn't have this legal protection? Wasn't Minna Specht incarcerated for being German in Britain at the time?

Dieter Krohn: Those Germans whom you talk about—I think it was in 1940, when the British saw a spy in every German. So all the Germans were interned, but those who were British citizens, of course, were not. That was another reason to marry—for the covering.

Rene Saran: Most of these political marriages were dissolved later, because the people concerned had a real marriage in front of them and wanted to be able to enter it.

Giulia Paparo: But she never had a real marriage and a family.

Rene Saran: No, she never had a family.

Giulia Paparo: It was total devotion to—probably the party was her family in a way.

Rene Saran: If by family you mean a husband and children, personally as a member of the group I can say that I had a husband in the later period of my life but I didn't have any children.

Patricia Shipley: Would you say marriage was discouraged by the Nelsonian group?

Rene Saran: Having children was discouraged in the group through the Nelsonian influence. Politics was much more important than having a commitment to a family, which means a financial, and an ethical and all kinds of other responsibilities.

Sally Redfern: Does that not mean that the group would die out?

Rene Saran: You could say that I am an example—because my mother had me. It needn't have happened, but it did happen, that I became a member of the group. But that did not always happen, and some children of members of the group went their own way. I went my own way also, but in a different way. In my earlier life I was politically very active in the Labour Party in Britain; but because as I developed as a person, I was more interested in education and I did something which I didn't want to do when I was at school, I didn't want to be a teacher. Something changed and I became very motivated to be a teacher, and then I became a teacher, initially in adult education and later in university education. So, I was a late developer.

It is a very interesting point about this business of family and politics, because another aspect of the group was that everyone who applied to join the group had to have sponsors. So I had to have two sponsors in order to join the group when I was about 20 or 21. I had a male and a female sponsor. My female sponsor said in her testimonial, which I read several decades later when I committed the archives from the group to Warwick University Modern Records Centre—but I read them before they went. This testimonial about me said: 'Rene is still very young and we have to be careful not to overload her with organisational responsibilities because she is good at organisation, because she may still develop in other directions'. That was when I was 20 or 21. For me, that illustrated the awareness, as part of the ethos of the group, that young people need to be encouraged and their own faculties need to have the opportunity to develop freely. The group, although it had this rigid or disciplinarian side to it, with certain demands which you then as a young adult accepted—nevertheless the rational faculty, the ability to be a self-determining being, as a grown-up person, that was something also in the school. That faculty had to be nurtured in the young, and it was also practiced when people were adults.

Gregor Schiemann: Was it an explicit rule not to have children?

Rene Saran: It was not a rule.

Dieter Krohn: For the officials. Not for the normal members.

Gregor Schiemann: Not for the normal members?

Dieter Krohn: They were not encouraged to marry and have children, but it happened.

Gregor Schiemann: How explicit is it? Was it: you want to marry? No problem; but please, don't have children. I think the idea is that without children we are able to educate more children. So education is very important. Having one's own children does not allow us to educate other children so effectively. So my other question is: maybe this is also the background?

Rene Saran: It is the background. For me personally I have not had children of my own; it wasn't because I was unable to have them, but because I chose not to have them. It had to do with this political culture and ethical culture in which I grew up. Because in order to make your contribution to fighting for social justice—it was argued to me and I felt it—having the responsibility of a child would make me less capable. Once you have children you have a big weighty responsibility, because you can't neglect your children and just go out every evening and do politics; you have to see to the children. That was, I think, the main reason for this emphasis in the group: preferably no children, although people had children, of course. Some people did and some didn't, and I was one of those.

Patricia Shipley: May I just briefly read a bit from her essay on 'Conquering Chance', with respect to this? She is talking about Nelson's moral philosophy and his principles—it is a kind of Kantian-like commandment: 'The first criterion is the commandment of justice, which has the character of duty [...]; the second is the ideal of rational self-determination' (Henry-Hermann 1991, p. 38). Then on the next page: 'But on the other hand there is the impression that this requirement demands something superhuman, that no one can do justice to it'. Then she goes on to fill this point out in much more detail and, for example, with reference to where you put the limits—where the boundaries to this are: 'the question what we are to regard as the interests of our fellow human beings [...] the extent of what it is to be taken into account grows immeasurably' (ibid., p. 40). So it is a different take on Kantian conditions of possibility, isn't it—the constraints and limits on action and reason?

Dieter Krohn: Well, I am eager to read the letter because you might be able to answer the questions I have in connection with that, and it might give you some pieces of information you can use. It is a letter written on 17th December 1933 by Gustav Heckmann to Grete Hermann.[6] The first sentence: 'Ich danke Dir herzlich für die Zusendung Deiner Arbeit'—'Thank you very much for your work you sent me'. I don't know exactly what it was, but it has got something to do with the quantum mechanics work, I'm sure. Now some words about Heisenberg's judgement about it: 'I talked to him the day before yesterday for more than an hour'. You remember Heisenberg was a friend of Gustav's in Göttingen while they were both students there. 'Sie nehmen Deine Arbeit voll und ganz *ernst*'—'They are taking your work absolutely *seriously*', 'und noch in den Tagen seines Hierseins wollte Heisenberg zusammen mit Bohr und einem Schüler Heisenbergs, Weizsäcker, die Antwort an Dich gemeinsam abfassen'—'During their presence here Heisenberg wanted together with Bohr and one of his pupils, Weizsäcker, to formulate an answer to you'. This letter was written, you need to know, in Copenhagen. Gustav Heckmann was in Denmark together with the school and he went to Copenhagen to see

[6]An English translation of this letter can be found in this volume, Chap. 13, while the original will appear in Herrmann (2017, Part III, Letter 2).

Weizsäcker, and Bohr was there and Heisenberg was there. So they talked about Grete's work.

Fernando Leal: I think you should emphasise that the three of them—Bohr, Heisenberg and Weizsäcker—wanted to respond to Grete together.

Dieter Krohn: Yes, together to formulate the answer. Then there is a name, and I don't know who it can be, so I don't know whether it is... Dirac? 'An Dirac, der auch hier war, habe ich im Moment nicht gedacht'—'I didn't think about Dirac at the time, who was here as well'; 'vielleicht ahnten aber Heisenberg und Bohr nicht, dass auch er Deine Arbeit bekommen hat'—'But maybe Heisenberg and Bohr did not know that you had sent him your piece of work as well'. Now some quotations—that must have been Heisenberg who said: '"Sachlich hat sie bestimmt unrecht", aber "eine fabelhaft gescheite Frau"'.

Gregor Schiemann: Is that a quotation in the letter?

Dieter Krohn: It is a quotation in the letter. I suppose it was Heisenberg who said so, or the others together: '"On the matter she is absolutely wrong", but "she is a fabulously intelligent—smart, clever—woman"'.

This reminds me of another little anecdote, which came to my mind when Rene talked about Peter Nemenyi. In one of the discussions in the late '20s, Grete was under attack from some people who discussed things with her—teachers—and she had an answer and could refute everything. Then one of the teachers said helplessly: 'If I were as clever as you I would see where the mistake is!'

And then another quotation (in the letter from Heckmann): '"Als bei meinem Vortrag eine Frau aufstand und mit etwas scharfer Stimme zu reden begann, dachte ich: um Gottes Willen, was wird das werden"'. Maybe it was Heisenberg again: '"When during my lecture a woman stood up and in a kind of sharp voice began to speak I thought: oh, by God, what's coming now"'. '"Aber ich war ganz überrascht über die Klarheit, mit der sie alles auseinander setzen konnte"'—'"But I was really surprised with the clarity with which she could analyse everything"'. 'Im übrigen meint er',—he means Heisenberg—'wie auch Du, Du müsstest noch mehr Physik lernen'—'He thought, as you do, you need to learn more physics'. 'Die 4 Aufsätze von Bohr studieren'—'Study the 4 papers by Bohr'—'Er hätte sie Dir genannt'—'He told you about them'.

There is a wonderful judgement about Heisenberg: 'Heisenberg spricht mit einer so unpräzisen philosophischen Terminologie, dass ein Friesianer die Wände hochgehen möchte'—'Heisenberg uses such an unclear philosophical terminology that a Friesian'—you remember Fries?—'would climb up the wall'. 'Wenn er den transzendentalen Idealismus verstünde, dann würde er wohl die Möglichkeit erkennen, auch für die Auflösung der philosophischen Schwierigkeiten, die aus der Quantenmechanik sich ergeben, den Schlüssel zu finden'—'If he understood transcendental idealism he would see the possibility to find the key for the solution of the philosophical problems which are raised by quantum mechanics'. 'Ich erzählte ihm vom transzendentalen Idealismus'—'I told him about transcendental idealism'. 'Er sagte, ich sollte doch mit Weizsäcker über diese Sachen einmal sprechen'—'He told me I should talk to Weizsäcker about those things', 'er habe sich stark mit Naturphilosophie befasst'—'Weizsäcker—he had been dealing with philosophy of nature'. 'Nun, das musst Du dann machen'—'Well, you have to do that' (then he gives her his address:) '(Dr Weizsäcker, Kopenhagen-Hellerup, A.N. Hansensallé 21). Weizsäcker arbeitet jetzt bei Bohr'—'Weizsäcker works together with Bohr at the moment'. 'Du und Dora' (Dora, that is Minna Specht) 'habt jedenfalls bei Heisenberg immer eine offene Tür und ein offenes Ohr'—'You will find an open door and an open ear at Heisenberg's all the time'. 'Benutze es noch, ehe Du an die schwerere Tür aus dem älteren Holz klopfst: die von Bohr'—'Use it before you knock at the harder, at the older door, that of Bohr'.

'Heisenberg sprach noch von einer gewissen schwer zu erarbeitenden Resignation der Physik gegenüber'—'Heisenberg also talked about a certain resignation which is difficult to deal with in regard to physics': 'man müsse einmal darauf verzichten, über die *wirklichen Vorgänge* im Atom etwas aussagen zu wollen—dann würde man fruchtbarer arbeiten'—'you would have to resist the intention to try to say something about the *real events* in an atom, then you will be able to work on that more fruitfully'. 'Auch Deine philosophischen Überlegungen würden fruchtbarer werden, wenn Du Dich der Atomtheorie gegenüber erst zu diesem Standpunkt durchgerungen hättest'—'Also your philosophical contemplations would be more fruitful if you had reached this different point of view regarding the atomic theory'. Now a quotation again. Apparently Heisenberg said: ' "Es kostet viel Kraft, erst dahin zu gelangen" '—" 'It's really strenuous to get there" '. ' "Grete Hermann scheint noch nicht ganz dahin gelangt zu sein" '—' "Grete Hermann seems not to be quite there" ', ' "auch Schrödinger ist noch nicht bis dahin gekommen" '—' "also Schrödinger hasn't really got there" '. 'Ich verstehe nicht diesen Standpunkt'—'I do not understand this point of view', Gustav Heckmann said, 'vielleicht übersiehst Du, was er meint'—'maybe you can understand what he means'.

This is from the end of '33; that means that they talked about it a lot with Gustav Heckmann and they knew about her work and also the others knew about it. They carried on, by the way, after the war. May I just add: after the war, when in the late 1950s the question of atomic armament was discussed, Gustav Heckmann and Grete, and a third person, wanted to form a kind of working group on that problem. They worked on that. Gustav Heckmann was strictly against atomic armament. He was one of the founders of the Ostermarsch, the Easter March movement in Germany. Grete was willing to work with them and they discussed things again and again with Weizsäcker. So they met him, they invited him for talks, and discussed things very openly. And we find some interesting ideas of Weizsäcker on who of all the physicists in Germany could possibly be interested in working together with them on this matter. He mentioned Bechert from Mainz, Holthusen, Rietzler (Bonn), Gerlach, Dr von Hoerner. He has good evaluations or judgements about all those people in a private letter.

Giulia Paparo: I want to ask something else. Do you have any idea what they are referring to with the work she has sent them?

Dieter Krohn: I suppose it is the beginning of her work on her 1935 paper.

Giulia Paparo: It is not clear whether there was already the von Neumann critique in there or not.

Dieter Krohn: They seemed, both Heckmann and Hermann, to be a bit critical of Heisenberg. Heckmann said we never know whose side he is on. There is one passage I wanted to quote in *Politics and Ethics*, a text that was published in '45 written by Grete Hermann (Hermann 1945). She talks about resistance against Hitler and the Nazis in Germany:

> But is it necessary to oppose them? Time and time again people who were perfectly aware of the nature of the Nazi regime have tried to hold aloof from the political struggle for its overthrow in order to serve aims which they considered had a value independent of the social conditions of the time. Scientific institutes in which they carried out research, artistic activities to which they were devoted, personal relations in the family or between friends, all became islands in the political stream of events where they dwelt secure from the political upheavals and unmolested by the conflicts of the world. (Hermann 1947, p. 66)

When I read that, I thought: aha! what is she thinking of? Is this her judgement about those German physicists who did their research? She continues:

> This attitude of escapism from a threat which affects the whole fate of society is rejected by those who have weighed up the claims of their individual lives against the bigger issues affecting humanity. The so-called independence of such a life of seclusion is an illusion. Devotion to art or science, to the creation of relatively free human relations which is possible in such a protected environment has helped to mislead the world about the real state of affairs in society. Indeed it has been assiduously exploited for this very purpose. Those who adapt themselves to a régime such as Hitler's and close their eyes to the political happenings around them for the sake of things which in themselves are valuable, support and strengthen the system in taking up such an attitude. There can be no neutrality when people stand face to face with the moral and cultural decline of a corrupt social order. Those who do not struggle against it grant it their support. However fine and noble the achievements otherwise obtained, they are rendered worthless by the share in the social injustices with which they are burdened. (ibid.)

I think this is quite clear.

Patricia Shipley: So what do you do in a situation like that? This weapon of mass destruction—the blueprint was there. Who was going to develop it? What do you do with it?

Rene Saran: That is what the whole play *Copenhagen* was about (Frayn 1998). There is fascinating dialogue in that play. On the basis of having seen that play three times, which is very rare with me, and having read it, I asked Dieter: what is the answer about Heisenberg? Because I wasn't informed. He told me it has been established that he did support the Nazis.

Fernando Leal: I think they planted microphones in the barracks where all the physicists were being held.

Elise Crull: That was Farm Hall, when Otto Hahn, Heisenberg and others were interned in Britain for a while. You can read the transcripts of those recordings—including their reactions when they received news over the radio of the bomb being dropped (cf. Bernstein 1996).

But I thought there was still some speculation as to Heisenberg's role—there was a basic equation that Heisenberg made a mistake on, regarding how much unstable material you would need to get a pile going. He miscalculated it and thought: 'This is impossible for us'. But when they hear the news of the Americans dropping the bomb, one of them does a quick back-of-the-envelope calculation for the amount of uranium needed, and gets it right. Post-war, Heisenberg would say he deliberately miscalculated as an act of sabotage. It is worth reading these transcripts, because you can hear the physicists saying: 'What are we going to tell people when we get out of here, about how far along the Uranium Club was?' It is an interesting moment of revisionist history.

Guido Bacciagaluppi: Thank you for sharing with us all your insights, all your special knowledge of Grete Hermann.

Rene Saran: I wanted to say at the beginning, but as usual I forgot. I was really, in a way, very delighted when I received the invitation to this workshop, because it is not the sort of thing which would have come across my way at all; and the trustees of our charity (SFCP) were also delighted. The fact that we have been able to come here and participate in things that certainly I don't understand much of—we have been able to make a contribution from a different perspective, and you invited us to do that. We are very appreciative of that.

References

Bernstein J (1996) Hitler's Uranium Club: the secret recordings at Farm Hall. AIP Press, Woodbury, NY

Frayn M (1998) Copenhagen. Anchor Books, New York City

Heckmann G (1985) Vorwort. In: Henry-Hermann Grete, Die Überwindung des Zufalls. Kritische Betrachtungen zu Leonard Nelsons Begründung der Ethik als Wissenschaft. Verlag Öffentliches Leben, Frankfurt, pp xi–xx

Henry G (1973a) Die Bedeutung der Verhaltensforschung für die Kritik der Vernunft. Ratio (German edition) 15(2):197–209

Henry G (1973b) The significance of behaviour study for the critique of reason. Ratio (English edition) 15(2):206–220. Translation of Henry (1973a)

Henry-Hermann G (1950) Ethik und Naturwissenschaft. Ein Beitrag zu der von Carlo Schmid angeregten Diskussion über die Voraussetzungen eines sozialistischen Programms. Geist und Tat 5:383–389

Henry-Hermann G (1991) Conquering chance: critical reflections on Leonard Nelson's establishment of ethics as a science. Philosophical Investigations (1):1–80, Translation of Henry-Hermann (1953) by Winch P
Hermann G (1936) Zum Vortrag Schlicks. Erkenntnis 6:342–343
Hermann G (1937) Die naturphilosophischen Bedeutung des Übergangs von der klassischen zur modernen Physik. In: Travaux du IX. Congrès international de philosophie, Paris, 1–6 Août 1937, pp 99–101
Hermann G (1945) Politik und Ethik. Renaissance Publishing Co., London
Hermann G (1947) Politics and ethics. International Publishing Company, Bombay. Translation of Hermann (1945)
Herrmann K (ed) (2017) Grete Henry-Hermann: Philosophie–Mathematik–Quantenmechanik. Springer, Berlin
Nelson L (1927) Die bessere Sicherheit: Ketzereien eines revolutionären Revisionisten. Verlag Öffentliches Leben, Manchester
Nelson L (1928) The better security: being the heresies of a revolutionary revisionist. Verlag Öffentliches Leben, Manchester. Translation of Nelson (1927) by Kumleben G
Nelson L (1932) System der philosophischen Ethik und Pädagogik. Verlag Öffentliches Leben, Göttingen
Nelson L (1970–1977) Gesammelte Schriften. 9 volumes. Meiner, Hamburg
Saran R, Neisser B (2004) Enquiring minds: Socratic dialogue in education. Trentham Books, Stoke on Trent
Shipley P, Mason H (eds) (2004) Occasional working papers in ethics and the critical philosophy, vol 3. Society for the Furtherance of the Critical Philosophy

Chapter 12
General Discussion

Mélanie Frappier: My question is very historical: was Hermann's work in ethics forgotten like her work in quantum mechanics, or did it have a bigger impact?

Fernando Leal: I can only give you my personal answer. I don't know, I cannot prove it, but my impression is that it was forgotten. Dieter told us that during the war she worked on a manuscript that was published in 1945 called *Politics and Ethics* (Hermann 1945); English translation (Hermann 1947), and this is still a very Nelsonian book; beautifully written and very interesting, but completely Nelsonian.

But from 1945 to 1953 something happened to her and she found what I think was her answer to Nelson. Finally she had got the arguments. The Nelsonians published a volume to honour Nelson, they published it in 1953 (Specht and Eichler 1953). Grete's contribution to that memorial volume was this paper called 'Conquering Chance' (Henry-Hermann 1953). That is the first time I think that she actually developed the whole critique to Nelson.

What happened to that paper? Who read it? What did they do with it? As far as I understand things, there was not much written about it and no particular reaction. I can tell you that the essay was translated into English by Peter Winch, a famous philosopher in his time. I read a letter in the archives by Peter Winch and he says something like 'much better than Wittgenstein, much more profound'. He writes: 'The only problem is that she keeps quoting this strange chap Nelson. If she kept him away it would have an impact on modern philosophy, because her arguments are right on target'. That was Peter Winch in a letter.

The following is a transcription of a general discussion held on the afternoon of Sunday, 6 May 2012, at the conclusion of the Grete Hermann Workshop, University of Aberdeen, Scotland. The initial transcription was made by Thomas P. Scott and edited by Guido Bacciagaluppi and Elise Crull. All footnotes in this chapter are editorial.

E. Crull and G. Bacciagaluppi (eds.), *Grete Hermann - Between Physics and Philosophy*, Studies in History and Philosophy of Science 42,
DOI 10.1007/978-94-024-0970-3_12

Dieter Krohn: The translation of 'Conquering Chance' was published in the journal *Philosophical Investigations* in 1991 (Henry-Hermann 1991). In the Editorial Note, D. Z. Phillips says: 'It could be argued that her most distinguished work is in the philosophy of science, but the essay on "Conquering Chance" should be of interest to readers of the journal with reference to what she says, in her own voice, in criticising Nelson's ethical system. The essay should be related to issues raised in contemporary moral philosophy concerning moral luck' (ibid., no page number). And then he goes on: 'I was introduced to Grete Henry-Hermann's work by Rush Rhees, who was a friend of hers. He thought that her essay was one of the finest things he had read in ethics and found himself returning to it again and again'. So in 1991 Phillips thought it might be of interest to those who discussed ethics in those times. Rush Rhees, just as a footnote, was one of the first trustees of the Society for the Furtherance of Critical Philosophy. So in those years, late 1930s, they must have met—Grete and Rush Rhees. You know that Rush Rhees was Wittgenstein's friend and the editor of his works after he died.

Giulia Paparo: The fact that in 1991 we had already a translation of her works on ethics and we still don't have a translation of her work in quantum physics—for me that shows that the former had more influence or impact than her work on quantum physics. I was looking around in German and Dutch libraries, and her comments on ethics are more easily available than 'Die naturphilosophischen Grundlagen der Quantenmechanik' and all her other works. The entire Netherlands had only one version of 'Die naturphilosophischen Grundlagen', while Amsterdam, Utrecht and the main university libraries had her works in ethics.

Dieter Krohn: You have stimulated us to comment on the publication of this essay, the 1935 essay. You need to take into account that the *Abhandlungen der Fries'schen Schule*—in which it appeared in 1935—had no great circulation; it was an insider journal. So it was published by Erich Irmer. Erich Irmer was a member of the ISK. He was a printer and he had a publishing firm, Verlag Öffentliches Leben. It was the publishing firm which published all the Nelsonian things. And, well, not all but most of the readers were ISK members and they, at the peak of their development, were 300, and not philosophers. This is another reason why it was not really taken into account by those who discussed these matters.

Fernando Leal: I want to contradict you a little bit, because when I was a student in Germany I made it my task wherever I went to visit the philosophy department library, and there was always a set of this journal. So it was perhaps not so unknown. It was, of course, not a popular journal; I agree with that.

Elise Crull: You were mentioning [in your presentation this afternoon] how the idea of complementarity was embraced by Hermann as her philosophy matured, and ultimately played a really important part in her critique of Nelson. But I am wondering to what degree her considerations of quantum mechanical results—the idea that the world really is only understood in segments and by crossing over from one context to another—are incorporated in her other thinking. Is she saying that the world, in all its aspects, truly is complementary? That these physical results translate all the way to, say, psychology? Or is she using quantum mechanical complementarity as analogy in a Kantian sense, as in complementarity bridges the gap in our understanding between unintuitive things and intuitive things? Or does she mean it in a weaker sense—that the concept of complementarity is a powerful way of describing a lot of different relationships in the world and in different areas of human understanding?

Fernando Leal: I will say that you have to remember that she was a Kantian. That means the label Ding an sich—the world in itself, in reality—is just what we would call a limiting concept. It is, as such, inaccessible. That is why it is Ding an sich, because it is defined as independent from us. So about it we cannot say anything. Basically we are organisms that have a particular cognitive apparatus. All we can do as philosophers is to enquire into how we can make sense of Ding an sich.

Patricia Shipley: You would have to dig pretty deep in stuff like this. She does talk about new physics in the preface to 'Conquering Chance', but only in broad terms. She says: 'The ideas in this work have grown out of my own experience with the practical consequences which Leonard Nelson's ethics gives rise to. Their philosophical clarification has been extensively enriched through the concern with aspects of philosophy of natural science, which has emerged in modern physics and which has thrown new light on crucial issues and doctrines in critical philosophy' (Henry-Hermann 1991, p. 1).

I have tried to pin that down more in the text. When she gets on to social factors she gets more into questions of relationships. It is indirect, I think; not enough to actually identify that as an illustration of this general point which she makes in the preface, is really what I am saying. I would have thought it was bound to have some influence.

Gregor Schiemann: Yesterday I raised the question whether in her work on the philosophy of physics the notion of chance appears. The answer was that it doesn't appear. There is a causal discourse, but it is not explicitly said whether there is chance or not. It is a causal network of relations, but chance doesn't play any role. Probably she is convinced that it is a deterministic scenario.

Elise Crull: I don't think so. I don't think you should equate indeterminism with chance. She definitely seems to understand that the world is indeterministic from the new physics. Chance seems to be a more anthropocentric idea.

Gregor Schiemann: Heisenberg is convinced new physics has proved absolute chance in nature. This is explicit in his writings.

Giulia Paparo: Is it chance or is it indeterminism?

Gregor Schiemann: He uses the notion chance, Zufall. He wants to make it explicit that indeterminism could also mean it is a problem of knowledge. So the question of hidden variables is still open, but for Heisenberg it is clear that he also uses the notion of absolute chance.

Elise Crull: What time period? Is this post-World War Two?

Gregor Schiemann: It is a manuscript, 'Ordnung der Wirklichkeit', I am referring to (Heisenberg 1942/1984). As far as I see it, the problem for Grete in the philosophy of physics is the notion of causality and not the notion of chance. But here in practical philosophy it is the notion of chance which is in the title. As far as I see these are two different discourses, so they are not very closely related. So there is a special notion of causality which she uses in quantum mechanics and its interpretation, and it is not the same notion when she speaks here of causality and chance. It is a different discourse. This is part of your complementarity ansatz. In this complementarity also causality is used in the context of practical reason and not in the context which she has used in the interpretation of quantum mechanics. The notion of causality she uses to interpret the new physics is a very narrow notion. It must be appropriate to the task of upholding Kantian philosophy against the conquering new physics, so you need a very special notion of causality which fits there; whereas here you have causality in the context of action. So there are reasons for your thoughts, and reasons for how you may behave. I think it is a completely different context. For me there is the problem of how she combines both, but my main impression is that she is very careful.

Patricia Shipley: There are many other groups of philosophers other than, shall we say, Kantians. There is Hume. There are pragmatists. I am thinking of the pragmatist school of American philosophy. Many early psychologists were pragmatist philosophers. I am thinking here, for example, of George Herbert Mead's work. And his idea of the self as 'me', a kind of third-person notion of selfhood, a kind of 'self in the past': the 'me' I can look back on as having done something in a certain way. Maybe one could actually attribute causes to what one did, as opposed to the self as 'I' in the first person: somebody here and now who has to think and act in the future and who has purposes which are not yet determined because I haven't actually acted.

One thing I have tried to do to pin this thing down a bit is to look—Elise, we were talking earlier about the comparison between Nelson and Grete Hermann on the concept of a scientific ethics, what their different thoughts are about what science is—and I think there are very real differences. I have tried to convey what I think science means for her, and it is not the same for Nelson. She is younger; she is immersed in a wider body of literature and scientific research and so on; she has explored way beyond the boundaries of Kantian philosophy. Her science is much more empirical, I think, much more observational in *this* world about what happens to people in their concrete experiences. She would value making observations of people, she would value empirical research.

Sally, there is this work by the American nursing researcher, Patricia Benner, I am sure you are familiar with this—some time ago (Benner 1982). She was using help from the skill researchers, the Dreyfus brothers, on the analysis of skilled behaviour (Dreyfus and Dreyfus 1980). In Benner's work she distinguished between what she called ordinary nurses and expert nurses. Do you remember this study? What was it that characterised these expert nurses? Well, they were behaving in this particular way as if they had free will. You know more about this study, Sally, what the settings were that these nurses were operating in and how they were distinguished methodologically. But these expert nurses, according to Benner, felt they could make a difference in their work. They were much more ethical in their practice. They would challenge doctors and other people in the system in the right kind of way, to benefit the patients.

Sally Redfern: To add to that, they worked in the intensive care field, so this was highly medically technical, even for nurses. These expert nurses reached a high level—Benner, through the Dreyfuses, had these stages of expertise—Benner's experts reached the top level which was based on intuition. They didn't have to think twice.

Patricia Shipley: Were they all female?

Sally Redfern: They probably were in those days but I don't think they had to be. So they didn't have to think twice about how to react in an emergency. Whereas if you are at the lower level you probably have got to go through a standard set of rules, maybe look them up; the same with the pilot when he is about to have a near miss, or something, or a crash, if he is less expert he will look through his manual, whereas if he is very skilled he will know exactly what to do, which would always be the right thing without thinking twice about it.

Patricia Shipley: It is a question of what underlies this skilled behaviour. The point I am trying to make is that to study that kind of real life behaviour can be very illuminating. It doesn't mean that you have solved the problem of free will, but you can say this is what seems to be happening in practice, in that kind of setting.

Rene Saran: I'm not sure if I am being fair, but according to your perceptions, Grete was closer to knowing about, and taking into account, the actual practice.

Patricia Shipley: Yes. And this kind of empirical research methodology, this approach to science. Yes, I think so.

Thomas Filk: I would like to come back to speculating about the question of where quantum mechanics might have influenced her general philosophy. Maybe we shouldn't emphasise so much the concepts of complementarity and causality in this context, but maybe more the concept which she herself introduced: the relative character of quantum mechanics, not the complementarity character of quantum mechanics.

I think what you realise in quantum mechanics is that there is no system which is isolated and determined by itself. The truth about the system is always in relation to its environment. Forget about measurement, we shouldn't talk about measurement. It is not the system by itself but its relation to the environment, this determines what we can say about it. Maybe that is what has influenced her thought.

It reminds me a little bit—Nelson takes the character of Newton: that something is absolute, there is an absolute truth in the system, and that's it. Then she recognised that, no, in quantum theory the system has always to be considered relative to the environment and only in that sense is there truth. Maybe this is a slightly different aspect than complementarity but I would like to keep this concept of complementarity out of it and emphasise more the relative sense.

Gregor Schiemann: For me this is a good reason not to make a close connection between the context of physics and the context of practical reason, because the relation to the environment in quantum mechanics is entanglement. This means losing the identity of objects. You cannot separate the different objects, and for practical reason it is constitutive that you must be able to identify the objects, that you have a responsible subject who is a person you can speak to. It is not admissible to have the sort of entanglement that you have in quantum mechanics. This for me is a categorical difference. So it is not possible to apply the analogies in a wider sense to the discourse on practical matters.

Thomas Filk: I wasn't saying that one should take the mathematical formalism which expresses for instance entanglement, only the insight that we cannot consider a system by itself but only in relation to its environment.

Gregor Schiemann: That is so general that we can deduce everything from it.

Martin Jähnert: That is also my impression. This idea of perspective is so general: why do you need the very special field of quantum physics for this? That is a quite general mode of thought, I would say. I don't know if she needed quantum physics to get to that.

Giulia Paparo: It wasn't a general mode of thought before quantum mechanics, at least in Newton's time. Whatever you measured, the result would have been the same, for Newton.

Martin Jähnert: Definitely, but this idea of looking at things in different ways; you don't need quantum theory or physics for that; you can look at art, for example.

Elise Crull: But she is more specific about it. She is critiquing Nelson in a neo-Kantian vein, using modern physics. She's not simply saying, hey, everything is relative. Yes—that concept has been around for a long time. But she is grounding her specific neo-Kantian approach in particular aspects of this new physics.

Giulia Paparo: No, I see more that she uses the new physics as an inspiration for looking at this—we don't even have to take it literally. It is clear that she is thinking about two different things: one is physics and the other one is ethics. Still, it is the human relating to nature and the human relating to other humans. The way we learn things and look at things might be similar; this is part of what she derives from the analysis of quantum mechanics. For me it is really clear that this is what she is doing in the last chapter of 'Die naturphilosophischen Grundlagen'. We are not speaking about the formalism at all; we are speaking about more general concepts.

Gregor Schiemann: But you have no intuition in modern physics any longer. And practical reason is dependent on intuition, on observation, on direct experience—so I think there is a very clear cut, so to say, between these two worlds. The challenge which modern physics puts to philosophy is to conquer this new situation. So I think one of her insights is that we must be careful not to suppose a close relation. The scientific worldview, which makes a very fast analogy between these two worlds, is in error.

Patricia Shipley: I'm also very interested in this question of social relations. There is a multitude of thoughts in there, with respect to the phrase 'social relations'.

To go back to pragmatist George Herbert Mead, the pragmatist philosopher and social psychologist: I think he made the suggestion that mind is not something which is encapsulated in the individual. It is a relationship between minds. It has lifted the level of analysis from the isolated, individual mind to a social level. I don't know if Benner's nurses were working in teams, for example, whether some kind of group dynamics were involved. But one can slide too much into some kind of speculation.

Guido Bacciagaluppi: This is just a small question, maybe: I wonder whether the physicists with whom Grete Hermann associated in the '30s kept up an interest in the other side of her work. It may be pure coincidence just because complementarity was Bohr's hobbyhorse and he thought it was very widely applicable, but I am surely not the only one who is thinking of Bohr's pronouncements about applying complementarity to justice. I think he actually uses the language of complementarity between justice and—is it 'love'?

Mélanie Frappier: It is, love and justice.

Guido Bacciagaluppi: Is it pure coincidence?

Mélanie Frappier: The example is this: you have a child, you love your child but she has been naughty so you have to punish her. So here you have to apply justice but it is going to be hard for you to punish your child because you love her. So you have to strike the right balance between justice and love.

Elise Crull: This is from Bohr when? Is it from the 1950s, his unity of knowledge stuff?

Thomas Filk: As far as I know Bohr got his conception of complementary from William James, right? William James explicitly mentions this word 'complementarity' in his writings. And he gives an example of what he means by complementarity, and the example is knowledge and belief. So knowledge and belief are two concepts which he thinks are complementary. One is not the opposite of the other, but the more knowledge you have the less belief, and the other way round. It is not orthogonal, but complementarity more in the sense which I think Bohr later uses.

Elise Crull: Was Hermann engaging with William James, with the pragmatists?

Thomas Filk: Well, I think Bohr was…

Elise Crull: *Bohr* was, but is James a common source, maybe?

Thomas Filk: I've no idea.

Elise Crull: Fernando, do you know, concerning her ethics, who she was reading?

Fernando Leal: All I can tell you is that Nelson was against James.

Elise Crull: That doesn't surprise me.

Fernando Leal: She doesn't seem to quote James.

Patricia Shipley: She would have known, surely.

Elise Crull: Would she have read a lot of James?

Patricia Shipley: She had read so much, it is obvious. We are talking about early twentieth century with James and also experimental psychology, and Freud—and she did psychoanalysis.

Elise Crull: She saw a psychoanalyst?

Fernando Leal: Yes, she did.

Dieter Krohn: She started it in Britain already.

Giulia Paparo: With whom?

Patricia Shipley: Who knows? It might have been Melanie Klein with her objects relations theory. That might have appealed to her very much. If you just want to know about psychoanalysis in a theoretical way you read books, but if you voluntarily undertake psychoanalysis it is a major commitment, if you see it through, even to embark on it.

Dieter Krohn: Gustav Heckmann had it done to himself as well.

Elise Crull: Heckmann is a really interesting connection, because he is speaking with Hermann throughout, and he is also connected to the physicists via Born and others. Can you tell us more about the role he played?

Guido Bacciagaluppi: Before we do that, just to clarify the question of psychoanalysis: there are two typical reasons for undergoing analysis. One is that you have got a problem. But analysts, especially orthodox ones, undergo analysis as part of their training—to actually learn the theory in a practical way. Do we know what the reasons were for Grete Hermann to undergo analysis?

Dieter Krohn: I suppose we have mentioned that problem a lot of times, and it's Nelson.

Several: Nelson?!

Dieter Krohn: Yes, it is true for both Heckmann and Hermann; they had to really free themselves from Nelson.

Patricia Shipley: We can blame it all on Nelson. But we don't want to set this woman on a pedestal—she was human. She could stand on her own feet, couldn't she? She was clearly a woman of considerable conviction and self-assurance, wasn't she? But at the same time, she was only human and what she was doing was courageous. There must have been a personal cost, don't you think? And then you, Dieter, told me the story about her mother and her regret that she felt she couldn't visit her mother and then never saw her mother again because it was too late, and how this was an indication to her of the inhumane nature of Nelson. I wouldn't be too surprised if she thought psychoanalysis might help her.

Fernando Leal: Well, there is another possibility—to come back to the question of whether there may have been theoretical interest on Grete's part: there may have been, because in Nelson's philosophy the idea of there being some kind of unconscious knowledge deep inside human beings was important. So maybe she was trying to find out from psychoanalysis whether she could get there. But Paul Branton told me once that she had this phrase: 'Es gibt keinen dunklen Keller'—there is no dark cellar. That was after psychoanalysis. Now we must remember that Nelson believed in the existence of some deep, originally obscure knowledge—it is originally obscure because it is not intuitive, it is not given, and this knowledge was the ultimate basis of all human knowledge and action. So by saying 'there is no dark cellar', Grete, as reported by Paul Branton, was setting herself against this Nelsonian thesis. She was in fact claiming that there was no such basis, no such Nelsonian obscure knowledge. What the role of psychoanalysis was in this shift of Grete's position about what was really a fundamental tenet of Nelson's philosophy, is anybody's guess.

Dieter Krohn: I think that some of the Nelsonians were deeply impressed by Freud and his findings. This could be interpreted as a kind of criticism of Nelson. He expected people to come to clear decisions and then do what they wanted to do. He did not take into account all those 'dark powers' within the individual. And then they realised: there is something we haven't seen before; there is something new. And they were really impressed by that idea. That was the case with Gustav Heckmann, to be sure, and with Grete as well. And when they felt they still had problems with Nelson—let's say, about 40 years after his death—they had to do something about it. They decided to choose psychoanalysis as a tool to help them.

You had a question about Gustav Heckmann.

Giulia Paparo: Could you also tell us something about Heckmann as a physicist? He studied with Born, you said.

Dieter Krohn: I don't know very much about it. I knew what his dissertation was about, but I forgot; I am not a quantum physicist myself. I wrote something about it when I wrote his biography and published it. It is important that he seemed to be quite a successful physicist. He did his dissertation under the supervision of Max Born. Then he worked in this field—I mentioned that he did some work together with Pascual Jordan, who turned out to be a Nazi afterwards. Everybody thought that Heckmann was one of the future physicists. And then in December of 1922, Nelson gave a lecture to the Pedagogical Society in Göttingen on the Socratic Method. The situation with Gustav Heckmann was that he had had to take part in the First World War. He had worked in the medical service there, and he was quite young when he had to do so. So when he came back from the First World War, he did not know of any values or ideas that were really worth following—he was at a loss. Then he listened to this lecture by Nelson telling them that there is a method: you can think together in groups and you can really find answers to your philosophical questions. That fascinated Gustav Heckmann. So he decided: I want to learn to facilitate those dialogues, I want to take part in those dialogues. He gave up his work in the field of physics and went to Nelson.

Nelson of course demanded all those awful commitments we have already heard about. Heckmann's family was Protestant, and thought it very important to be a member of the church. So it was hard for him to leave the church, but he did all this. It was decided he should first go and complete his training as a grammar school teacher in mathematics and physics. He did that, and then came back to the Walkemühle as a teacher. He taught mostly adults at the Walkemühle, and that was it—he never came back to work in the field of physics. But he still kept in contact with the physicists. He had met Weizsäcker; he met Heisenberg a lot—they were friends when they were both students. So after the Second World War he approached them again, to win them over to his political activities against atomic armaments. So there have always been connections, and even connections between Grete Hermann, Weizsäcker, Heckmann, and others—Max Born was among them. So they worked together, not in the field of physics, but in the field of politics.

Martin Jähnert: Where was Gustav Heckmann in the time, say, 1923 to '27?

Dieter Krohn: In '32 he was among those who produced the daily paper—he was in Berlin. But you said 1920s?

Martin Jähnert: 1922 to 1927, after he left physics.

Dieter Krohn: He first went to complete his training as a teacher, north of Berlin; he was in one of the grammar schools there. Then he went to the Walkemühle—Nelson's boarding school—where he taught.

Martin Jähnert: And the letters you mentioned are from this period?

Dieter Krohn: I have got letters from after the Second World War between Grete Hermann and Heckmann—and the one from December 1933, which was one of the early ones. I might have more information in all of the papers I still have.

I remember somebody asked the question why Grete Hermann did not take up a post as a physicist again, after the Second World War. I know that especially Weizsäcker again and again approached her, to convince her that she should take a chair at one of the German universities. But she didn't want to do that. I think you, Giulia, told me that she had nearly done it. Maybe you can help us by explaining that.

Giulia Paparo: Hermann wrote to Weizsäcker, I think it was in 1960, asking to collaborate with him again and to visit the Max Planck Institute. And she arranged everything: she had leave from the institute where she was working to go there, and arranged the dates and everything, but then apparently nothing happened.[1]

Martin Jähnert: But Weizsäcker's institute was not a physical sciences institute.

Gregor Schiemann: It was the Max-Planck-Institut zur Erforschung der Lebensbedingungen der wissenschaftlich-technischen Welt [Max Planck Institute for the Study of Living Conditions in the Scientific-Technical World]—the one in Starnberg, which he co-directed with Habermas. Before, he had a professorship in Hamburg in natural philosophy, but this was in the philosophy department. I think he had not worked as a physicist after the Second World War.

Thomas Filk: He did in the '60s and '70s and the beginning of the '80s—definitely a lot. He was always interested in philosophical questions also, but there were other people in his group working with him, who were really working on physics—in unified theories and so on. I know many people who worked on physics with him in the Starnberg Institute.

[1]On this episode, see above, Chap. 1, p. 14.

Giulia Paparo: And this was Hermann's purpose—she wanted to work on physics again.

Dieter Krohn: And then she didn't go?

Giulia Paparo: I wrote to the Max Planck Society and actually got a reply one week before getting here. They told me they had done an extensive search in their archive but couldn't find anything about her, so she probably hadn't gone there and they don't know the reason. I guess it was because Minna Specht was ill at the time, and then she decided to take care of her.

Dieter Krohn: That is a very good reason. It shows the spirit of the community, really, because Minna Specht and Grete Hermann lived together in Bremen. Minna Specht became quite ill and Grete Hermann cared for her. And if this happened when she had the chance to go to Starnberg, I am sure it was enough reason for her not to go, but to stay with Minna Specht.

Gregor Schiemann: You can also ask the question, why had she not started after the war directly, and asked Heisenberg or Weizsäcker? The war was over. Why did she go to this pedagogical institution? She could also have done good work with Heisenberg and Weizsäcker after the world war. It is very late when she eventually writes to Weizsäcker.

Dieter Krohn: I think there is a reason for that, an explanation why she did not do that. First, remember my quotation from *Politics and Ethics*. Her attitude towards the physicists in Germany was very critical. Then, in the correspondence between Grete and Gustav Heckmann I found some remarks on that—that she couldn't stand the atmosphere at German universities, German professors. That was the reason for her not to go. So I was a bit surprised that nevertheless there was a time when she thought about it.

Gregor Schiemann: Are these remarks in the '30s?

Dieter Krohn: No, after the war.

Gregor Schiemann: So it might have been a problem for her that they took part in the development of the atomic bomb—Weizsäcker and Heisenberg?

Dieter Krohn: Possibly.

Patricia Shipley: Did she say what she didn't like about the German professors? Were they wrinkled old men with gray beards or something? Or was it the kind of work they were doing?

Gregor Schiemann: For their taking part in the development of the atomic bomb?

Elise Crull: No, her criticism was that their *inaction* was enough to implicate them in moral wrongdoing.

Gregor Schiemann: Inaction in the context of the atomic bomb?

Elise Crull: For not even leaving Germany. Einstein criticises Schrödinger for sending in a paper to *Naturwissenschaften* after they'd kicked Berliner out.

Guido Bacciagaluppi: The Schrödinger cat paper (Schrödinger 1935).

Elise Crull: Einstein writes, 'Why are you still publishing in these journals?' Einstein was very vocal about this: don't even 'carry on'—make an active protest. And that sounds in the vein of Hermann's criticism as well. So maybe it wasn't even their direct involvement, just that they remained in Germany or didn't leave the universities there.

Mélanie Frappier: I just wanted to say that even in her language, she uses the word 'island'—she points out that people are like islands. When Heisenberg recalls his discussion with Planck about whether or not he should stay in Germany under the regime, he puts that exact word in Planck's mouth: we will need islands during the world war to protect the students, and then we can relaunch German science after the war. And so Heisenberg sees himself as this neutral island in the midst of Nazi Germany.

Gregor Schiemann: What is the relation of this to Hermann?

Mélanie Frappier: I was just pointing out that in her criticism of those scientists she uses the word 'islands': they act as islands that are immune to whatever is happening.

Gregor Schiemann: She said they had better leave the country?

Mélanie Frappier: I didn't say that she said they should leave the country, but not act as islands—this idea that you are isolated from the political situation.

Dieter Krohn: Just another incident, may I add, as you mentioned Einstein. I told you something about 1932 when they had the newspaper—the daily paper in Berlin. They also published a so-called Dringender Appell—an urgent appeal to everybody, especially to left parties, to unite and fight Hitler. They needed signatures from famous people. Of course Gustav Heckmann was one of those who collected these signatures. So he went to Käthe Kollwitz, the artist in Berlin, and her husband, and he went to Einstein. We used to ask him to tell us the story of how he went to Caputh—where Einstein had lived—and Einstein wasn't there when he came, but his wife was there. And his wife showed him—and was very proud about it—the new Frigidaire she had in that house. Then Einstein came and Heckmann explained what they wanted, and of course Einstein signed this urgent appeal.

Giulia Paparo: For me this fact that she wrote to Weizsäcker is interesting. But what is interesting is not that it came so late, but that it *came*. From her biography you get the feeling that she is not interested in physics anymore after the war—that she is just dealing with politics and education. For me this is showing that she kept her interest in physics and she was still thinking about these matters.

Gregor Schiemann: For me it is clear that she has this interest, but the situation seems to be different. There is this criticism, possibly against the German physicists. Then of course she has the task to rebuild Germany, and her position in Bremen was directly rebuilding the institutions. So she is very practically engaged. For me it seems very probable that she is still interested in this field, as she always was. She has to leave Germany, because she is a member of this Kampfbund [ISK]. She could not stay in Germany. What would she have done if she did not have to leave Germany? Perhaps she might have stayed with Heisenberg. What do you think?

Dieter Krohn: I think the problem is that she had contacts with all the resistance groups.

Gregor Schiemann: So she *had* to go.

Dieter Krohn: She had to leave Germany when the first groups were—how to put it—when the Gestapo found them and they were put on trial. The danger was that somebody would mention her name—that she had been with them, discussed things with them—and then it would have been too late to leave. So she had to leave exactly then, in 1936.

Patricia Shipley: She obviously makes some very major decisions, some very big choices. Was she compelled by an inner conviction?

Elise Crull: It is interesting to compare the anecdote of when she is meant to give a lecture and she goes, instead of staying with her sick mother, and when she stays with Minna Specht instead of visiting the Max Planck Institute. Or when she decides her work in Germany should be rebuilding after the war instead of pursuing physics again. It is interesting to see that she is carrying out this practical aspect of her convictions.

Mélanie Frappier: Giulia, do you have any clue what she wanted to do with Weizsäcker? Because I had assumed up until now that she was going to talk about physics, but she could have wanted to go to talk about nuclear armament and the politics of that.

Giulia Paparo: No. I have the quotations somewhere, and she says really clearly that she wants to put her hands on her early thoughts on quantum mechanics again. This is pretty clear.

Elise Crull: Specifically quantum mechanics?

Giulia Paparo: Yes, specifically quantum mechanics, and also to get up-to-date on the newest changes. It is really clear that this is what she wants to do there.

Dieter Krohn: She was born in 1901. This would have been after her job at the Bremen College of Education? It would be characteristic of her—to first finish something, and then start a new thing, even at that age. Of course it isn't old, is it?

Guido Bacciagaluppi: She was actively involved with the PPA from that time onwards?

Dieter Krohn: She was involved with the PPA all the time, either as the chairperson or as a member.

Giulia Paparo: I just found the letter—I was wrong, it is dated '56 and not '60—the letter to Weizsäcker (Henry-Hermann 1956; Herrmann 2017, Part III, Letter 23). And she says: 'Mir geht es darum, die Ansätze, die ich in früheren Aufsätzen gewonnen habe, zu vertiefen und damit grundsätzlich zu überprüfen'.[2] Minna Specht dies in '61. She was ill for several years, so it could be that's the reason why she might not have gone.

Weizsäcker says something about being at the Max Planck Institute and not being sure whether he would be in München or Stuttgart.

Unidentified Scotsman: I wonder what the feelings are in the room concerning Grete Hermann's duty to do more to publicise her writing. Or whether there was a duty?

Elise Crull: I feel like it would have been Heisenberg's duty to recognise it... or even for Weizsäcker to say, 'Look, Heisenberg—she gets it. Come on!'

Guido Bacciagaluppi: I agree. We did say that in the context of her own work—Giulia gave us a nice analysis of that—in the context of her work, certain aspects—the critique of von Neumann in particular—didn't play such a big role. So, no, I wouldn't think she had some particular duty to publicise it more widely. But others should have recognised it more.

[2] 'My concern is to deepen the approaches I obtained in earlier essays, and thereby to reassess them thoroughly'.

Martin Jähnert: Plus what we heard from the letter before: she did send her 1933 work not only to Bohr and Heisenberg and so on, but also Dirac. That Dirac does not delve into these interpretational questions is not that much of a surprise, really.

Guido Bacciagaluppi: He never writes back more than two lines.

Dieter Krohn: I suppose she sent it to Friedrich Hund as well, because Hund was also a friend of Gustav Heckmann. In some of the papers you can read that she attended one of Hund's lectures in Leipzig, and they had some discussions (Herrmann 2017, Part III, Letter 7). So that was another contact.

Martin Jähnert: Hund is a fascinating figure.

Dieter Krohn: A fascinating figure. When I took him from the station in Hannover to the celebration we had on the occasion of Mr Heckmann's 85th birthday, he was also a very fascinating character: very very small, very very thin, very very old. He was a very good friend of Gustav Heckmann's.

Martin Jähnert: Hund was more an experimental physicist, and not so much into the theoretical side.

Mélanie Frappier: I have two questions that are linked, but are very different. The first one is: do you have the obligation to try to publicise the work of another, especially if you don't think it is important or useful or right? And linked to this: I cannot recall any of those physicists doing that kind of thing. Not very often, and only for very specific, gifted people. This is the second question. Am I wrong in thinking that they didn't do that as often as we might see it now?

Guido Bacciagaluppi: Maybe they did. The paper in *Die Naturwissenschaften* (Hermann 1935)—*Die Naturwissenschaften* was a very prestigious journal. By October 1935 it wasn't even Arnold Berliner any more who was editing it—it was in the hands of somebody else more in favour with the regime. Would it in fact have been difficult for Grete Hermann to publish there? Did she need some help from Heisenberg or people like that to publish in *Die Naturwissenschaften*? Or would it have been relatively easy to do so?

Giulia Paparo: I think she definitely had some help, and also for the Avenarius Prize. Heisenberg writes her a letter telling her, 'Yes, so you are going to get a good surprise'. It is clear that he helped her in getting the Avenarius Prize. This is clear from his letter—that he personally intervened. And he is also speaking about the other two articles awarded the prize, and says that the one by Thomas Vogel—I think—he agreed with that, while the last one he wasn't so happy with. So if he did it in this case, why shouldn't he have done it in other cases?

Dieter Krohn: Heisenberg was on the jury, wasn't he? In this role he could write the letter. As he knew about the political background of Heckmann and Grete, he was afraid that, had he openly supported her in other contexts, people would have asked: 'Oh, what's behind it? Does he sympathise with those leftish circles?' So being a member of a jury is in itself a different thing.

Giulia Paparo: Yes, but the critique of von Neumann's proof is not such a leftish thing. I understand your point, but I think it was also about content.

Mélanie Frappier: My question is, did physicists support students who were up-and-coming scholars in the way they do now? But to come back to your point, there are two big differences between Heisenberg and Weizsäcker. The first one is Weizsäcker's dad, who is quite able to protect his son. Heisenberg doesn't have this luxury of having a well-placed father in the government.

The second thing that I think is very different is that Heisenberg actually risks losing a very important job, but Weizsäcker is not quite in that position. And Heisenberg is already starting to stretch himself to help people here and there, writing letters to help the parents of a student or supporting a Marxist student in his lab, to the point that in 1938, he spends a weekend with the Gestapo as their 'special guest' in a basement in Berlin, because they have questions for him. And he cannot travel that year to Poland to this big meeting on quantum mechanics, and Bohr reads his paper then. He seems to be walking a very thin line here.

Gregor Schiemann: For me it seems to be very risky for Heisenberg to give this support, because the ISK is already forbidden. It is very clear that Hermann is very politically engaged. It is also very risky for her to stay in Germany still. For Heisenberg, this is an example of an 'island': he supports the people in his island of physics, and this is a risky line.

The other thing for me that is still unclear is how Grete managed this work, because she is politically so engaged at this point of her life—she is writing for newspapers, she is into the education stuff, she mentions all these new situations. And while she is doing this she starts to do this work in the philosophy of physics. For me it is incredible. Es geht nicht mit rechten Dingen zu![3] It takes an incredible intellectual power to manage this. The physicists are concentrating on their questions—they think of nothing else—and she is travelling around, fighting here and fighting there, and then solving von Neumann…

Guido Bacciagaluppi: Actually, there are two questions we are considering about this support. One is personal support for Grete Hermann—her career maybe. The other thing is, should Heisenberg and others have taken on board more her criticism of von Neumann, rather than coasting along with the cult status of von Neumann's proof? These are two questions.

---

[3] 'It isn't natural!'

Elise Crull: In the correspondence that Heisenberg is having with everybody at the time, he's writing about all these questions and never mentioning Hermann. Yet he is talking about Einstein and other people who would have been politically out of favour.

Giulia Paparo: For me, he could have been mentioning von Neumann's critique without mentioning Hermann, since it is not so connected—Hermann's work is mainly connected with Friesian philosophy and Kant and so on. I think it would have been easy to just mention von Neumann's proof without having to risk anything concerning Grete Hermann. It doesn't mean supporting her here.

Mélanie Frappier: Just one question: when does von Neumann go to the States, and why?

Elise Crull: Von Neumann is at Princeton already in 1930. Then his book is published in 1932.

Mélanie Frappier: Here's my reason for bringing up that question: is not talking publicly about the problem in von Neumann's proof another way of supporting someone who is in disfavour? Just not attacking someone, by saying: 'Well, it's not that important. Yes, it's about physics, but we don't think that overall it's a big problem for our position anyway. Let's just not attack someone who is really an ally'. I'm just wondering if it could be another of those weird ways of supporting someone by silence.

Elise Crull: But she's making an *intellectual* criticism—and they're all freely doing this. Pauli is personally criticising people all the time in his letters and interactions.

Mélanie Frappier: But isn't he the one during the war who is writing all those papers under other people's names, so that the Germans don't know that they are actually working on the bomb? Who is doing that? There is one famous physicist doing that during the war. I am pretty sure it's Pauli. He published under other people's names.

Martin Jähnert: Pauli is already in Zurich by then.

Guido Bacciagaluppi: I think there may be a wider question about the reception of von Neumann's proof. We all know that in the folklore, there is this cult status. That certainly is enough to silence the masses of physicists—that might be enough for the average physicist who hasn't thought much about it, and so on. But what do the big guys—Bohr, Heisenberg, Pauli and so on—actually think about von Neumann's proof? It might have had that effect of being this myth, and the average physicist who works in quantum mechanics has this in the background of their minds that 'Oh yes, von Neumann showed that there are no hidden variables'. But I'm not aware of seeing explicit statements by Bohr, Heisenberg, Pauli and so on, in *support* of von Neumann's results.

Martin Jähnert: I think there wouldn't be, given way they argue about quantum mechanics all along, because what they are thinking is: 'Oh, we have the formalism, it works, and we do not yet have a problem that forces us to consider an alternative like hidden variables. So according to them you do not need a *proof* that hidden variables are impossible; the burden of proof would rather consist in finding a problem which demands them.

Elise Crull: But that comes about in 1935, with EPR. All of a sudden there is another reason to think about hidden variables.

Martin Jähnert: But as you see, for example, from the Pauli letter [to Heisenberg, 15 June 1935], about the EPR paper (Pauli 1985, pp. 402–405)—he says, 'Oh, Einstein finally got it. *We* knew this all along; *we* have no problem.'

Elise Crull: But he's the only one who feels that way—the rest of them are pretty confounded by it and take quite seriously this criticism. Pauli's the only one who is flippant about it; the rest of them seem to take it pretty seriously.

Martin Jähnert: But until 1935 I would explain it as I suggested before.

Giulia Paparo: For me, to support this idea that they knew about it and took for granted that von Neumann's proof wasn't so good, there is also Bell's paper, in a way. For he is not stopping there, he is going so much further than that. His criticism is just at the beginning of the paper, then he goes on to speak about Gleason, and Kochen and Specker. It is as if among the big physicists it was already clear that von Neumann's proof wasn't so perfect and as universal as we had the feeling it was. But it's just an idea I was thinking about the other day—that they all kind of knew about it.

Martin Jähnert: Plus maybe one thing: the research interests—at least of Pauli and Heisenberg—have shifted radically. They are working on quantum electrodynamics by then, which is of course troubled far beyond lacking a proof against finding hidden variables.

Giulia Paparo: For me another indication of the fact that they might all have known about the flaw in von Neumann's proof is the fact that it took two years for Bell to get the paper published. He handed it in and he didn't hear anything about it, and it got lost. It seemed to me that he didn't worry that much about getting the paper published, as it didn't have anything so new in it, or so astonishing. And after two years they asked him: 'What happened? You told us you were sending us a paper, and it never got here'. And he said, 'Yeah, I handed it in—I thought you lost interest in it'. That is why they published it two years after.

Guido Bacciagaluppi: We shouldn't forget that each of these—Bohr, Heisenberg, Pauli—they all had their own reasons for believing that hidden variables were impossible. Bohr centred this on the doctrine of complementarity—that values were not well defined outside of an experimental context. Heisenberg has this idea that if you've got hidden variables, that will suppress interference effects that you do see experimentally. Pauli, in the letter to Schrödinger on the 9th of July 1935 (Pauli 1985, pp. 419–422), has a neat argument, which in fact mathematically is related to von Neumann's argument but is much more direct and simple. If you assume values for position and momentum—say you have an ensemble with sharp momentum, and you subdivide it into sub-ensembles in which also the position has a sharp value—what are the values, say, for the energy, or the angular momentum? You get a continuous distribution. But we know that the angular momentum or the energy of the harmonic oscillator are quantised in quantum mechanics, so you get wrong predictions. If one analyses the logic of that—mathematically, the way one calculates the energy, or the way one calculates the angular momentum—that really uses the linearity of the assumed dispersion-free expectation values. The background assumption is the same as von Neumann's, but it's such a direct and plausible argument as stated by Pauli.[4]

Each of those guys have their own reasons why it was obvious to them that you couldn't have hidden variables. So maybe they just didn't think that von Neumann's proof was that important to them. They might have thought: O.K., there is also this argument, maybe there are some problems with the assumption, but we know anyway what the truth is.

Giulia Paparo: Yes, but that is Hermann's attitude as well—she also thinks that it doesn't work. But in fact, she *published* that it is wrong, that it has a problem.

Elise Crull: She 'put it into practice'.

Danny McShane: I guess David Bohm put it into practice as well. I think John Bell—part of his motivation was that he had seen the impossible done, so he had to set about re-examining the proof. When David Bohm came up with a theory that *was* hidden variables, then John Bell had to revisit the maths.

Thomas Filk: Maybe I am seeing things sometimes too psychologically, but when I talk to physicists—I mean the everyday physicists in the Physics Institute—and I mention Bohmian mechanics, I very often get the reaction: 'Oh, we don't want to go back to *that*. We all struggled a lot until we thought we understood quantum mechanics. And now somebody comes and says: "Well, I have a Newtonian ontology which can explain everything". We don't want to go back to that stage'. And this reaction I hear quite often; you don't really hear scientific arguments made so often against Bohmian mechanics, but much more often this intuitive 'We don't want to go back'. I can quite well imagine people like Heisenberg and Bohr and Pauli, who went

[4]Cf. the discussion of these points in Bacciagaluppi and Crull (2009, Sect. 4.2).

through the struggle much more than we did when we learned quantum mechanics—they went through this for years, for many years—and finally came to a point where they think they have a story to tell. If then at that stage somebody comes and says that maybe hidden variables could be the case, they have this reaction: 'Well, we don't want to go back'. Although they don't say it this way, it could've been a reaction like that.

Giulia Paparo: But I find it even more dangerous because they were the fathers of the quantum theory. I understand today when we just decide 'this' and not 'that'—but they were actually *creating* 'that'. So not looking into other possibilities in their case is for me much more intellectually worrying.

Gregor Schiemann: I'm interested in how you, Thomas, see the situation. I agree completely with your description of physics, but I think in the philosophy of science it's quite the reverse: there is no philosopher of physics—nearly no philosopher of physics—who is not very interested in hidden variables. Is this your impression, too, Guido? That the attitudes are quite reversed in philosophy and physics?

Guido Bacciagaluppi: Yes, in the philosophy of physics it's an open question—an open ball game and who knows: hidden variables, spontaneous collapse, many worlds...

Elise Crull: But in a sense people come to the field because they are interested in pursuing these sorts of questions to begin with. So maybe there's already a sort of self-selection.

Does anybody have any last thoughts—tying together the different aspects of her life? Obviously she was a remarkable woman.

Patricia Shipley: Thank you for setting up such an interesting programme. I found it very stimulating. I didn't know what to expect—I came with some hypotheses which didn't hold up. I had some fun this morning. Including that there's really no discernible connection between quantum physics and Kantian philosophy. That's a big one. Are there two separate fields and two sets of discourses and the twain will never meet? On the other hand, the main subject has been this very interesting woman, a very remarkable woman, clearly intellectually highly competent. She achieved a great deal in her life and a lot of it has been related to her background in moral philosophy. Can we say that?

Sally Redfern: Does that mean another workshop needs to be set up in two, three, four, five years' time? Where we can review some of these points?

Guido Bacciagaluppi: Why not? We had high expectations for the workshop, and our expectations were exceeded. We met some fabulous people whom we didn't know. I am all for continued forms of interaction, and it may well be that we feel at some point we need to touch base again—that maybe we've got more things to say to each other. If we have another workshop like this, I certainly wouldn't be disappointed. Thank you all for coming.

References

Bacciagaluppi G, Crull E (2009) Heisenberg (and Schrödinger, and Pauli) on hidden variables. Stud Hist Philos Mod Phys 40(4):374–382

Benner P (1982) From novice to expert. Am J Nurs 82(3):402–407

Dreyfus SE, Dreyfus HL (1980) A five-stage model of the mental activities involved in directed skill acquisition. Tech. rep., UC Berkeley Operations Research Center, no. ORC-80-2

Heisenberg W (1942/1984) Ordnung der Wirklichkeit. In: Blum W, Dürr H, Rechenberg H (eds) Werner Heisenberg, Gesammelte Werke, Abteilung C: Allgemeinverständliche Schriften, vol I. Piper Verlag, München, pp 217–306

Henry-Hermann G (1953) Die Überwindung des Zufalls: Kritische Betrachtungen zu Leonard Nelsons Begründung der Ethik als Wissenschaft. In Specht and Eichler (1953), pp 25–111

Henry-Hermann G (1956) Letter to C. F. von Weizsäcker, 30 June 1956. In: Archiv der sozialen Demokratie der Friedrich-Ebert-Stiftung, Bonn: Grete Henry-Hermann Collection, Box 1/GHAJ000009

Henry-Hermann G (1991) Conquering chance: critical reflections on Leonard Nelson's establishment of ethics as a science. Philosophical Investigations (1):1–80. Translation of Henry-Hermann (1953) by Winch P

Hermann G (1935) Die naturphilosophischen Grundlagen der Quantenmechanik. Die Naturwissenschaften 23(42):718–721

Hermann G (1945) Politik und Ethik. Renaissance Publishing Co., London

Hermann G (1947) Politics and ethics. International publishing company, Bombay. Translation of Hermann (1945)

Herrmann K (ed) (2017) Grete Henry-Hermann: Philosophie–Mathematik–Quantenmechanik. Springer, Berlin

Pauli W (1985) Scientific Correspondence with Bohr, Einstein, Heisenberg a.o., vol II, 1930–1939, Springer, Berlin. Edited by Meyenn K von

Schrödinger E (1935) Die gegenwärtige Situation in der Quantenmechanik, I–III. Die Naturwissenschaften 23:808–812; 823–828; 844–849

Specht M, Eichler W (1953) Leonard Nelson zum Gedächtnis. Verlag Öffentliches Leben, Frankfurt am Main/Göttingen

Part IV
Translations

Chapter 13
Letter from G. Heckmann to G. Hermann, 17 December 1933

Gustav Heckmann

Copenhagen
17/12.33.

Dear Grete!

I thank you heartily for sending me your paper. I am unable to *study* it given my current daily schedule. And since it is futile to comb over it, I shall let it lie until January; then I have some time. For now, just a few words about Heisenberg's judgement. I spoke to him the day before yesterday for over an hour.

They take your paper absolutely and completely *seriously* and in the days while he is still here, H[eisenberg] together with Bohr and a student of H[eisenberg]'s, Weizsäcker, wanted to jointly draw up an answer to you. At the moment I did not think about Dirac, who was also here; but maybe H[eisenberg] and Bohr did not suspect that he had also received your paper. [Quoting Heisenberg:] 'In substance, she is certainly wrong', but 'a fabulously clever woman'. 'When during my lecture a woman stood up a[nd] began speaking in a rather sharp voice, I thought: for Heaven's sake, what will this come to. But I was quite astonished by the clarity with which she could analyse everything'. Otherwise he thinks, as you do, that you must learn still more physics; study the 4 papers by Bohr—he has told you about them. Hei[senberg] speaks with such imprecise phil[osophical] terminology that it would make a Friesian climb the walls. If he understood transc[endental] idealism, then he would surely realise the possibility of finding the key also to solving the philos[ophical] difficulties arising from quantum m[echanics]. I told him about transc[endental] id[ealism]. He said I should really talk with Weizsäcker about these things: he has thought a lot about

Translated from the German by Elise Crull and Guido Bacciagaluppi. From Nachlass Gustav Heckmann, private collection Dieter Krohn (also Nachlass Grete Henry-Hermann, Archiv der sozialen Demokratie der Friedrich-Ebert-Stiftung, Bonn, file 1/GHAJ000006), forthcoming (in German) in (Herrmann 2017, Part III, Letter 2). This letter was introduced to those present at the Grete Hermann Workshop by Dieter Krohn. During the conference's panel discussion (cf. Chap. 11, pp. 191–193), Dieter read aloud excerpts from the letter and translated them into English on the spot.

E. Crull and G. Bacciagaluppi (eds.), *Grete Hermann - Between Physics and Philosophy*, Studies in History and Philosophy of Science 42,
DOI 10.1007/978-94-024-0970-3_13

natural philosophy. Well, you had better do that then (Dr Weizsäcker, Kopenhagen-Hellerup, A. N. Hansensallé 21). Weizs[äcker] is working with Bohr at present. You and Dora[1] will always find an open door and an open ear with Heisenberg. Use it yet, ere you knock at the harder door made of older wood: that of Bohr. Hei[senberg] spoke also about a certain difficult-to-acquire resignedness in the face of physics: one just has to give up the wish to state something about *real processes* in the atom—then one is able to work more fruitfully. Also your philos[ophical] considerations would be more fruitful once in regard to atomic theory you had managed to fight your way to this st[an]dp[oin]t. 'It requires a great deal of strength to get there in the first place. Grete H[ermann] seems not to have quite got there yet, also Schrödinger has not yet arrived there'. I do not understand this st[an]dp[oin]t, perhaps you can see what he means.

It makes me very happy that you have acquired the considerable regard of Bohr and Hei[senberg] with your paper. Get onto them with the transc[endental] id[ealism]!

Niels Bohr's address: Institut for teoretisk Fysik, Kop[enhagen], Blegdamsvej 15.

Dirac looked radiantly cheerful.

Hopefully you too.

Warmly
Gustav.[2]

Reference

Herrmann K (ed) (2017) Grete Henry-Hermann: Philosophie–Mathematik–Quantenmechanik. Springer, Berlin

[1]Dieter Krohn believes that 'Dora' is a reference to Minna Specht (*eds.*).

[2]A postscript to this letter appears to be written in a different hand, and concerns mainly social arrangements (*eds.*).

Chapter 14
Determinism and Quantum Mechanics

Grete Hermann

> Although the new theory seems thus well established in experience, one can still pose the question of whether *in the future*, through extension or refinement, it might not be made *deterministic again*. In this regard one must note: it can be shown in a mathematically exact way that the established formalism of quantum mechanics allows for *no* such completion. If thus one wants to retain the hope that determinism will return someday, then one must consider the present theory to be *contentually false*; specific statements of this theory would have to be refuted experimentally. Therefore, in order to convert the adherents of the statistical theory, the determinist should not protest but rather test.[1]
> Born (1929)

> Declaring a fundamental surrender in epistemology—that indeed goes too far. Who shall stop a scientist, if he is attracted irresistibly to the open questions of the individual process? Who shall prophesy that nothing ever will come of this?
> (von Laue 1932)

1. The Purpose of the Following Considerations

This work is devoted to the concerns that are levied on the part of quantum mechanics against the law of causality.

The interest that I bring to the law of causality is a philosophical one. It arises from the intuitions of the critical philosophy as they have been developed in the school of Kant, Fries and Nelson. According to the investigations of this school,

[1] Original emphases omitted by Hermann (*eds.*).

Translated by Elise Crull and Guido Bacciagaluppi. Manuscript originally titled 'Determinismus und Quantenmechanik', sent to Dirac with cover letter, and located in the Dirac Archive, document DRAC 3/11, University Libraries, Florida State University, Tallahassee, FL. A copy of the Dirac Archive is located in the Churchill Archives, Churchill College, Cambridge, who kindly supplied us with a copy of the manuscript. The original German text of the manuscript as well as of the cover letter will appear in Herrmann (2017).

Translators' note: where Hermann gives references, these have been completed and put in the References section.

E. Crull and G. Bacciagaluppi (eds.), *Grete Hermann - Between Physics and Philosophy*, Studies in History and Philosophy of Science 42,
DOI 10.1007/978-94-024-0970-3_14

human rationality exhibits a priori knowledge of the validity of the law of causality—knowledge that finds application in experience but that can be neither justified nor refuted through observation and experiment.

As the statements by Born and Laue given above show, examining this claim nowadays leads immediately to questions of physics. The point of view of the critical philosophy appears to have been refuted by physics on two counts: in the understanding of the majority of its exponents, quantum mechanics has discovered natural processes that are not determined strictly causally; and this refutation of the law of causality has occurred on the basis of observation and experiment.

In the following, the grounds for these physical claims shall be examined. But first, two remarks for the purpose of making the hereby formulated task clearer, and of paring away pseudo-solutions.

1. One merely apparent solution is the often attempted defense of the law of causality that assumes that quantum mechanics has directly challenged only the predictability of certain measurement results, but not the existence of causes for these measurement results. One argues more or less in the following way: Heisenberg's uncertainty relations have demonstrated that in-principle unsurmountable limits are set to our observation of atomic processes. It follows from this that, no matter how precisely we may observe a process, we cannot know the causes for certain changes, and accordingly cannot predict these changes either. However, since this proof of the unpredictability of certain measurement results relies on the unknowability of their causes and thereby presupposes these causes are real, one cannot then further infer from unpredictability to the absence of causes and thus to the failure of the law of causality.

As to this criticism, which is in fact invited by several presentations of quantum mechanics, the rejection of the stated circularity is indeed called for. However, the position of the law of causality is thereby not yet secured. Indeed, when it is the case, on whatever grounds, that some future event is unpredictable in principle, thus that a cause that determines it sufficiently cannot be found in principle, then the claim that such a cause nevertheless exists is mysticism, not a proposition about nature. (I speak here and in what follows of the possibility of prediction even where the margin of error in the prediction is non-zero but can still be reduced beneath any margin different from zero.)

This consideration makes it possible to eliminate from the present investigations the highly contentious philosophical question of the correct formulation and interpretation of the law of causality. If quantum mechanics provides evidence that the result of some specifiable physical measurement cannot be predicted with arbitrarily small—although perhaps not vanishing—error, however advanced the knowledge of the initial conditions and the laws of nature may be, then the law of causality is done for as a principle of natural knowledge; then not determinism but indeterminism holds in nature. We can therefore restrict ourselves in what follows to the question (which no longer contains the concept of causality): *Has quantum mechanics determined an unsurmountable limit in the predictability of natural processes in this sense?*

2. Born's advice to the determinists not to protest but to test opens up an opportunity for the law of causality that the advocates of indeterminism generally concede: even for quantum mechanics—despite the splendid confirmation that it has found to date—the day could come when, on the basis of new experimental evidence, one of its fundamental statements is given up as false or at least must be modified in essential points. Such a correction could bring about the downfall of indeterminism.

Nothing is gained for the proponents of critical natural philosophy with such an experimental refutation of indeterminism, since it too leaves the decision about the validity of the law of causality to experiment and observation. Hence in the following I rule out this escape allowed from side of the physicists, by assuming that the formalism of quantum mechanics developed on the basis of the physical experience to date, will prove itself with respect to future experience also. Then only the question remains: *Is the claim made by numerous exponents of quantum mechanics, that the results of certain specifiable physical measurements are unpredictable in principle, a necessary theorem of quantum mechanics, or can it be detached from this theory without affecting in the slightest its value for the explanation of the available physical experiences?* I maintain that such a detaching is possible.

2. Heisenberg's 'Uncertainty' Relations

The claim about the impossibility of predicting the results of certain measurements with arbitrarily small error relies on the so-called Heisenberg uncertainty relations.[2] In Heisenberg's description, these relations find their theoretical grounding in the duality of wave and particle picture that has come to light in certain experiments. Since every atomic process, as is apparent from the discussion of these experiments, must be representable in both the wave as well as the particle picture, yet these two interpretation[s] contradict one another, provided one assumes the physical quantities that are characteristic for each to be perfectly precise, it follows that the physical process in question cannot, at least with respect to all of these physical quantities, be characterised by unique numerical values each fixing exactly a given quantity. Heisenberg obtains his relations through a mathematical discussion of these circumstances: let q and p be two canonically conjugate variables in a physical problem denoting, say, an electron's position and momentum along a given spatial direction, and let further dq and dp be the precision to which these quantities are determined, then at all times $dq \cdot dp \geq h$, where h represents Planck's constant of action.

Now, what do these relations actually mean? According to their derivation from the duality of wave and particle picture, the following: the description of the motion

[2] The German term 'unbestimmt' can arguably be translated both as 'undetermined' (that which has not been determined but has a value) and 'indeterminate' (that which does not have a determinate value in the first place). Hermann clearly treats the German terms 'unbestimmt' and 'Unbestimmtheit' as (misleadingly) suggesting the first reading. We have therefore translated them here, respectively, as 'undetermined' and 'uncertainty' (the standard English term and not as unwieldy as 'undeterminedness') (*eds.*).

of an electron as the motion of a corpuscle has only the bearing of an analogy. This analogy fails in certain places, specifically in that a corpuscle, but not an electron, has an 'exact' position (i.e. characterisable in each direction by a single numerical value) and an equally 'exact' momentum. Position and momentum of the electron are characterised in every direction by intervals on the position or momentum axis, respectively; thus, at least the four numbers q, dq, p, dp are necessary to fix them—the interval lengths dq and dp satisfying Heisenberg's relations. Any statement that ascribes to an electron 'more exact' position and momentum values is, according to a statement of Heisenberg's, 'just as vacuous as the application of words whose meaning has not been defined'.[3]

In this interpretation of the relations only one thing remains unintelligible: the name under which they have become known. In what sense are we dealing with 'uncertainty' relations here? What remains undetermined? To answer that the exact position and the exact momentum are undetermined or unknowable is out of the question given Heisenberg's considerations. Whoever argues in this way assumes in so doing precisely that the electron has also an 'exact' position within the interval of length dq, and thereby gets tangled in contradictions with the physical starting point of the argument—that the motion of the electron must *also* be interpretable as a wave process.

To this consideration the response is occasionally given that, the Heisenberg relations notwithstanding, the concept of 'exact' position for the electron remains meaningful; for it is always possible through a suitable experimental arrangement to measure the position of the electron with arbitrary accuracy, i.e. with arbitrarily small dq'. But it does not follow from this that already *before* the measurement the electron possessed a more exact position than was determined by the interval (q, dq) given at the time by q and dq. Physicists agree nowadays that, as a rule, a position measurement alters the state of the electron. Manifestly this change partly consists in the fact that the position interval (q, dq) is transformed into another one of potentially shorter length. But this does not mean there was an undetermined position within the interval before the measurement.

Alternatively one can interpret the 'uncertainty' of Heisenberg's relations as an uncertainty in prediction. If position and momentum of the electron are given by the intervals (q, dq) and (p, dp) with $dq \cdot dp = h$, and if one performs now an exact position measurement, its result cannot be predicted from q, dq, p, dp. But does it then follow—with this we arrive at the crucial question of this investigation—that this result is unpredictable in principle? By no means! It is only established that position and momentum do not suffice for the calculation of the results of a future position measurement; the possibility remains, however, of discovering other, as yet unknown physical quantities, such that the knowledge of their values together with the position and momentum of the electron are sufficient to predict the result of the position measurement. These quantities need not necessarily be determinants just of the observed electron itself; they could also pertain to the measurement apparatus

[3]The quotation is from (Heisenberg 1930, p. 11). This same sentence is also quoted in Hermann's 1935 essay (see below, Chap. 15, p. 246) (*eds.*).

used for the position measurement, since this indeed exerts a demonstrable influence on the electron.

The task of exactly predicting the result of the position measurement is thus vastly more complicated than classical mechanics assumed. It is at present still completely unsolved. But it has not been shown to be insoluble.

Here one may well reply that experience and theory have indeed shown that the position interval of Heisenberg's relations represents the range of possible results of a position measurement; a number is assigned to every point of this interval that corresponds to the probability of finding the electron exactly there. Since now for a given momentum interval the length of the position interval cannot be reduced beneath a given limit, it is impossible to predict the result of the position measurement with probability 1, i.e. with certainty, or even with a probability that differs arbitrarily little from 1.

The answer to this is that the relevant probability statement can be properly applied only if one considers merely already-known physical quantities. At any rate, experience has managed to show no more than that in cases that agree with each other merely with respect to already-known physical quantities, the results of a position measurement are distributed in accordance with the given probability values. That one will always obtain the same distribution—even if in the progress of physical research one brings the examined electrons into agreement with respect to ever-new physical parameters—is a claim that far exceeds experience so far, and cannot be derived from it.

Let us assume, say, one performs position measurements on a large ensemble of electrons; before the measurement let the state of all the electrons be characterised by the same four numbers q, dq, p, dp. Then measurement results will be distributed over the interval (q, dq) according to the probability coefficients calculated by quantum mechanics. But why should it be impossible now, in the cases where the position measurement gives the same value q' (with arbitrary small deviations), to find a trait that pertained to the electron or the measurement apparatus or both together already prior to the measurement, and that was not present in the other cases that led to different measurement results—and then to use the presence of this trait as a basis from which to predict the occurrence of the result q' (or of a value very near it) in future measurements? The objection that with the help of such a property one could determine the position of the electron already before the measurement more precisely than through the interval (q, dq), does not apply. *Before* the measurement, the electron does not yet have the position q' but rather is spatially determined by the interval (q, dq), which finds physical expression in the fact that the Schrödinger wave function characterising the state of the electron is non-zero over the whole interval (q, dq), and not only in an arbitrarily small neighbourhood of q'.

Heisenberg justifies the statistical interpretation of quantum mechanics, and with it the in-principle impossibility of predicting particular measurement results, by pointing out that as a rule an observation disturbs the observed system in an uncontrollable way. The phrase 'in an uncontrollable way' is obviously significant for his argumentation; for a disturbance that is controllable with respect to kind and magnitude can be precisely traced calculationally, and keeping it under control is only a mathematical problem—albeit perhaps complicated; no in-principle incalculability results from such a disturbance. Now, the disturbance that an electron experiences by being observed is, according to the preceding considerations, in no way uncontrollable. For instance, if the position interval of the electron is shortened significantly in a position measurement, the momentum interval is thereby correspondingly lengthened. One tends to see the uncontrollable disturbance in this change of momentum. If, however, following Heisenberg, one insists that it is meaningless to speak of a 'more precise' momentum for the state after the measurement than is given by the new momentum interval, then the change of momentum consists exclusively in the transition from the shorter to the longer momentum interval. Nothing is left undetermined here, and consequently no reason exists for inferring from such an uncertainty to the impossibility of predicting, say, precise [results for] momentum measurements.

3. Dirac's Construction of Quantum Mechanics

In order to avoid the risk of limiting the discussion to isolated, perhaps arbitrarily selected quantum mechanical arguments, and thereby overlooking that the overall structure of the theory might indeed rule out the predictability of certain measurement results, I turn in the following to a closed theoretical presentation of quantum mechanics. I choose for this purpose the beautiful and clear construction of Dirac. The mathematical doubts that, e.g., Neumann raises against Dirac's δ-functions play no role in the context of the question of whether and in what way this theory rules out the predictability of certain measurement results. Their discussion is therefore unnecessary.

Dirac's own presentation and interpretation of his formalism undeniably includes indeterminism. Hence the question is only whether it is a necessary component of his theory, such that in giving it up one loses essential elements of the theory, or whether indeterminism can be detached from the theory without thereby affecting its explanatory value.

If we take Dirac's entire axiomatically-constructed apparatus—i.e. his state and observable calculus—as an instrument that has proved itself for ordering the facts of experience, then it turns out that this instrument bears the stamp of indeterminism in a place that is exactly specifiable. It receives it through the fact that Dirac interprets $\phi \alpha \psi$ as meaning the average value of the observable α in the state ψ. (ϕ and ψ are symbols for the same state of a system, defined through a 'maximal observation'; α is an observable, that is, a symbol assigned to a physical quantity.)

This interpretation leads to indeterminism. As Dirac shows, in the first place it is mathematically equivalent on the basis of the postulated calculational rules to the following statement: let $\psi = \sum c_i \psi_i$ be the unique decomposition of a normalised ψ into normalised eigen-ψ_i of the observable α; then $|c_i|^2$ is the probability that the eigenvalue belonging to ψ_i is found in an eigenvalue measurement. Now, in the framework of the Dirac formalism, one has the theorem that for each state ψ there is [an] observable α, for which ψ is not an eigenstate but a superposition of eigenstates.

Let α be such an observable, for which thus multiple c_i are non-zero. Then it follows that the result of the eigenvalue measurement of α in the state ψ cannot be predicted with certainty, rather only probabilistic statements about it are possible, no matter how precisely the state ψ and the method used for the eigenvalue measurement may be known.

So the question is: what significance does this interpretation of $\phi \, \alpha \, \psi$ or the equivalent one of the c_i have for the explanatory value of Dirac's theory? (In order to simplify the arguments, I limit myself in the following to the interpretation of the decomposition coefficients.)

The probabilistic interpretation of the decomposition coefficients states the following. Let there be n physical systems, all given in the same state ψ (n very large), and let an eigenvalue measurement of α be performed on each. Then, in $n|c_i|^2$ [systems] one will obtain the eigenvalue belonging to ψ_i. Since more than one c_i is different from 0, it is always the case that $n|c_i|^2 < n$. Therefore it also follows: for systems having the same state, the same eigenvalue measurement leads to different results.

Evidently the description of this purported experiment reaches significantly beyond experience to date. Indeed, however these n systems may be given in experience, no experience rules out that through careful scrutiny one might have discovered already before the measurement new, as yet unknown traits in which the n systems differ. As we have already seen, such distinctions could provide the basis for predicting the different measurement results.

Dirac excludes this possibility, specifically by limiting himself to those states that are defined through 'maximal observations'. A maximal observation here is a system of mutually compatible observations that cannot be supplemented by any further observation compatible with the previous ones. (Two observations are compatible, when neither of them changes the state with respect to the feature of interest to the other observation, so that all knowledge of the outcome of one—regardless of whether it predicts this outcome with certainty or only with a certain probability—is left unaffected by whether or not the other one was performed shortly earlier.)

The concept of maximal observation so defined lacks only one thing: the applicability to the facts of experience. For in order to apply it, we need a criterion to the effect that the given state may include no further trait of any kind that has not yet been captured by previous observation, but which might have been observed without thereby modifying the state with respect to the traits already observed. This is not the place to show the impossibility of such a criterion in general; for the present investigation it is sufficient to note that physics up to now has provided no such criterion. For it follows from this already that the explanatory value of Dirac's theory cannot be affected by dropping the, at least up to now, unrealisable condition that ψ be introduced through maximal observations. On the contrary: only after giving up this condition does it become possible to apply Dirac's symbols to the facts of experience and interpreting these with the help of that formalism.

From the perspective of Dirac's theory one can surely reply, that, notwithstanding this criticism, the concept of maximal observation contains something that is indispensable in the overall structure of the theory. This is correct, as the following consideration shows. If one places no completeness condition at all on the observation defining a state, then two kinds of superpositions of states become conflated—ones that objectively characterise the physical system, and ones that are attributable merely to a deficiency in the observation. Indeed, the fact that a state ψ is not an eigenstate of an observable α but a superposition of its eigenstates can either be due to the fact that the observation that has led to the determination of the state ψ was aborted before what could have been observed in terms of the observables had in fact been observed; on the other hand, it can have its basis in the uncertainty relations. When, say, the position and momentum of an electron are defined by intervals of length dq and dp, then one can ask whether the interval length could have been shortened through more detailed observation without thereby modifying the state. Quantum mechanics shows that the observation was certainly maximal *with respect to its determination of position and momentum* if $dq \cdot dp = h$. Analogous uncertainty relations can be derived from the commutation relations for any two physical quantities with non-commuting observables. They supply at least a sufficient criterion for whether an observation, as I want to put it, is maximal with respect to such a pair of non-commuting observables. An observation is called maximal with respect to certain observables if it is compatible with no further observation that determines one of these observables in terms of a narrower range within its eigenvalue space than the original observation does.

It seems reasonable to replace Dirac's requirement that the ψ occurring in the system be defined through maximal observations with the weaker condition that these ψ, or rather, the observations underlying them, should be maximal with respect to all observables that play a role at all within the framework of quantum mechanics according to the present status of physical research. The crucial difference between quantum mechanics and the classical theory then presents itself clearly: according to classical mechanics, this condition leads to the restriction to states that are eigenstates of all occurring physical quantities; according to quantum mechanics, even under this condition, there are for each ψ certain observables for which ψ is not an eigenstate but a superposition of eigenstates. The range in the eigenvalue space of the relevant observables involved in this superposition (for which, that is, the corresponding eigenstates ψ_i in the decomposition of ψ are assigned coefficients c_i different from 0) is uniquely defined by the observation maximal with respect to these observables that determines ψ. Later observations compatible with this observation, which thus do not change the state with respect to the traits that are the object of the observations, will similarly not change what determines this range. The same is valid—as also follows from the commutation relations—for the decomposition coefficients c_i, and thereby also for the expression $\phi\,\alpha\,\psi$.

If accordingly the c_i are uniquely determined through an observation that is maximal with respect to the physical quantities recognised as fundamental up to now, then one may assume that also the physical meaning of these symbols is determined by the data given in such an observation. Applied to Dirac's interpretation of the decomposition coefficients, this means: the ensemble of physical systems that in an eigenvalue measurement of α display a distribution over the eigenvalues of α corresponding to the $|c_i|^2$, obeys this and only this condition, that they all have the state ψ, and that means here that they must agree with one another and with ψ with respect to those physical quantities that play a role in the present state of physics.

In this version the interpretation of the c_i does not lead to indeterminism. For it is not required that the systems in the ensemble agree with one another with respect to all controllable traits. The possibility of finding traits that distinguish them and of predicting with their help the result of the eigenvalue measurement thus remains open.

Perhaps someone can object that the discovery of such a trait would still conflict with Dirac's interpretation. For if one incorporates the observation of this trait—which by assumption is compatible with the observation that defines ψ—into the definition of ψ, then the c_i no longer have the old probability interpretation, since in fact the representation $\psi = \sum c_i \psi_i$ is not affected by this incorporation, while on the other hand the possibility of predicting the result of the eigenvalue measurement changes.

To this one has to say that according to the interpretation of the c_i introduced here, these probability statements may be assumed *without further confirmation* to hold only if the definition of the ψ takes into account merely the quantities today available to quantum mechanics. The experiment described is changed in an essential respect if one adds to the previous conditions obeyed by the systems in the given ensemble the requirement that these systems must agree with one another also with respect to any newly discovered trait. It will not do to first link the interpretation of the c_i to the domain of the physical quantities known to date, and then modify it with every expansion of this domain. One would then indeed arrive to a new interpretation of the quantum mechanical formalism with each such extension, and would ask again in each case whether this newly interpreted formalism agrees with the facts of experience! That it should no longer do so the instant it is possible to predict the result of the eigenvalue measurement of α, speaks neither against the possibility of such a discovery nor against the fact that in the current interpretation the quantum mechanical formalism does justice to the facts of experience in an admirable way.

He who wishes, may however add to the interpretation rule with the restriction to the system of physical quantities thus far recognised by quantum mechanics also the claim that the numbers $|c_i|^2$ will always represent the given probability distribution no matter what further conditions the systems in the ensemble—in the given experiment—may be subject to. He will not thereby get tangled in contradiction—unless he is refuted in the future by a solution to the problem of predicting eigenvalue measurements—and from this assumption he can draw the necessary conclusion of indeterminism. Only one thing he cannot claim for his assumption: that it is required by any of the experiences to date or by any formal posit in the calculus of quantum mechanics. And therefore it is possible to remove this assumption, and with it indeterminism, from quantum mechanics without entering into conflict with the physical experiences of quantum mechanics and without altering anything in the formalism developed by Dirac. Neither his axiomatic connection of the symbols nor the interpretive association of these symbols with the facts of experience underlying quantum mechanics is affected. Eliminated is only an anticipation of future experience that groundlessly extrapolates beyond the experience to date, or better: the groundless ruling out of a possible discovery not excluded by any experience to date.

Summarising, I can describe the result as follows: if one gives up on the notion of a maximal observation, which lacks a criterion for being applied to the facts of experience and insofar has to be indeed disregarded in the application of the theory, Dirac's interpretation of the expression $\phi\,\alpha\,\psi$ or the equivalent one of the c_i consists of two parts: an interpretational rule that makes reference to the physical quantities considered in the state of the theory up to now, and an anticipation of future experiences that drops this reference to the quantities occurring up to now. *Only the interpretational rule is essential to Dirac's theory. The further-reaching claim instead is not supported by experience and does not contribute to the explanatory value of the theory. But it is this claim, through which indeterminism enters the theory in the first place. If one drops it, one thereby detaches indeterminism from the theory without compromising its explanatory value.*

4. Neumann's Proof of Indeterminism

In his presentation of quantum mechanics, Neumann claims to have proved in a mathematically exact way that the purely statistical character of quantum mechanics cannot be overcome through the discovery of new, previously unnoticed physical quantities; rather, the formalism of quantum mechanics stands in 'forceful logical contradiction' to the assumption of strict causality. It is thus necessary to consider whether this demonstration offers new grounds for indeterminism that I have overlooked.

The very clearly presented proof relies solely on the following assumptions, which in the formalism of quantum mechanics highlight the essential difference of this theory with respect to classical mechanics:

I. There is a bijective correspondence between physical quantities and certain Hermitian operators.
II. If the operator r corresponds to the physical quantity R, the operator s to the physical quantity S, then for any function f the physical quantity $f(R)$ corresponds to the operator $f(r)$, and $R + S$ to the operator $r + s$.

Apart from these assumptions, the following definitions are essential for the proof.

1. A function that assigns values to the physical quantities is called an $Exp(R)$, if
   a) for every R, which by its nature cannot be negative, $Exp(R)$ is ≥ 0, and if further
   b) for any numbers a and b and any physical quantities R and S one always has

$$Exp(aR + bS) = a \cdot Exp(R) + b \cdot Exp(S) .$$

2. An $Exp(R)$ is called dispersion-free if for all R one has

$$s_R = Exp(R^2) - (Exp(R))^2 = 0\,.$$

Neumann's proof leads to the following result: there are no dispersion-free $Exp(R)$. Moreover, it follows in the context of the whole theory that for every $Exp(R)$ the measure s_R of the dispersion satisfies the uncertainty relations.

The proof is mathematically unexceptionable. Hence, in what follows I assume his claim to be established without further discussing the proof. The inevitability of indeterminism for quantum mechanics results from this claim if one applies it to the so-called 'expectation value functions'. Neumann does this, based on the assumption that every expectation value function is an $Exp(R)$.

By an expectation value function $E(R)$ one understands here the following: let A be any ensemble of physical systems; $E(R)$ is the function that assigns to every physical quantity R the expectation value it has in A; it thus denotes the average value of the measurement outcomes one obtains if one carries out eigenvalue measurements of R on the elements of A.

If each such expectation value function is an $Exp(R)$, then it follows from the stated theorem that any arbitrary ensemble of physical systems, by whichever conditions may it be determined, 'disperses' with respect to any pair of physical quantities that have non-commuting operators, and that means here that the eigenvalue measurements of these quantities on the elements of the ensemble lead to different values; and this dispersion cannot be reduced beneath a fixed limit in any subset of the ensemble. It readily follows that under these circumstances, the same unsurmountable limit is set to the prediction of measurement results.

Therefore all depends on the question of whether every expectation value function satisfies the two conditions defining the class of $Exp(R)$. For classical mechanics this is self-evident; in quantum mechanics there arise difficulties regarding the second condition, specifically in the case of physical quantities with non-commuting operators. If one is dealing with physical quantities R and S that obey the laws of classical mechanics, or with quantities in quantum mechanics to which correspond commuting operators, then the value of $aR + bS$ is defined as the sum of the values of aR and bS. From this definition condition b) follows for any expectation value function. However, this definition fails for quantum mechanical quantities that have non-commuting operators, because these quantities are not 'simultaneously measurable' on one and the same physical state. The sum $R + S$ in this case can be defined only indirectly as the quantity corresponding to the sum $r + s$ of the operators belonging to R and S. As Neumann shows in an instructive example, the eigenvalues of $R + S$ in no way need to be sums of those of R and S; this, however, would be necessary to ensure that the proof of condition b) for expectation value functions could be carried over from the classical theory.

Neumann thus needs another proof for quantum mechanics. He finds it in the following consideration: in the formalism of quantum mechanics, the expectation value of a quantity R in the state $[\phi]$ is given by the symbol $(r\phi, \phi)$, where r is the operator belonging to R. Since $((r + s)\phi, \phi) = (r\phi, \phi) + (s\phi, \phi)$ holds for this symbol, the expectation value of a sum is indeed equal to the sum of the expectation values.

Here we stand at the crucial juncture in Neumann's proof of indeterminism, and it turns out it is the same juncture where one needed to apply the critique to Dirac; for Neumann's interpretation of the expression $(r\phi, \phi)$ proves to be formally identical to Dirac's approach to $\phi\alpha\psi$. Hence, the considerations of the foregoing section provide the key to assessing also Neumann's argumentation.

The symbol ϕ of the Neumann formalism can be interpreted through the Schrödinger wave function for physical systems. The expression $(r\phi, \phi)$ is then a number that is uniquely determined by the operator r and this function. If this expression denotes an expectation value, one thus has to be referring correspondingly to systems that are characterised by the fact and only by the fact that they are assigned the same Schrödinger function. This means: for an ensemble of physical systems that are identical in this respect and are subject to no further conditions, $(r\phi, \phi)$ is the average value of the results one obtains through eigenvalue measurements of R.

If one restricts oneself to this interpretation, then one can no longer infer that for any arbitrary ensemble the expectation value of a sum is equal to the sum of the expectation values. Rather, this statement holds only of ensembles whose definition (i.e. the condition imposed on its elements) is based solely upon those physical quantities that play a role in the present state of quantum mechanics, specifically that of determining the Schrödinger function. For these ensembles—but only for these—has Neumann proved the inevitability of dispersion. The question, however, was whether at some time in the progress of research a new hitherto unknown physical trait might not be found through which the dispersion (for at least some physical quantities) could be reduced beneath the scale fixed by the uncertainty relations. Such a discovery, which would provide a cue for predicting the result of eigenvalue measurements of these quantities, is not excluded by Neumann's proof. Indeed, for ensembles of physical systems agreeing with one another besides in the wave function also in terms of such a newly discovered trait, it has not been shown that the expectation value function has the form $(r\phi, \phi)$ and is thus an $Exp(R)$.

Therefore, in terms of the predictability of measurement results we have the following: as Neumann's proof shows, a physicist who *only* knows a given system by its Schrödinger function is bound to limits conforming to the Heisenberg relations in predicting measurement results. More is not proven. In other words: *while this proof shows that the problem of exactly predicting certain measurement results is at present unsolved; it does not show that this problem is insoluble.*

Whoever wants to extract more from Neumann's proof must already assume that $(r\phi, \phi)$ represents the average value for the eigenvalue measurements of R for any ensemble whose elements, besides with respect to ϕ, agree with one another also with respect to arbitrary further conditions. But that all these ensembles have the same average values is an assumption justified neither by previous experience nor by the hitherto confirmed theory of quantum mechanics. Without it, the proof of indeterminism collapses.

5. The Statistical Interpretation of Quantum Mechanics

However logically correct the here described separation of indeterminism from quantum mechanics may be, yet it will give the impression that it is based on an at least aesthetically objectionable idea. It introduces into the interpretation rule for one of the theory's important symbols, Dirac's expression $\phi\alpha\psi$, a reference to the domain of physical quantities to be taken into account *given the current state of physics*. How can the physical interpretation of an expression that evidently represents an objective feature of a given observed system be dependent on the momentary state of physical theory?

Of course it would be easy to eliminate the reference to the present state of research by providing a complete survey of the domain of the quantities to be considered. This domain might potentially also be determined differently depending on the particular observable considered. So for instance it suffices for interpreting the position interval that maximally determines the spatial location of an electron, to consider an ensemble of electrons that agree with the given one with respect to the four quantities q, dq, p, dp. However, this does not change the unsatisfactory situation that the interpretation of the position interval, which according to the entire framework of the theory is objectively determined by the state of the given electron, can only be displayed in a large ensemble of electrons, whose states need not even agree in every respect. If the case obtains, which is not ruled out by anything in the theory, in which a trait is found that allows an exact prediction of the position measurement, then for an ensemble of electrons agreeing with one another not only in their position and momentum intervals but also regarding this property, the position interval (q, dq) ascribed to them would have no intuitively apparent meaning.

Therefore, difficulties of understanding stand in the way of conceiving of a quantum mechanics freed from indeterminism. They do not in fact pertain to the possibility of detaching indeterminism from quantum mechanics, but they raise specific questions on whose answer depends an also intuitively satisfactory interpretation of the theory.

Addressing these questions goes beyond the aim of the present investigations. In what follows, I want to discuss them only inasmuch as the preceding considerations provide a cue that might potentially lead to their solution. I limit myself here to the consideration of an electron whose position and momentum data are maximally specified by the intervals (q, dq) and (p, dp) in accordance with the Heisenberg relations.

In connection with the experience-based conception of the duality of wave and particle pictures, on which the uncertainty relations are grounded, we had been led to the assumption that within the position interval, the electron would not possess a 'more exact' position. The simplest intuitive picture that positively conveys this notion consists surely in imagining that the whole interval (q, dq) is filled with the smeared-out electron. Similarly one arrives at the notion that this electron has no single momentum, but rather is an object diffusing like a wave-peak, whose momentum values fill the interval (p, dp). Then the expressions $|c_i|^2$ of Dirac's theory give the density, as it were, with which a definite point of this position or momentum interval is occupied by the electron.

In this approach, the probabilistic statements drop out *of the interpretation* of the uncertainty intervals and decomposition coefficients, and appear—only insofar as they are justified by experience—as theorems of quantum mechanics. Thus the difficulties mentioned above do not arise here. This approach coincides with the one originally advocated by Schrödinger for the interpretation of the wave function $\psi(q)$, according to which $e|\psi(q)|^2$ represents the charge density of the electron at the point q (e total charge of the electron).

Schrödinger himself has subsequently raised doubts against this view. It feigns, so he argues, a return to the concepts of classical mechanics, while it is clear that the motion of the electron cannot be described as the motion of a mass smeared out over the electron's position interval and following the laws of classical mechanics.

The objection is correct; however, it merely means that the conception of 'smearing out' the electron over its position and momentum intervals is insufficient to recover the quantum mechanical approach. An assumption must be added that makes clear that the motion of a smeared-out electron is not that of a correspondingly extended mechanical system. The direction in which one has to search for such an assumption emerges from the analysis of the question: what does it really mean to say that physical systems are limited in terms of, say, their position and momentum data, by Heisenberg's uncertainty relations? Why is it not allowed, e.g. to treat a system distributed over the spatial interval (q, dq) as subdivided into two parts, each of which comprises half the interval? The answer is simple: since neither of these subsystems could exhibit the experimentally proven duality of wave and corpuscle properties, it cannot itself be the subject of a physical observation. This evidently means that only the whole system, satisfying the Heisenberg relations with respect to its position and momentum intervals, can enter into interaction with other physical systems. In any observed physical effect of a system, thus, the system is always participating as a whole. Thus, e.g., one may not describe the force that is exerted by one system on another by integrating over the forces issuing from arbitrarily small parts of the system. The classical assumption whereby interaction between masses can be reduced to interaction between point masses, is thus supplanted by a more complicated assumption in quantum mechanics: interaction between physical systems presupposes that each of them exhibit an extension corresponding to the Heisenberg relations.

Since every observation of a system constitutes an effect of the system on the observer, it follows that the object of an observation can only be a system that is subject to these relations. Now if one performs an eigenvalue measurement on it—say of position (defined by an interval before the measurement)—then this measurement transforms the system into another that has one single position but is distributed over the entire eigenvalue space of momentum. The laws according to which this transformation takes place in detail, which therefore determine which position is then observed, are at present unknown. However, it is not impossible that one of these days physics will get onto their trail.

References

Born M (1929) Über den Sinn der physikalischen Theorien. Die Naturwissenschaften 17(7):109–118

Heisenberg W (1930) Die physikalischen Prinzipien der Quantentheorie. S. Hirzel, Leipzig

Herrmann K (ed) (2017) Grete Henry-Hermann: Philosophie–Mathematik–Quantenmechanik. Springer, Berlin

von Laue M (1932) Zu den Erörterungen über Kausalität. Die Naturwissenschaften 20(51):915–916

Chapter 15
Natural-Philosophical Foundations of Quantum Mechanics

Grete Hermann

Preface

The fruitful progress of investigations in natural philosophy has always depended on working out the philosophical difficulties that arise during the development of physical theories in close contact with physical considerations—a condition that is ever harder to satisfy today given the widely established estrangement between physics and philosophy, as it presupposes a mutual understanding between representatives of both fields of research. In the physics institute at Leipzig I had the opportunity to pursue the problems in natural philosophy raised by quantum mechanics by engaging with the physics circle there, and I thank therefore here above all Professor Heisenberg for his willingness to discuss the foundations of quantum mechanics, which was crucial in helping the present investigations.

March 1935 Grete Hermann

Introduction

In its bold and successful advance, modern physics has shaken positions that in the classical theories were still considered unassailable foundations for any natural science—a judgement that critical philosophy, in its doctrine of the a priori principles of experience, had given a natural-philosophical interpretation and justification.

Translated by Elise Crull and Guido Bacciagaluppi. Originally published as Hermann (1935a); this paper was reprinted separately for Verlag Öffentliches Leben (Inh. Erich Irmer, Berlin 1935), with the addition of a Preface by Hermann (included here), and a table of contents (not included) (Hermann 1935b). Excerpts of Sects. 1, 2, 3, 8, 9, 10 and 12 (with minor variants) were also published in Hermann (1935b) under the same title. The original version of the paper will also be included in the forthcoming volume by Herrmann (2017).

Translators' note: we have translated the following Kantian terms in accordance with Eisler (1930): Anschauung as intuition; Einheit as oneness; Erkennen as perception; Erkenntnis as knowledge; mannigfaltig as manifold; Spaltung as splitting; unmittelbar as direct; Wahrnehmung as sensation. The format of the references has been mostly modernised, and a list of References has been created.

E. Crull and G. Bacciagaluppi (eds.), *Grete Hermann - Between Physics and Philosophy*, Studies in History and Philosophy of Science 42,
DOI 10.1007/978-94-024-0970-3_15

It is said that experience has decided against this standpoint; one celebrates the physical achievements of this century as the great victory of experience triumphing over all preconceived opinions, as the delivery from prejudices that empirically obtained conjectures had dressed in the luminous but deceptive garb of eternal truths. Einstein declares that, forced by the facts, he has got down the concepts of space and time 'from the Olympus of the a priori in order to overhaul them and return them to a usable state'. The exponents of quantum mechanics endorse analogous proposals for amending the law of causality.

As inappropriate as it would be to give up as obsolete the philosophical discussion of those a priori principles in the face of these judgements based on purely physical considerations, so would it hardly do justice to the problems arising here were the philosopher to respond to the *physical* critique only with a critical recapitulation of the *philosophical* deduction of a priori principles. For, even if the physical development of the theory is not sufficient to put the foundations of the thus achieved knowledge of nature into the sharp light of awareness, still the scientific progress that has been obtained in these theories precisely through the willingness to abandon or revise old familiar concepts provides the guarantee that new and fruitful points of view have been introduced here into research. Only their philosophical interpretation and elaboration will produce clarity concerning both the philosophical arguments for the a-prioricity of natural-philosophical principles and the objections to them arising from the side of physics.

Viewed in this light, the attack from physics on traditional philosophical opinions represents a valuable stimulation and fertilisation of philosophical work—a welcome opportunity to put new lessons from experience to use for the old investigations of the foundations of natural knowledge.

The aim of the present work is to take on this task inasmuch as it is posed by the development of quantum mechanics, that is in the first place to scrutinise the revision of the law of causality announced by the theory.

The course of the investigation is already fixed by the preceding: to be discussed, starting from the lessons of experience, are the natural-philosophical implications of the physical achievements. The route of quantum mechanics, from the observations and experiments that have led to the crucial new questions and difficulties, to the results that have strictly physically proved useful in answering these questions and resolving these difficulties, thus forms the subject of our investigations.

We are not concerned with a *physical* critique of this route. The way out frequently suggested by physicists to the friends of the law of causality—that of finding a fault with quantum mechanics in the physical domain through experimental verification and extension, whose correction, if luck will have it, will at the same time clear up the concerns about the law of causality—[such a way out] is not of interest for the present investigations. For—besides the fact that the hope for such corrections would only have its basis in a shrinking away from the philosophical difficulties springing from quantum mechanics and not rather in physical indications of the necessity of modifications—even if such a revision were justified, it would only uphold the validity of the law of causality by empirical means. Anyone who thereby claims that the upset that came from quantum mechanics is overcome and the law of causality is trustworthy again, even in so doing has left the decision concerning this law to the mercy of experience, of observation and experiment, and has let the thesis of its a priori character die without so much as a whimper. Since in the following we are concerned precisely with subjecting this thesis to a critique based on the quantum mechanical results, and since this critique does not depend on whether and how the ideas of quantum mechanics are reshaped by further progress, we will proceed on the assumption that drastic corrections are no longer required to the presently accepted edifice of quantum mechanics.

This does not mean that the portrayals of the theory given by physicists should be taken on without scrutiny. For these portrayals—understandably—far exceed what has directly proved itself reliable through successful predictions or the correct rendering of observed data. This direct empirical corroboration pertains strictly speaking only to the formalism of the theory and its correspondence to the data of observation. The conceptual interpretation of the formalism that must necessarily accompany it in every physical theory only receives direct physical corroboration in as far as the corresponding formalism proves to be an appropriate means of describing observed events. Where this interpretation is ambiguous, then, one can only make a physical-empirical decision between the different views if they give rise to different extensions of the formalism.

Chapter I. The Limits of Predictability

§1. Causality and Predictability

The physical result from which the unsettling of the hitherto dominant causal concepts has arisen states that a sharp, insurmountable limit is set to the predictability of future natural processes. The idea of Laplace's demon—who completely knows the present state of nature, has complete insight into all natural laws, and because of this knowledge can foresee the future course of events in full detail—loses thereby every application to nature. It is no longer the unattained ideal that is nevertheless to be aspired at beyond all limits, but proves to be a phantom that has been rejected through the progress of experience as a construct not corresponding to nature. And yet this idea was only the expression of the certain conviction that every natural event in all its features has been caused by preceding events, and thus must be predictable from these causes for one knowledgeable of the natural laws. Together with the faith in the unlimited possibility of such predictions, also the conviction in the pervasive causal connection of natural events thus comes to falter.

Defenders of the law of causality have occasionally sought to secure its validity in the face of this scepticism, in that they have disputed the connection between the discovered limits of predictability and those of the law of causality, roughly on the basis of the following argument: Heisenberg's uncertainty relations have shown that in-principle insurmountable limits are set to the observation of atomic processes. From this it follows that the physicist, no matter how precise the observation of a process may be, cannot fully perceive the causes for certain changes, and accordingly cannot predict these changes with certainty. This proof of the unpredictability of certain measurement results is based upon that of the unknowability of its causes, and precisely because of that presupposes these causes are real; hence it is absurd to subsequently infer from this unpredictability to the *non-existence* of causes and thus to the failure of the law of causality.

As much as this criticism is prompted, even necessitated, by some expositions of quantum mechanics—especially popular ones—very little is won by it for the status of the law of causality. Removing the indicated circularity does not resolve the conflict that arises for the assumption of pervasive causality from the fact that—according to the claims of quantum mechanics—the *criterion* for causal connection affords only limited usefulness. This criterion is the ability to predict future observations. For on and only on this basis can one judge whether a given physical hypothesis correctly reproduces the appearances as connected through natural law, that events can be predicted from it whose occurrence can be checked through observations. What then, if, for whatever reason, limits are set to the possibility of such predictions? One who wished to brush this off with the excuse that, while the *knowledge* of the causes determining the processes is limited, the *existence* of such causes is not put in doubt, removes the law of causality from the realm of the principles governing natural knowledge into that of mysticism. Where it is impossible in principle to decide what falls under a given concept in nature, the statement *that* anything falls under it also loses its meaning.

Physical research has itself long embraced this principle, and handles it with absolute assurance; instead it has entered philosophical considerations almost only in the positivistically distorted form, according to which all physically relevant concepts are taken from observation and should accordingly be unrestrictedly applicable to it.

The fate of the law of causality therefore indeed depends on whether and how far—according to the claims of quantum mechanics—an in-principle limit is set for possible future predictions. To be discussed in the following are thus the meaning and implications of the observations underlying this claim and of the rules of the formalism that are based on them.

§2. Duality Experiments and Uncertainty Relations

The experimental catalyst for considerations culminating in the claim that there are insurmountable limits to predictability was given by a series of experiments that were entirely unintelligible within the framework of the classical theory.[1]

The classical theory distinguishes between radiation processes that consist in the rapid movement of small material particles, of corpuscles, and those in which a wave propagates. Corpuscular rays are recognisable by the discrete paths of the individual particles, by their deflection in electric and magnetic fields, from which the mass, velocity and electrical charge of the particles can be calculated. In contrast, criteria for wave processes are provided by the phenomena of interference, which are only intelligible under the assumption of a wave motion extended over a wide region of space, and from which the wavelength can be ascertained.

Compelled by experiment and observation, quantum mechanics has given up on this separation between the two kinds of radiation processes. It has found characteristic features of both pictures in the same processes, indeed both in those processes that according to the classical theory have corpuscular character and in those that were previously considered wave motions. Matter rays, like those emitted by radioactive elements—which leave linear traces behind when passing through saturated water vapour, thereby showing the discrete character of the particles here in motion, and which furthermore display in electric and magnetic fields all the features of corpuscular processes—lead on the other hand, when they seep through a grating or are reflected by it, to definite interference phenomena, and thus force the researcher to assume one is dealing with a wave process. In a similar way, light rays, which since the discovery of interference phenomena were unequivocally interpreted as a wave motion, have displayed properties that allow one to infer to their corpuscular nature. One example is the photoelectric effect: electrons that are emitted from a metal plate when it is irradiated with ultraviolet light present quite the appearance as if the plate has been hit by a shower of discrete particles—Einstein's 'light quanta'—and not by a wave train. The velocity of these electrons and thereby the kinetic energy they carry, is independent of the intensity of the light, thus independent of whether strong or weak light, much or little energy has hit the plate; it is solely determined by the number of oscillations of the light—by its frequency. The intensity of the light instead affects the number of ejected particles. And this means: the intensity does not appear to determine the strength of the impact on a point of the plate, rather the number of discrete impacts sustained by the plate—and thus the number of impacting particles in the light ray—while the energy carried by a single such particle appears to be independent of the intensity of the light and determined by its frequency. Therefore the experiment necessitates a corpuscular interpretation of the propagation of light.

Quantum mechanics does justice to these experiments—which allow one to distinguish discrete corpuscles in the propagation of a wave and demonstrate the wave-like character of processes that initially appear as corpuscular—through the assumption

[1] In the following presentation, I essentially follow the treatment by Heisenberg (1930).

that each atomic process must be representable *also* in the wave picture, each wave process *also* as corpuscular. Since, however, given that the two pictures are contradictory, it is impossible that one and the same process can have all the traits of a propagating wave as well as all the features of the motion of a corpuscle—otherwise it would, for instance, have to be spread out over the whole of space as well as being constrained to definite discrete points—compatibility of the two pictures is only possible through the applicability of each one limiting that of the other. A process that displays traits of both pictures can be neither a wave motion nor a corpuscular ray, but only in certain respects behave like a wave, in others like the motion of a corpuscle.

Heisenberg has managed to give mathematical expression to this thought, and to determine exactly in the oft-cited uncertainty relations the limitations that wave and particle picture, applied to the same physical process, impose on each other. The best-known of them forbids the simultaneous sharp determination of the position and the momentum of particles in applying the corpuscle picture: if Δq is the precision with which the position, say of an electron, is determined, and Δp is the precision with which its momentum is determined, then the relation $\Delta q \cdot \Delta p \geq h$ holds, where h is Planck's constant. A similar limitation holds in the wave picture for the simultaneous determination of the electrical and magnetic field strengths. Similarly, other uncertainty relations pair together two variables in each picture in such a way that the measurement of one limits the precision of the measurement of the other, and even completely removes the possibility of such a measurement in the limiting case.

In contrast to the classical theory, the formalism of quantum mechanics precludes thus that the different physical quantities of a same physical system may be determined with arbitrary precision independently of one another; the determination of one curtails the possibility of precise determination of the other and completely cancels it out in the limiting case. In other words: the determination of one quantity implies a statement about another—not in fact a statement about its value, but indeed one about the limit within which it is still determinable. The independence of the measurability of the two quantities that holds in the classical theory is thus upset. Whereas classically the state of a system can be expressed through a mere enumeration of the values of all occurring physical quantities, the quantum mechanical formalism employs novel symbols in the description of the state that express the mutual dependence of the determination of different quantities.

These symbols, the wave functions of physical systems, and the mathematical formalism that prescribes the correct rules for their combination, follow the classical theory closely because of Bohr's *correspondence principle*. The classical description is compatible with the quantum mechanical one insofar as its quantities remain undetermined to such a degree that the uncertainty relations are fulfilled. All predictions that can be derived classically from such classical data are preserved in quantum mechanics; conversely, any prediction obtained from the quantum mechanical formalism can be interpreted through the classical concepts of the wave or corpuscle picture, whose application, however, finds its limits in the uncertainty relations. The quantum mechanical formalism is legitimised through this correspondence as the natural extension of the classical theory imposed on it by the duality of the wave and the corpuscle description.

On the other hand, this correspondence entails that the quantum mechanical formalism does not allow one to predict the result of a measurement with arbitrary precision, but only supplies more or less far-reaching probabilistic statements depending on the wave function characterising the physical system before the measurement; for, precisely because of the correspondence principle, these predictions can go no further than those that can be derived classically from the classical description of the process in the wave or in the corpuscle picture, respectively. But since both pictures are only of limited applicability, so, too, the predictions derived from them are necessarily limited; they leave the result indeterminate within an interval corresponding to their own uncertainty.

In the formalism of quantum mechanics this is expressed by the fact that the symbols peculiar to it correspond to the data of observation through probabilistic statements. The modulus of the Schrödinger wave function at any one point in space determines the probability with which a corpuscle can be found at that point.[2] For any physical quantity other than position, the Schrödinger function can be converted through a suitable transformation into another function whose modulus determines the probability that a measurement of this quantity yields a certain value.

§3. Uncertainty Relations and Predictability

How far are we hereby into the investigation of the quantum mechanical formalism? We were seeking in it grounds for the thesis that *in-principle insurmountable limits* are imposed on the prediction of the future. We have shown *that this formalism itself* provides the basis *only for limited predictions*, that, to the question about the result of a future observation, it answers as a rule only with probabilistic statements, not by exhibiting this result with certainty. The given task is not completed with this proof: the insurmountability of the indicated limits is neither guaranteed nor disproved by it. Even if—as we assume here—the formalism will prove itself also in the future as it has so far, what prevents us from assuming that through an extension of physical knowledge it may be augmented with new formulas and rules that, together with the current formal approach, make precise predictions possible again?

Everything here depends on answering this question. Not because searching for such a completion would be of physical interest; the fact that in the existing material of experience one does not find the smallest hint for such a development, rejects as fruitless any attempt to advance in physics by such hypotheses, at least for the time being. However, the understanding of the novel natural-philosophical situation featured in quantum mechanics, heralded precisely by the thesis about the in-principle limitation of possible predictions, can only be achieved by one who is able to fully piece together the reasons for this thesis from the structure of the formalism, and to assess their weight.

---

[2]If the correctly normalised wave function of an electron is $\phi(x, y, z)$, then $|\phi(x, y, z)|^2 \Delta x \cdot \Delta y \cdot \Delta z$ is the probability of finding the electron in the interval $\Delta x \cdot \Delta y \cdot \Delta z$.

To this purpose, it lies at hand first of all to consider the points of the formalism at which the imposed limits appear—the uncertainty relations and the probabilistic interpretation of the Schrödinger functions based on them—and to examine whether they already show more than the merely provisional limitation of future predictions.

The uncertainty relations appear to say more, in fact. If position and momentum of a particle both cannot fundamentally be measured with arbitrary exactness, how then shall one obtain definite statements about the future motion, which is determined precisely by the instantaneous position and momentum of the body?

But this argumentation is based on the notion mentioned once already that irrespective of the uncertainty relations, the classical conception that describes the electron as a corpuscle remains valid in full detail. According to this conception, the electron has at each time an exact position and an exactly determinate momentum which—up to external disturbances—determine its future motion. The uncertainty relations state in this understanding only that *this cause* of the forthcoming physical development is unfortunately always hidden from the searching gaze of the physicist. Thus these relations are only subjectively interpreted, and appear to say nothing about the nature of electrons and other physical systems.

However, this subjective interpretation is incompatible with the derivation of the uncertainty relations from the duality of the wave and particle conception: the required subsumption of every atomic process *also* under the features of the wave picture is only possible by limiting the application of the corpuscle picture, i. e., in that not all features of point masses in motion—in the classical sense—may also be properties of the electron in flight. The uncertainty relations show where the applicability of the corpuscle theory ceases; they also state that an electron cannot have simultaneously something like a sharply defined position and a sharply defined momentum. In the words of Heisenberg: any more precise use of the words 'position', 'velocity' going beyond the uncertainty relations 'is just as vacuous as the application of words whose meaning has not been defined'.[3] What is *positively* determined herewith about the nature of the electron—whether and how, say, one can conceive of an electron that has no exact position—we must here still leave open. For the time being it is only definite that the assumption that an electron has simultaneously exact determinations of position and momentum, even though these are inaccessible to observation, contradicts the duality experiments.

But if according to these considerations the electron does not simultaneously have an exact position and an exact momentum, then how should its exact position and its exact momentum be determining its further motion? Dropping this assumption that the exact position and the exact momentum of an electron at a definite instant are crucial for its further motion thus opens the door to the question of whether or not one might find other features upon which the course of the motion depends and from which it can be calculated. The formalism of quantum mechanics knows no features that should make such a calculation possible. But with what right does it anticipate future research and declare the attempt to try to find such features moot from the start?

[3] Heisenberg (1930, p. 11).

The urgency of these questions comes out all the more sharply when one considers the general chain of thought upon which it is based. It is the thought that, given the openness of experience, the possibility of finding the explaining cause of a hitherto unexplained natural process never completely disappears. For as long as not all in-principle detectable features have been tested for whether they determine the process in question, one cannot rule out that this review of physical features might lead to success. But it is impossible to obtain an overview of all in-principle detectable features in nature. And so one who, in such a case, nevertheless declares further research into explanatory causes to be fruitless in principle, is faced with the difficult task of proving the impossibility of such explanatory features independently of the future lessons of experience. The argument from the uncertainty relations that the position and momentum of an electron do not fully determine its further motion in no way satisfies the requirements of this proof.

§4. The Failure of the Statistical Arguments[4]

A different, commonly advanced argument meant to prove fundamentally mistaken the attempt to obtain arbitrarily exact predictions of the future fails at exactly the same point. (However, this argument uses, besides the uncertainty relations and the formal approach depending on them, also further physical results—thus already for this reason cannot be used to clarify the natural-philosophical consequences of the duality experiments and the uncertainty relations.)

Certain physical experiences make it necessary to posit that the wave functions of physical systems consisting of a whole ensemble [Schar] of particles be symmetric or anti-symmetric with respect to these particles; symmetric in the case of systems of light quanta, anti-symmetric in the case of systems of electrons. These are the experiments that led to the Pauli exclusion principle (according to which no two electrons in an atom can coincide in all quantum numbers) and those that led Planck in the calculation of his radiation law (which represents the capacity of black bodies to emit light as a function of the absolute temperature of the body and the wavelength of the radiated light).

Accordingly, for systems of electrons—or respectively, of light quanta—only such wave functions are allowed as remain unchanged (possibly up to a factor of -1, which is negligible for the characterisation of the physical state of a system) under the exchange of two elements. In other words: there is no change in the state of such a system if the states of two electrons or two light quanta, respectively, are interchanged. This means, however, that these corpuscles are indistinguishable from one another—that it is meaningless to treat them as individuals within this system, any of which one could, on the basis of any features whatsoever, follow in its individual fate and re-identify again and again within the ensemble of particles. This absence of individuality finds its mathematically sharpest expression in the statistical laws that determine the probability of the states of such systems: the formalism prescribes Einstein–Bose statistics for systems of light quanta, Fermi statistics for systems of electrons. Both have in common that the different possibilities that formally result from an arbitrary state of the system by a permutation of the particles are not treated as different cases, but only as a single case in the counting of possible cases.

[4]This section, as with later passages in small print, contains discussion of physically difficult arguments. These passages are not necessary for the comprehension of the subsequent considerations.

However, if the particles of such systems possess no individuality in this sense, then they obviously also cannot exhibit features that have hitherto escaped research and by which they can be more closely determined than the quantum mechanical formalism already does. For indeed, such features would directly provide the opportunity for distinguishing the particles from one another and tracing any of their state changes within the system. The hope of being able, through the discovery of such features, to exactly predict the result of future measurements on the electrons or light quanta in such systems is therefore already shown by current experience to be misplaced.

As correct as this reasoning may be, on closer examination however it says nothing against the possibility of predicting on the basis of newly discovered features the measurement outcomes not yet predictable today. What is proven is only that systems of electrons and light quanta may not be interpreted as ensembles of corpuscles in the strict sense of the word, and this because their alleged elements are not individuals distinguishable from each other. Also, if such a system has originated from ensembles of electrons or light quanta, or if in later experiments such particles can be extracted from it, nevertheless it evidently cannot be interpreted as a side-by-side of independent and distinguishable particles. Thus e.g. the photoelectric effect shows how, upon the impingement of ultraviolet light on a metal plate, individual light quanta become effective. Nevertheless, the decomposition appropriate to the nature of light rays is not into individual corpuscles, but into monochromatic components. Although the intensity of a component represents the number of light quanta corresponding to this wavelength, one cannot therefore trace single corpuscles within such a component, each possessing an individual fate.

Although it follows from this that future observations of light quanta or electrons—extracted from such systems through some experiment and about whose future fate exact predictions apparently are impossible—certainly cannot be predicted from provisionally still unknown features of *these particles*, this is so just because the ray strictly speaking is not a system of individual particles at all. If it is not, however, then one cannot conclude from the absence of further distinguishing features of the particles that their different later fates should have been in-principle unpredictable. The decomposition of the ray into a system of light quanta or electrons has proven an analogy of only limited use that does not allow one to recognise already now within the system some particle that is perhaps later extracted from the system. But this prompts the question—again thanks to the openness of experience—whether hitherto unknown properties of the whole system could not prove to be determining for what happens later to such an extracted light quantum or electron. What should exclude the possibility of discovering such features? The appeal to the peculiarity of statistical approaches yields no answer to this.

§5. The Probability Interpretation of Wave Functions

One is led to similar considerations in discussions of the other point, the probability interpretation of the wave function: let $\phi(x, y, z)$ be the wave function of an electron, attributing no precise location to it but rather only fixing an interval within which the electron is found upon a position measurement. The modulus of the function at a point in this interval determines the probability of finding the electron exactly there.

Now, such a probability statement is always a statement about a whole *ensemble* of physical systems. It claims that for a large number of systems whose states are characterised throughout by the same wave function, the results of sharp position measurements are distributed over the position interval in question in proportion to the given probabilities. These measurements thus provide *different* results for the elements of this ensemble; expressed differently: this ensemble of physical systems exhibits a 'dispersion' with respect to the position measurement, whose measure is fixed by the wave function.

If one takes the probability interpretation of the wave function to be that *every* set of physical systems in the state ϕ disperses under position measurements to the extent corresponding to this function, then the possibility of exactly predicting the result of a position measurement is quite excluded. For if one could discern in a system, on the basis of any features whatsoever, already before the position measurement what result it will lead to, then an ensemble of systems that agree, besides in the wave function, also in these other features would exhibit throughout the same result in a position measurement; contrary to the requirements of the wave function it would be dispersion-free with respect to these measurements.

A curious proof! Supposed we had investigated an ensemble of electrons that all have the same wave function. Position measurements made on every element of this ensemble have yielded the distribution over a spatial interval corresponding to this wave function. In the age of classical physics, this result would have given grounds for thinking that the electrons of the ensemble were only apparently in the same state before the measurement, despite their agreement with respect to the wave function. The physicist would have made the attempt to find new features in them, already at a time before the measurement, that had escaped previous enquiry, with respect to which they could be distinguished and which would have provided the basis for predicting the different results of the measurements. Based on what experiences does quantum mechanics reject this attempt as futile from the start, without performing it?

The probability interpretation of the wave function, which apparently requires this renunciation, is an expression of the correspondence between the quantum mechanical statements and those classical statements allowed within the scope of the uncertainty relations. Because the application of the wave function to the data of experience is mediated solely through the classical conceptions of the wave or corpuscle picture, only those predictions that can be extracted from the interpretation of the wave function given a limited applicability of these pictures will result. Concerning the outcome of the position measurement of an electron, on the basis of its wave function one can only say what one would know of it were it a corpuscle in the strict sense of the word but with only an incompletely known position: one could only give with certainty an interval within which the electron is to be found; however, it is only possible to state probabilistically where it is within this interval. And this means: for a large ensemble of such corpuscles whose positions are subject *to the condition but only to the condition* that they are located within the interval in question, one has to expect a distribution over the whole interval corresponding to these given probabilities. Instead, one should already no longer expect this distribution for an ensemble that, in addition to this one, is subject to further conditions—say that its members have entered the prescribed spatial interval a short time earlier through a narrow slit.

The consequences for the probability interpretation of the wave function: It also can only mean, in order to be applicable to the current facts of experience, that position measurements on a large ensemble of physical systems selected *by and only by the criterion* that they all possess the same wave function ϕ, exhibit the distribution determined by ϕ. In this precise version, however, the probability interpretation of the wave function is no longer able to rule out the possibility of thoroughly exact predictions for such measurement results. For it stipulates something only for those ensembles of physical systems that are already open to analysis in the present state of research, and defers the question of whether, on the basis of any hitherto unknown features, from these ensembles one might extract sub-ensembles in which position measurements would lead to a different distribution of results—unless one may rule out the possibility of such features once and for all through some conclusive proof.

§6. What Are 'Maximal Observations'?

In his closed development of the formalism,[5] Dirac has cut off the prospect for the discovery of such features by virtue of the fact that his quantum mechanical characterisations refer only to so-called 'maximal observations'. By this he means observations that cannot be sharpened or extended without the influence of the measurement apparatus disturbing the observed system and thereby rendering obsolete the results of existing observations.

That observations can disturb each other in this way is a simple consequence of the uncertainty relations. They show that with certain pairs of physical quantities, as e.g. with position and momentum of an electron, the exact measurement of the one precludes that of the other. Thus a measurement of such quantities has taken place with maximal exactness, when the indeterminacy that still inheres in them after the measurement satisfies the Heisenberg relations exactly.

In contrast to classical physics, which assumed in principle the possibility of completely sharp observations of all physical quantities—thus for which the maximal sharpness of observation was achieved only when each quantity was determined univocally and by a definite numerical value—in quantum mechanics there are inevitably determinations of quantities that specify only an interval in which the quantity in question lies. In this case it is necessary to distinguish whether the specification of this interval is only the expression of ignorance of the situation, thus is based on an inadequacy of the observation and could have been replaced by more exact specifications under more exact measurements, or whether it is the result of an observation of the quantity in question carried out with maximal sharpness, and thus represents in itself a physically significant statement about the system.

In this sense, however, one can only say that the characterisation of the physical system is maximal *with respect to a pair of physical quantities*, i.e. that in the measurement of these quantities it has achieved the highest sharpness of observation

[5] Dirac (1930).

allowed by the uncertainty relations; but it is quite another matter to require unqualifiedly maximal observations, in the sense that an observation extending without disturbance what was previously observed is absolutely impossible—either through the sharper observation of the quantities already taken into account for the characterisation of the system, or through the discovery and measurement of new quantities, hitherto overlooked by research, that determine the state of the system as well. It is this requirement, however, that Dirac needs in order to reject, invoking the probability interpretation of the wave functions, any possibility of predicting future processes arbitrarily exactly on the basis of newly discovered physical features. But one may not directly infer the one interpretation of maximal observations from the other: as important as it may be for the physical theory to start from maximal observations with respect to all investigated physical quantities, the condition thereby imposed does not exclude in the least that ongoing research may discover new physical quantities not considered hitherto. If a physical system is maximally determined with respect to specification of its position and its momentum in the manner described here, then it is indeed impossible to make the position measurement more precise without thereby upsetting the result of the earlier momentum measurement, and vice versa; but it is not yet proven impossible by the previous considerations that no new features of the system might be observed without thereby rendering obsolete the specifications of position and momentum.

Dirac's requirement of maximal observations thus prompts the question of the *applicability* of this concept: where lies the criterion for deciding whether the variables of a physical system are maximal, not just with respect to the observed quantities—whether they are so can be decided on the basis of the uncertainty relations—but rather maximal in an absolute sense, and so may not be expanded through new discoveries? Dirac does not provide such a criterion.

§7. The Circle in Neumann's Proof

In fact, there is no lack of efforts to prove the impossibility in principle of such discoveries that might again provide a possibility of exact predictions of all measurement results. This proof is worked out particularly thoroughly in Neumann's mathematical development of the formalism.[6] But a detailed assessment shows here, too, that this mathematically otherwise faultless argumentation introduces into its formal assumptions, without justification, a statement equivalent to the thesis to be proven. It is contained in the following consideration. Let there be given some ensemble of physical systems, and let $\mathfrak{R}$ and $\mathfrak{S}$ be physical quantities that can be measured on the systems of this ensemble: under the expectation value of $\mathfrak{R}$, $Exp(\mathfrak{R})$ shall be understood the mean value of the measurement results that arise from an $\mathfrak{R}$-measurement on all systems of the ensemble, thus the value that is to be expected as the probable result of an $\mathfrak{R}$-measurement on any not further specified element of the ensemble. For the expectation value function $Exp(\mathfrak{R})$ thus defined by means of an ensemble of physical systems, which assigns a number to every physical quantity, Neumann assumes that

[6]Neumann: 'Mathematische Grundlagen der Quantenmechanik'. Berlin, 1930 [sic] [von Neumann (1932)].

$Exp(\mathfrak{R} + \mathfrak{S}) = Exp(\mathfrak{R}) + Exp(\mathfrak{S})$. In words: *the expectation value of a sum of physical quantities is equal to the sum of the expectation values of the two quantities*. Neumann's proof stands or falls with this assumption.

For classical physics this assumption is trivial. So, too, it is for those quantum mechanical features that do not mutually limit each other's measurability, thus between which there are no uncertainty relations. Because for two such quantities, the value of their sum is nothing other than the sum of the values that each of them separately takes, from which follows immediately the same relation for the mean values of these magnitudes. The relation is, however, not self-evident for quantum mechanical quantities between which uncertainty relations hold, and in fact for the reason that the sum of two such quantities is not immediately defined at all: since a sharp measurement of one of them excludes that of the other, so that the two quantities cannot simultaneously assume sharp values, the usual definition of the sum of two quantities is not applicable. Only by the detour over certain mathematical operators assigned to these quantities does the formalism introduce the concept of a sum also for such quantities.

However, for the so-defined concept of the sum of two quantities that are not simultaneously measurable, the formula given above requires a proof. Neumann carries it out in two steps: since each ensemble of physical systems can be decomposed into sub-ensembles whose elements agree with each other in terms of their wave functions, then it follows, first, that the theorem in question needs to be proved only for ensembles whose elements satisfy the condition of equal wave functions. But for these ensembles Neumann relies on the fact that, in the context of the formalism, the rule $((R + S)\varphi, \varphi) = (R\varphi, \varphi) + (S\varphi, \varphi)$ holds for the symbol $(R\varphi, \varphi)$, which represents a number and is interpreted as the expectation value of the quantity $\mathfrak{R}$ in the state φ. (Here R and S are the mathematical operators assigned to the quantities $\mathfrak{R}$ and $\mathfrak{S}$; φ specifies the wave function of the systems under consideration.) From this rule Neumann concludes that for ensembles of systems with equal wave functions, and therefore for all ensembles generally, the addition theorem for expectation values holds also for quantities that are not simultaneously measurable.

The interpretation of the expression $(R\varphi, \varphi)$ is here crucial for the whole proof. To posit that it indicates the expectation value of the quantity $\mathfrak{R}$ on systems in the state φ, essentially—as the formalism shows—amounts to the same as the probability interpretation of wave functions. The considerations relating to this interpretation can thus be carried over without further ado: until the proof of the impossibility of new variables—which has yet to be given here—the expression $(R\varphi, \varphi)$ may denote the expectation value of $\mathfrak{R}$-measurements only for such ensembles of physical systems on which *this but only this condition* is imposed—of being in the state φ; to remain applicable, this [posit] must instead leave open whether this expectation value is also the same in all subsets of such ensembles that are selected from them on the basis of any new features. But if one leaves this open, then one can no longer infer, from the asserted addition rule for $(R\varphi, \varphi)$, that also in these subsets the expectation value of the sum of physical quantities is the same as the sum of their expectation values. In this way, however, an essential step in Neumann's proof is missing. If instead—like Neumann—one does not give up on this step, then one has implicitly absorbed into the interpretation the unproven assumption that there can be no distinguishing features, of the elements of an ensemble of physical systems characterised by φ, on which the result of the $\mathfrak{R}$-measurement depends. However, the impossibility of such features is precisely the claim to be proven. Thus the proof runs in a circle.

On the other hand, from the standpoint of Neumann's calculus one can argue against this, that [in this calculus] it is an axiomatic requirement that all physical quantities are uniquely associated with certain Hermitian operators in a Hilbert space, and that through the discovery of new features invalidating the present limits of predictability, this association would inevitably be broken. Indeed, any discovery that is representable in the operator calculus would have its contents specified only through the form of a wave function, which for quantities not simultaneously measurable exhibits the smearing out required by the uncertainty relations, and which finds application only by way of the probability interpretation.

By this consideration, however, the crucial *physical* question of whether the progress of physical research can attain more precise predictions than are possible today, cannot be twisted into the impossibly equivalent *mathematical* question of whether such a development would be representable solely in terms of the quantum mechanical operator calculus. There would need to be a compelling physical reason, if not only the physical data known to date, but also all the results of research still to be expected in the future are related to each other according to the axioms of this formalism. But how should one find such a reason? The fact that the formalism has so far proven itself, so that one is justified in seeing in it the appropriate mathematical description of known natural connections, does not mean that the as yet undiscovered natural law connections should also have the same mathematical structure.

§8. The Fundamental Difficulty of Such a Proof

The arguments considered thus far, which rush in from all sides of quantum mechanics, to prove the fundamental limitation of possible future predictions, lead—according to the above—indeed to the uncovering of tremendous difficulties lying in the way of the attempt to overcome the present limits of prediction, and onward beyond any limit, to approach the ideal of classical physics of accurately reading the future from the present state of nature. They show that the intuitive concepts of classical physics fail in this attempt; but how should one find a new intuitive approach that makes the coexistence of wave and particle picture understandable, and that provides the basis for unlimited prediction? They show that the mathematical formalism of quantum mechanics does not help in this attempt, rather that if a chance at all exists for its success, it must seek a new mathematical path—but where should there be hints as to the direction in which progress here is possible?

Notwithstanding these difficulties, the hitherto examined arguments do not show what they purport to demonstrate: the *in-principle* insurmountability of the presently existing limits. On the contrary: the assessment of these limits has brought out equally sharply the temerity of these claims and the deep difficulties of the proof to be provided. In addition, these difficulties appear to be insurmountable.

What is the situation?

For any quantum mechanically characterised state of a physical system there are measurements whose results cannot be predicted on the basis of knowledge of this state; in such measurements one gets sometimes this, sometimes that result.

The otherwise usual path in such a situation—to look for new features, to refine through them the determination of the states of physical systems, and to find in them the reason for the difference in measurement results—this approach is supposedly blocked here. But how so? The problem of whether and how with this approach we can again make exact predictions, is not one of those pseudo-problems a more detailed examination of which proves that whichever answer one might give to them, an empirical check of its correctness or falsity is inconceivable. Here it is different: the search for new features of physical systems, on which their different behaviour in some measuring experiment depends, obeys the requirements of physical verifiability: whoever pretends to know such features, ought to prove the supposed correctness of his assertion, by deducing from it correct predictions about the outcomes of measurements. Scientific, physical methods thus provide the criterion for the legitimacy of his answer.

Anyone who, despite this possibility of verification and the security it provides against [the accusation of] idle fantasy, denies altogether the possibility of the features in question, comes into conflict with the principle of the openness of experience, as has been shown for all arguments examined so far. There is no other criterion for having captured all the essential features and circumstances in a natural domain, than the possibility of understanding the regularities of the processes occurring in this domain on the basis of the knowledge of nature [thereby] gained. Whether one has identified these natural-law connections, in turn comes from being able to derive from them predictions that are confirmed empirically.

Hence there can be only one sufficient reason for abandoning as fundamentally useless the further search for the causes of an observed process: *that one already knows these causes.*

Accordingly, quantum mechanics, with its claim to be forever limited in the predictability of measurement results, stands before the following dilemma: either it names itself the causes that completely determine these measurement results—but then how shall it prevent the researcher from determining these causes in the individual case and predicting the measurement result from them? Or it does not name these causes—but then how shall it, without arbitrarily anticipating the investigation of still unknown areas of nature, exclude the possibility of future discoveries of these causes?

§9. The Solution: The Relative Character of Quantum Mechanics

The quantum mechanical formalism contains a way out of this dilemma. Having taken it is the essential achievement of this magnificent theory, and at the same time the step by which the construction of this edifice has gained its great natural-philosophical significance.

The direction in which the solution to the difficulties is to be found is indicated by Bohr's correspondence principle. This principle permits and requires that, within the domain of application of the classical concepts allowed by quantum mechanics, every consequence that arises classically from the characterisation of the present circumstances, should also serve as a basis for the quantum mechanical posits.

Now this consideration in terms of correspondence gives, in certain cases of quantum mechanically unpredictable events, accurate information concerning how they arise and about those physical variables on which these processes depend in all their essential features. These are the cases in which such an event is an element of a measurement process and contributes to bringing the observed object into relation with the measuring instrument, so that from the state of the instrument after the measurement, the property of the object to be measured can be determined.

Let us take the simple case where the measuring instrument indicates the result of measurement by a pointer position—an electric or magnetic field strength, the weight of a body, or whatever it may be; then the step from the reading of this pointer position to the quantum mechanical posit for the state of the observed physical system presupposes a theory of interaction between this system and the measuring instrument. This theory, which makes the quantum mechanical posit possible to begin with, is based solely on classical concepts, and with their help shows that and to what degree the deflection of the pointer depends on the state of the measured object, and hence gives clues to its determination. The use of any electrical, any optical instrument, any balance, is thus based on an inference from the measuring instrument to the object of measurement, and in this inference the reading of the measuring instrument is explained as the necessary effect that the system to be measured has imposed on the instrument in the process of measurement.

In the case of a measurement whose result was not quantum mechanically predictable, the same evidently holds also for the pointer position of the measuring instrument that registers the result of the measurement. However, for this unpredictable event—the occurrence of this particular pointer position—the interpretation of the measurement process itself, as we saw, gives the reasons whereby it has come to be. It would thus be pointless to wish to seek the cause of its occurrence in new physical features hitherto overlooked by research. *The theory of measurement already contains a sufficient basis for explanation.*

But it is evidently no different for the state of the measured system. Because indeed it is contingent in the course of a natural process whether it so captivates the interest of a physicist that it becomes the object of his measurements and investigations, or whether it interests the researcher only as a means for measuring other processes. For no natural process is it completely excluded that in some context it may be considered and accordingly interpreted only as part of a measurement process. The classical causal reasoning that, in the interpretation of the observed measurement outcomes, leads from the measuring instrument to the observed system thus need not be broken off there, since it is always possible that this system in turn may serve as measuring instrument for some other system with which it has interacted. If, e.g., an elastic collision has taken place between two bodies, it is sufficient to measure the change in momentum that one has experienced, in order to determine also that of the other. If one of them is measured, it can thus be considered on the one hand as the object of this measurement, on the other hand as a measuring instrument for the determination of the other, in which case one then calls upon the explanations provided by the classical theory of elastic collisions for the occurrence of this particular change of momentum, and precisely in so doing one has proven the attempt to explain this process through new as-yet undiscovered features to be pointless.

The possibility of finding new features that strictly determine the result of a measurement is therefore, indeed, excluded in quantum mechanics by the only reason that is cogent in the face of the openness of experience: *the features that determine the measurement result are already given in quantum mechanics itself.*

At first glance this solution admittedly appears extremely strange. If quantum mechanics knows how to explain the measurement result completely after it has occurred, how then is it possible that it offers no handle on calculating it already before the measurement from the subsequently ascertained basis for its explanation? Furthermore: if the causes for the occurrence of an unpredictable measurement result are uncovered through the application of *classical* laws, does one not for this reason return to the criticised error of considering the uncertainty relations an expression of a merely subjective deficiency in our knowledge of nature, and secretly expecting that the physical system itself is exactly determined with respect to all physical quantities, though for the time being unknowable to the observing physicist?

The solution to these difficulties must, if at all, be desumable from the considerations that prove the impossibility of overcoming the limits in the quantum mechanical formalism. These considerations are along the lines of the consequences that result—again because the correspondence principle—from the uncertainty relations: the predictions that one obtains from the quantum mechanical characterisation of a system can never exceed those that are derivable from the classical conceptions that have only limited applicability. Thus, if after the reading of the measuring instrument one explains backwards the position of its needle through a theory of the measurement process, this [explanation] traces the process of measurement back to states of the measured system and the measuring instrument that were not, and could not, have been included in the preceding description of these systems. The description of these systems is thus not univocal in quantum mechanics. The process of measurement, through the interaction between object and measurement instrument, creates a new context for physical observation, in which both systems are presented to the observer in a new way that cannot be uniquely predicted from the previous one.

Herein lies the solution to the puzzle: the quantum mechanical description by which, on the basis of some observation, a physicist determines his system, does not characterise this system completely and absolutely, but (so to speak) reveals only one aspect of it—precisely the aspect that presents itself to the researcher on the basis of the observation made here. From the point of view of this observation—relative to it—the system has no sharp values with respect to certain physical quantities; hence—relative to it there are therefore also no features of the system from which the result of a sharp measurement of these quantities can be read off. However, if one makes such a measurement—which necessarily disturbs the system and brings it into a different state—then one obtains for this new state not only a quantum mechanical description that assigns this quantity a sharp value, but moreover in the context of this mode of description one can also find causes for precisely this unpredictable value having had to result. Nevertheless, those causes were not utilisable for a prediction of this result, since they also determine the system only relatively—exactly like the description given before the measurement—and, indeed, relative to the observation that was made in the first place with the measurement itself. They could therefore be available to the physicist only after this observation, and hence allow him no prediction of its result.

§10. Discussion of an Example

In this relative character of the quantum mechanical mode of description lies the new and amazing natural-philosophical aspect that is here introduced into the view of nature. What it consists of, can best be understood on the basis of a thought experiment treated by a pupil of Heisenberg.[7]

Let the position of an electron be known only to the extent that a plane is specified in which it lies; where it is within the plane instead is indeterminate. Then, according to the uncertainty relations, only the component of the electron's momentum in the plane can be given; in the direction orthogonal to it instead the momentum remains indeterminate.

Now a position measurement on this electron is to be undertaken, i.e. its location within the plane is to be determined. The measurement is carried out by illumination of the electron; the light scattered by the electron shall go through a microscope, and then be captured on a photographic plate.

In order to have simple conditions, we imagine the intensity of the light employed here to be so reduced, that the whole process involves only a single light quantum. The duality of the wave and particle picture, applied to the light hitting the electron, means that this light quantum on the one hand is to be considered as a corpuscle that collides with the electron according to the classical laws of elastic collision, and on the other hand as a wave that, deflected by the electron, penetrates into the microscope and there proceeds through the lenses according to the classical laws of optics.

The conservation of momentum is valid for the collision between the light quantum and the electron: both are deflected in the collision; the changes in their momenta are opposite and equal.

In order to obtain a sharp image of the electron, we place the photographic plate in the image plane of the microscope corresponding to the plane of the object, thus in the plane in which all wave trains proceeding from a point in the object plane are combined again in a single point after passing through the lenses of the microscope. On this plate one obtains a sharp image of the electron illuminated by the light, from which the position of the electron at the time of the collision with the light quantum can be inferred.

In this application of the wave picture, one evidently proceeds from the conception that, from the place of the collision, there spreads in all directions a spherical wave that, insofar as it strikes the aperture of the microscope, penetrates into its lenses.

[7] Von Weizsäcker (1931).

Therefore, in this process the total aperture angle of the microscope is involved, and therefore—now again in the corpuscle picture, on which we depend for the description of the collision of the electron and light quantum—it is meaningless to pick out a definite direction in which the light quantum has been reflected by the electron, and has penetrated into the microscope. From this, however, it follows that also the change in momentum that the electron has experienced through the collision cannot be determined exactly. One will thus have to characterise the state of the electron, immediately after the collision, by a wave function that determines a sharp position but a less sharp momentum as compared to the previous state.

One arrives at an entirely different description of the collision of the electron and the light quantum if, instead of placing the photographic plate in the image plane of the electron, one fixes it in the focal plane of the microscope. In the focal plane, all the light rays that have penetrated into the microscope from the same direction are combined in a single point. Also in this case the photographic plate will display a sharp image; the light quantum can only darken the plate at a single point, since it only has enough energy to excite a single atom of the plate. This point of the focal plane hit by the light quantum is characteristic for a definite direction from which the light has penetrated into the microscope. The conception of the wave picture that we must draw upon here for the interpretation of the observed event is thus entirely different from the first case. Here we have to work with the picture of a bundle of parallel rays, which through their refraction in the lenses of the microscope are combined into a single point in the focal plane. The direction in which the light quantum has entered the microscope is therefore fixed, but the position in the object plane from which the light has proceeded, and thus where the collision between light quantum and electron has taken place, remains indeterminate. If the momentum of the light quantum was known before the collision with the electron—which can be accomplished by an appropriate arrangement of the light source—then with the specification of its direction after the collision, the change in momentum experienced by the light quantum is also fixed. According to the conservation of momentum, one obtains from this the change in momentum experienced by the electron. In this case, the place of the collision remains indeterminate instead. So although in this case nothing different has happened than in the first one with the electron, one must now characterise its state immediately after the collision differently from before: now by a wave function with unsharp position and relatively sharp momentum.

> Finally, if one sets up no photographic plate at all, but allows the light quantum to pursue its path without detecting it, then one obtains yet a third—though not in the same way intuitive—description of the state after the collision. In this case, the physical system composed of the light quantum and the electron is assigned a wave function that describes a linear combination: each of its terms is the product of one wave function describing the electron and one describing the light quantum. Through this linear combination the light quantum and the electron are thus not described each by itself, but only in their relation to each other. Each state of the one is associated with one of the other.

The coexistence of these different possibilities now evidently means that, depending on how one procures one's knowledge of the observed system, or, as we can say for this, depending on the relevant observational context, one can obtain different wave functions for the same system and for the same instant—namely for the electron at the time immediately after the collision with the light quantum. Thus the quantum-mechanical characterisation, unlike the classical one, does not pertain to the physical system still somehow 'in itself', and this means here: independently of which observation one uses to procure one's knowledge of it.

At the same time, the example shows that the quantum mechanical formalism cuts off the question about new features to be discovered, upon which the outcomes of arbitrary measurements depend, by itself providing sufficient reasons for these outcomes, and that it nevertheless affords no clues for the prediction of all measurement outcomes. So, in the case under consideration, it is in principle impossible to predict at which point the light quantum will darken the photographic plate set up, say, in the focal plane of the system of lenses. Nevertheless, the inference from the observation of this point to the momentum imparted to the electron during the collision, allows one to identify precisely in this exchange of momentum the cause for the light quantum being found exactly at this point on the plate.

The causal claim that is expressed in this inference, however, cannot be strengthened in the sense of predicting the impact of the photon at this precise point on the plate from processes during the collision. Because in this case the observation of the effect (that is, the darkening of the photographic plate at this point) was what allowed one in the first place to find the cause (the entry of the light from a very definite direction). However, the posited causal claim can, indeed, be used indirectly for the prediction of an observation result, and be checked by carrying out this observation: it is sufficient for reaching a conclusion about the change of momentum of the electron, which for its part can be checked.

That in this case only the *indirect* checking of the causal claim is possible, and not the direct one consisting in the prediction of the effect, is understandable in view of the above. The cause from which this effect could be deduced—the definite momentum transfer in the collision between light quantum and electron—belongs to this process as a quantum mechanical feature, and that means it belongs to it only relative to an observational context, specifically the one the observer enters *after the fact*, only with the observation of the plate. The knowledge he had of the collision beforehand was valid relative to a different observational context, and thereby could not contain the cause of the process observed later.

The final result here is that the discussed limits of predictability are in fact insurmountable in principle—at least as long as the system of quantum mechanics retains its physical validity. For the precise prediction of a measurement outcome, through which the observer steps into a different observational context for the system under consideration, would only be possible on the basis of a theory of the measurement process that describes it objectively, thus independently of how the observer in turn gains knowledge of it. But such an objective description is only possible, as demonstrated by quantum mechanics, within the range of application of classical physics, and this is not sufficient for an exact prediction of the measurement outcome.

§11. The End of Laplace's Demon

The old ideal of physics—in the course of research to work one's way more and more beyond any limit towards the clairvoyance of Laplace's demon—has not been abandoned without a struggle. Too much did renouncing this goal seem to impose intolerable limits on physical research itself! Thus to this day, even in the ranks of physicists, the hope is still not extinct that the limits of predictability discovered by quantum mechanics should one day prove to be provisional and surmountable. M. von Laue and Schrödinger have but recently advanced the thought that—not the quantum mechanical formalism—but unjustified adherence to outdated classical conceptions should be what makes the limits of prediction appear insurmountable.[8] 'The imprecision relations set a limit', so Laue concludes, 'to any corpuscular mechanics, but not to all physical knowledge'. But this hope is treacherous; the limits to predictability that are set by the uncertainty relations can, as the preceding investigations show, be negated by no future physical approach—unless one negates these relations themselves, that is, one discovers a physical error in the current edifice of quantum mechanics. Neither do the arguments that Laue and Schrödinger bring against this consequence of quantum mechanics upset this result; they show only that it is not so simple to make way into the thicket of the known arguments for indeterminism through to the crucial turn in the quantum mechanical description of nature.

Thus it is indeed correct, as both researchers emphasise, that the fact of the unavoidable disturbance the physical object suffers through measurement is not by itself enough to prove the uncontrollability of this disturbance. But this consideration in turn is not enough to justify the hope that Laue and Schrödinger attach to it: that it be possible, through a future completion of the theory, to control the disturbance, and thereby surmount the provisional limits of predictability. Rather, the quantum mechanical formalism itself already contains the reasons that rule out such a completion. Control of the arising disturbance does not fail because the formalism is still defective with respect to the explanation of this disturbance and thus in need of completion, but because the explanations it provides—which are complete and therefore not liable to emendation—are valid, as are all quantum mechanical statements, only relative to a certain observational context. In fact, to the one in which the disturbance in question is considered, thus the one that comes about in the first place through the observation. The explanations for the disturbance provide a foothold for predictions only to one who has performed this observation, and thus finds himself in this observational context; the outcome of the observation itself consequently cannot be predicted with their help.

Now we have the data together that the quantum mechanical formalism provides on the question of the in-principle possibility for predicting future measurement results. We now need to understand them in natural-philosophical terms and draw consequences from them.

---

[8]Von Laue (1934), Schrödinger (1934).

Chapter II. The Natural-Philosophical Situation

§12. Causality and Quantum Mechanics

The unrest arising from quantum mechanics regarding the familiar foundations of the knowledge of nature has first of all affected the conception of the law of causation and its applicability. What revision is now actually to be made here? Two results of the preceding considerations characterise the position of quantum mechanics regarding this question: the limits of predictability of future events have turned out, indeed, to be insurmountable in principle, however there is no phenomenon for which one shall not find causes in the framework of the quantum mechanical formalism, from which it has necessarily followed.

The apparent contrast of these two results so closely related in content appears most sharply, when each of them is used as the basis for an assessment of the law of causation. For while the first of the two claims asserts that unavoidable limits are set to the application of causal inferences and the mastery of mankind over nature that they confer, the second emphasises the in-principle unlimited applicability of causal concepts to which fundamentally every natural process—indeed with regard to all physical features characterising it—can be subordinated.

To avoid contradictions we thus require a precise critique of what is meant with the assertion of the law of causality, which claims of lawfulness this law makes on nature. And this critique has to proceed from those concepts that play a crucial role in the quantum mechanical results mentioned: the concept of the predictability of natural events on the one hand and of the causal relations between natural processes on the other.

We have already once glanced at the relationship between the two concepts, touched upon the close link between them, which consists in that the explanatory value of a physical hypothesis can be verified only by the prediction of future natural events, and that without the possibility of such a verification, the assertion of causal connections loses the character of knowledge of nature.

This relation has often misled one to believe that in a strict sense one has identical concepts here, and that only the linguistic designation feigns a distinction. The causal dependence of one event upon another seems to indicate nothing other, than that the first one can be predicted inasmuch as the other is known. The claim of the law of causation, that the chain of natural processes is causally determined throughout and with respect to every physically detectable feature, thereby acquires the meaning that one can predict every observable natural event from others that are accordingly assigned to it as its causes.

On this construal, the two quantum mechanical claims—that of the in-principle limitation of predictions and that of thoroughgoing causal connection—come into indissoluble contradiction with each other. If the relationship of cause and effect consists in nothing other than that the effect can be predicted if the cause is known, then there are no causes for in-principle unpredictable events, and it would be meaningless to try to causally trace them back to such [causes].

The fact that quantum mechanics looks for and presupposes natural-law connections, even where the causal inference from the observed cause to the forthcoming effect is ruled out, therefore shows that equating the two concepts is based on a mistake—on a failure to recognise the difference that the linguistic expression already indicates: the causal link of two processes actually only refers to the necessary sequence of events itself; instead, the possibility of predicting these on the basis of insight into the causal relationships provides the criterion for the correct application of causal concepts.

If we disassociate the two from each other, thus formulating the law of causation at first independently of the criterion for its applicability, then we obtain in return the claim that nothing in nature occurs that is not in all physically determinable features, caused by (and this means: follows necessarily from) previous processes. In this sense, gapless, unlimited causality is not only compatible with quantum mechanics, but is even demonstrably presupposed by it.

This disassociation, which leads to a usable formulation of the law of causation, does not however exonerate us from asking by what means one can recognise the existence of a causal relation in the individual case, thus in particular how the claimed necessity of the succession can be verified. Also quantum mechanics requires a *criterion* for causality and takes it, like classical physics, from the possibility of predicting future events. In contrast to classical physics, however, it has broken with the hitherto almost self-evident appearing presupposition that each causal claim may be verified directly by the prediction of the effect. In all cases of in-principle unpredictable events, the causal explanation that quantum mechanics gives for them can only be verified indirectly by inferring backwards from these events to their cause, and by further deriving, from the assumption that this cause was present, predictions of forthcoming events whose occurrence can be verified empirically. So in the discussed example, the darkening of the plate is traced back to the processes during the collision of the electron and the light quantum, from which one can then infer forward to the state of the electron, which is still an accessible observation.

The inferences [Rückschlüsse] underlying these indirect predictions however must not be confused with those familiar inferences that apparently allow one to determine a physical system for a past time segment with accuracy exceeding the uncertainty relations.[9] Thus it appears as though one could for instance escape the limits on measurement accuracy by taking two position measurements on an electron in quick succession, or by first determining its momentum and shortly thereafter its position, in order then to calculate from the results of both measurements together the exact trajectory of the electron for the time interval between the two measurements, thus to determine with arbitrary accuracy its respective position and momentum for this time.

This posterior construction of an electron trajectory differs from the inferences [Rückschlüsse] that are necessary for the interpretation of a measurement in that it takes no account of the duality of the wave and particle picture, limits itself to the intuitive features of one of these pictures and, since each single observation makes reference to both pictures and hence admits unlimited use of neither, fixes these features through two different observations. In this way one apparently succeeds in combining the variables belonging to different observational contexts into a single description of the physical system.

---

[9]Compare Heisenberg (1930, p. 15).

In contrast to this, the causal inference that determines the state of the observed from that of the measuring instrument must take duality into account in the interpretation of a measurement. For it is essential for the process of measurement that every atomic process exhibits also a wave character, every wave motion also a corpuscular character. The *single* observation of the measuring instrument that grounds the inference to the object of measurement determines in which way the wave and particle picture delimit each other in the present interpretation. Thus in the example considered the propagation of light into the microscope is described in the wave picture, whereas for the interaction between light and electron one must draw upon the corpuscular conception of the light quantum. How both conceptions are made consistent with one another depends on the type of measurement: if the light is intercepted in the image plane of the observed object, then one is working in the wave picture with the conception of a spherical wave propagating from one point, and correspondingly ascribing a sharp position but a smeared exchange of momentum to the corpuscularly interpreted collision between electron and light quantum. If one carries out the observation in the focal plane of the microscope, then one deals with a parallel beam of rays, and accordingly in the corpuscle picture with a precisely determined exchange of momentum but an unsharp position. The single observational context that the physicist enters through observation of the photographic plate therefore determines which features of both pictures are used. The preservation of the uncertainty relations is thereby assured.

The significance of this contrast follows from the relative character of the quantum mechanical way of description. Since every physical description and explanation of processes is valid only relative to its respective observational context, so the calculation of, say, a corpuscle trajectory combining the variables of different observational contexts into one representation remains physically vacuous precisely to the extent that it exceeds the uncertainty relations: it is uncheckable and provides no grounds for future predictions.[10] The controversial representation Heisenberg has given of this situation—that it is purely a question of taste whether one should assign physical reality to such a calculation of trajectories—is therefore to be amended in the sense that this calculation disregards the quantum mechanically crucial relation each specification of a physical variable [physikalische Bestimmung] bears to the

[10]See Heisenberg (1930). One has attempted, however, to use such calculations of trajectories to derive predictions in a similarly indirect manner as happens in the interpretation and checking of measurements, and thereby to overcome the limits of the uncertainty relations. One such attempt is found sketched in Popper: 'Zur Kritik der Ungenauigkeitsrelationen', *Die Naturwissenschaften*, volume 22, issue 48 (Berlin 1934). The same issue contains a reply by Weizsäcker that uncovers the physical error in Popper's thought experiment [Popper and von Weizsäcker 1934]. The real reason for this error, apparent only from the more detailed discussions in Popper's book *Logik der Forschung* (Vienna 1935 [sic]) [Popper 1934], lies in a misjudgment of the duality experiments and their consequences. Popper is misled by the probability interpretation of the wave functions to apply these quantum mechanical state descriptions, and the uncertainty relations given with them, only strictly speaking to ensembles of physical systems, and for a single appropriately chosen system to assume instead no restriction through the uncertainty relations. In this he misunderstands that because of the duality experiments the applicability of the classical conceptions is limited according to the uncertainty relations already for every single elementary process, and that accordingly wave functions can in fact be used for describing the state of individual systems. That this use of wave functions is consistent with their probabilistic interpretation is based once again solely on the relative character of the quantum mechanical way of description: on the one hand, the wave function is completely determined by the values of those physical quantities that have a sharp value within the momentary observational context for the system. In this respect it characterises the system quantum mechanically relative to the observational context present. On the other hand, the probability interpretation of the wave functions yields those variables [Bestimmungen] that remain of the classical-intuitive description according to the correspondence principle and that fix which statements can be made for the passage from one observational context into another.

> respective observational context, and in this sense becomes physically meaningless. Conversely, the possibility of checking even only indirectly the causal claims that arise in the interpretation of a measurement process depends on the fact that this interpretation does justice to duality, and thus implicitly to the relative character of quantum mechanics.

The modification that quantum mechanics has made to the criterion of causality, consisting in allowing as criterion besides direct verification also the indirect derivation of predictions, cannot now itself, however, be inconsequential for the meaning of this principle. The classical requirement that in every case of a causal link its existence must be verifiable directly by the prediction of the effect, manifestly presupposes more about the law-like connections in nature than the law of causality, which requires one cause for each event. For, as quantum mechanics shows, this law is also compatible with causal connections in certain circumstances being liable to verification only indirectly through consideration of measurement processes. But which presupposition was it that forced classical physics to assume the further-reaching criterion? The answer is clear from the preceding: the in-principle limits of prediction become comprehensible in that and only in that the quantum mechanical description of nature proves to be relative—relative to the respective observational context in which the researcher is situated in relation to his object. This relative trait of the description of nature is foreign to classical physics; for it [classical physics], the characterisation of any system is unique and independent of the manner in which the observer procures knowledge of it. And hence it [classical physics] necessarily arrives at the assumption that with sufficiently sharp observation and sufficient knowledge of natural laws, the investigation of physical systems allows one to determine the causes of their further evolution with arbitrary distinctness and thereby to predict this further evolution.

Seen in this light, the difficulties into which the advocate of the law of causation is plunged by the discoveries of quantum mechanics are rooted in the fact that various principles have been merged together: the principle of causality in the narrower sense, whereby every event in nature has causes from which it follows with necessity, has been merged with the assumption that physical knowledge accounts for natural events adequately and independently of the observational context. This assumption has in fact not been introduced explicitly, but has crept into the criterion of the applicability of the principle of causality as an undisclosed and seemingly self-evident premise. It finds its expression in the presupposition that any causal link between processes gives rise to a prediction of the effect from the cause—indeed, that the causal link is in fact identical to the possibility of this prediction.

Quantum mechanics requires us to resolve this merging of different natural-philosophical principles—to drop the assumption of the absolute character of the knowledge of nature and to handle the principle of causality independently of it. Consequently, it has not refuted the law of causality, but has clarified it and freed it from other principles that are not necessarily connected with it.

In a perfectly analogous way the conflict between determinism and indeterminism that has broken out on the basis of quantum mechanics is now also settled. It finds its solution through a clarification of terminology, which turns out to be necessary given the distinction between natural-philosophical principles that has been drawn:

the judgement falls for or against determinism depending on whether one declares determinism to be the intuition that every process in nature is strictly causally fixed, 'determined' by prior states, or on whether one expresses by the doctrine of determinism the confidence that with sufficient research into nature, its future development may be predicted with arbitrary accuracy. For the result of the previous investigations is just this: the opinion prevailing throughout, and even plausible in, classical physics—that gapless causality and the possibility of in-principle unlimited future predictions are inseparably linked with one another—has proven to be false. It has been refuted by the demonstration of the merely relative character of the description of nature, which confirms afresh the assumption of thoroughgoing causality, but has broken once and for all with the hope of arbitrarily sharp predictions.

§13. Correspondence and Complementarity

Accordingly, even though under a proper understanding it is not the principle of causality where the radical break in the development of physical theory has occurred, the rejection of classical concepts is still prodigious. The effort of classical research to obtain an adequate account of natural processes through intuitive constructions has failed: in place of an intuitive description of natural events comes the formal assignment of a wave function, which already makes any intuitive interpretation difficult if not impossible by developing not in the usual three-dimensional space but rather in a higher-dimensional phase space. The break is even deeper with the classical notion that physical systems and the interactions existing between them may be grasped objectively, meaning: independently of the manner in which the observer obtains knowledge about them. In place of such a unified and objective description of nature come representations that are valid only relative to their respective observational context, lose their applicability with new observations, and are replaced by new descriptions.

The question of whether, with the development of quantum mechanics, experience shall have asserted its authority against the hubris of philosophical claims, which, after the fact and rashly, have presented the hitherto adequate physical maxims as unshakeable rational truths—this question thus still stands in all its poignancy.

To answer it, it will be crucial to envisage more clearly the connection between classical physics and quantum mechanics, but also the fracture between the two. The presentations that Bohr and Heisenberg have given of the principles of the quantum mechanical description of nature[11] contain all the essential lines of thought for this.

Characteristic of the relation of classical physics to quantum mechanics is the coexistence of reliability and failure of the classical concepts in the face of the typical quantum mechanical evolution. In Bohr's presentations these two aspects find their most succinct expression on the one hand in the correspondence principle, on the other in the complementarity relationships that pervade quantum mechanics.

[11] Bohr (1931), Heisenberg (1933).

According to the *correspondence principle*, the intuitive classical concepts still form the bridge between the data of sensation and the formulas of the theory. Without them one cannot obtain from a measurement a viable starting point for theoretical inferences; they provide the key to the interpretation of unintuitive quantum mechanical formulas and thereby make possible their application to experience.

However, as the duality experiments show, these concepts are limited in their application to experience. This limitation surfaces *in the feature of complementarity* that Bohr demonstrates in the quantum mechanical description of nature. There are essentially three mutually distinct but content-wise related relationships that Bohr combines under the idea of complementarity.

a) To begin with, the *wave* and *corpuscle picture* stand in a relationship of complementarity to each other, both of them being classical-intuitive constructions of processes in space and time. Because of the irreconcilable conflict that exists between them, the combining of both pictures in the description of one and the same physical process required by the duality experiments is only possible in that each [picture] specifies the limits unto which the other can find application.
b) These limits are in turn manifested in a relationship of complementarity that exists within each of these two pictures *between its characteristic features*. Thus, in the corpuscle picture the features of position and of momentum, whose exact determination is necessary for the full application of this picture, stand in complementarity to each other: the exact determination of one of them excludes that of the other.
c) The requirement to arrive, for the sake of a unified description of nature, to a formalism that does justice to these different complementary conceptions and at the same time states the limits of their respective domains of application leads to the third, the essential relationship of complementarity in quantum mechanics. Alongside the intuitive-classical mode of description, which retains its validity within the specified limitations, enters the quantum mechanical formalism, which combines into new symbols the various mutually constraining conceptions. These symbols—the wave functions of quantum mechanics, the operator calculus of the physical observables—escape direct intuitive interpretation; they bring together aspects of conflicting intuitive pictures and refer to a pseudo-space whose dimensions are determined by the degrees of freedom of the physical system considered. Through this departure from intuition, the quantum mechanical formalism succeeds in the step, where classical physics fails, of combining into a single state description the apparently mutually contradictory intuitive variables of the system, and of thereby capturing *the strictly causal course of natural-law relationships* for the states so characterised. On the other hand, through this departure from intuition it [the formalism] denies itself direct access to the data of experience, to the interpretation of the result of an arbitrary measurement as read off the measuring instrument. From the point of view of the formalism, this access can thus be preserved only indirectly, in that the old classical mode of description enters as complementary to this specifically quantum mechanical framework. It has the *intuitiveness* that the other lacks; it lacks instead

the completeness of the quantum mechanical characterisation, since after the breakdown of classical physics it is only fragmentarily applicable, and thus [lacks] the ability to interpret the natural processes causally throughout.

The demonstration that two modes of description so different as these seamlessly align and complement one another in the mastery of nature through physics is one of the most wonderful results of quantum mechanics. It relies, as far as the interpretation of *experience* is concerned, on the realisation that every observation is associated with a *disturbance of the observed system* due to the interaction of the system with the measuring apparatus and, given the only limited applicability of the classical concepts, is *uncontrollable* to a certain degree. In the example discussed earlier of the illumination of an electron: the interaction between electron and light leads, due to the corpuscular properties of the light quantum, to a transfer of momentum between the two; due to the wave nature of light that forces the physicist to interpret the further motion of the light quantum—depending on how he later observes it—either as the spreading of a spherical wave, or as the propagation of a parallel beam of light, the transfer of momentum cannot be determined in terms of [both] its location and the magnitude of transferred momentum. Experimentally, this finds its expression in that every determination of position obstructs the paths that could lead to a measurement of the transferred momentum, and vice versa.

From the point of view of *theory*, the classical and quantum mechanical modes of description are joined at the '*cut*' that is inevitable in the theoretical treatment of any measurement event. The calculation that tracks such a process—if it is not to lose itself in empirically inapplicable formulas, which furthermore have nothing to do with the measurement outcome present to intuition—must limit the quantum mechanical approach, i.e. the use of wave functions, to the state of the observed system. For the measuring instrument, for the state of one's own body—which likewise interacts with the observed system during measurement—the physicist is dependent upon the intuitive classical approach that makes intelligible to him the meaning of the registered measurement result. What in this division he considers to be the observed system and what the measuring apparatus, thus how far he follows in the quantum mechanical formalism the strictly causal development of the change in state, is largely left to his fancy. He can make the passage to intuition sooner or later; but he must make it somewhere if he wishes to incorporate the new observations into his considerations. And at the place where he makes it, he necessarily ceases to keep strict causal track of the quantum mechanical states in favour of the classical-intuitive—but fragmented and accordingly unsharp with respect to certain variables—interpretation of the interaction between measuring apparatus and system.

§14. The Necessity of the Complementarity Description

All these expositions of the complementary character pervading the quantum mechanical mode of description, however, still do not allow one to perceive with sufficient clarity whether or not they depict a truly new, characteristic aspect of natural processes that has escaped classical physics, and what that is.

After the failure of the wave and corpuscle picture, why should not the construction of a new unified and intuitive model of nature succeed that, like the natural processes themselves, combines traits of wave propagation and corpuscular motion?

Why should it not be possible to eliminate entirely the classical conceptions from the physical perspective? In doing so the advantage of intuitively understanding one's own observations might be—at least provisionally—lost, but the unambiguous association obtaining between the classical description of a physical system and the specification of its wave function (or, for non-maximally sharp observations, the specification of a mixture of such wave functions) provides the basis for letting the quantum mechanical formalism take the place of the incomplete classical description throughout, thus also in the evaluation of the measuring apparatus.

Wherein lies the criterion for the disturbance introduced by observation being uncontrollable *in principle*? Is it not so just as long as one still clings to the classical pictures of waves and corpuscles, which have demonstrably failed?

With what right does quantum mechanics forbid the physicist to ignore the 'cut', and to express and use within the quantum mechanical formalism everything he knows of the investigated physical systems, of the measuring apparatus and his own body, insofar as this interacts with the object during measurement?

All of these questions are very closely related to the old question of whether the quantum mechanical limits on predictability are insurmountable in principle or whether they represent a physical challenge that requires the completion of quantum mechanics through new physical discoveries. The answer that resulted for this question yields also the key to the correct understanding of the doctrine of complementarity: the quantum mechanical formalism is on the one hand *physically closed* in the sense that it completely states the natural-law relationships in what happens—apart from certain as-yet still unsolved physical problems, like that of the structure of the nucleus, the solution to which physics is investigating: it is in this respect incapable of *completion* on the basis of new discoveries, again except for those specific problems that are still open. On the other hand, it characterises physical systems only *relative to the respective context of observation* in which the physicist stands towards his object; so it is excluded from intuitive interpretation, which is possible only where physical processes can be unambiguously construed as motions in space and time. The hope of being able to preserve from the classical theory certain firm pillars upon which the construction of a new intuitive picture of nature could rest, is therefore moot. Despite this unintuitive character, the quantum mechanical formalism ultimately signifies *no detachment from intuition*; as the correspondence principle shows, in each interpretation of a sensation, in each passage from one observational context into another, it [the quantum mechanical formalism] seamlessly retains the connection to the intuitive space-time constructions of classical physics. To wish to eliminate these constructions thus means to obstruct the access to intuition and thus to a meaningful association between the data of sensation and the posits of a physical theory.

§15. Classical and Quantum Mechanical Description of Nature

Every explanation of nature is based on observation and must prove itself by it. But observation alone—that which directly presents itself to sensation—is not sufficient for the explanation of nature; indeed, on its own it is not even sufficient to justify the simplest judgement about the things that surround us every day. Observation becomes knowledge of nature only once it is ordered into the range of experience already acquired, and thus—whether consciously or unconsciously—is interpreted and processed from the standpoint of certain theoretical approaches. This interpretation of mere knowing through sensation underlies the practical handling of chairs and tables just as much as [it underlies] the physical utilisation of measurement results.

In this handling of the contents of sensation, both the experiences of everyday life and the explanations of classical physics presume to understand the data present in observation as excerpts from a natural process that plays out in terms of intuitive space-time relationships and in strict causal conjunction. Accordingly, every classical explanation of nature uses a model constructed from the forms of space and time that claims to depict the true conditions of nature and to exhibit in them the connections of natural law.

Despite all the failures of the classical theories, quantum mechanics also retains the method for proceeding from observation to the explanation of nature through the construction of such models and of the causal regularities valid within them. Forced by contrary experiences, it has only broken with the single assumption that these models describe the course of nature *objectively*, that is, independently of the observer and the manner in which he observes. The intuitive physical models, and indeed those of classical physics, appear to be still indispensable for the explanation of nature, even if they merely serve as *analogies* for it.

The duality experiments show first of all that a single model is no longer sufficient to characterise a physical system completely. The wave and corpuscle picture overlap, and restrict one another in their application to the observed objects. But the limit up to which one or the other model finds application is not itself an objective property of the object; rather, it depends on the respective observational process. In the example of the illuminated electron: depending on the manner in which the diffracted light is observed, the corpuscle description of the electron finds its limit in an imperfect specification of position or of momentum.

This relative character of the knowledge of nature is responsible for the fact that a new observation, as a rule, represents a break with the mode of description employed until then. For in specifying the object through the data of this observation, one enters into a new observational context with respect to the object and thereby changes the limits up to which the various classical pictures find application. This entails that the process of such an observation or also any of its sub-processes can enter the physical considerations in two fundamentally different ways.[12] On the one hand, like any

[12]Heisenberg (1930, p. 44).

natural process it can be followed mathematically within the present observational context. On the other hand, it can be considered as an element of the *measurement* and to that extent as a step that makes possible and intelligible the passage to a new observational context.

In the first case we are dealing with an application of just the quantum mechanical formalism. It formally combines the different complementarily related classical descriptions characterising the system on the basis of the preceding observations, and thereby provides an overview of the strictly causal development that the given system has so far undergone and will further undergo from its current state.

> Thus, the interaction between an object and the measuring apparatus used for its measurement can, in quantum mechanics, be completely followed and mathematically predicted, provided only that one restricts oneself to making all statements within the observational context that exists prior to measurement. The quantum mechanical formalism prescribes for this the multiplication of the two wave functions assigned to the object and the measuring apparatus prior to interaction, then the determination of the operator that is assigned to this interaction, and the application of this operator to the product of the wave functions. The result will generally be a linear combination whose individual terms each connect one state of the object to one of the measuring apparatus. These various possibilities, among which observation decides, are combined on an equal footing by a symbolic addition into a combined wave function that develops in a phase space determined by the degrees of freedom of both systems. But this means that one must forgo *within this observational context* the intuitive tracking of each of the two individual systems.

It is quite different when such a process is regarded as a measurement process, and is used for the purpose of presenting the observed object in a new observational context. Here, everything depends on making anew the transition from the data of observation to the explanation of nature, and this transition presupposes, in quantum mechanics as well as in classical physics, the classical-continuous pictures of processes taking place in space and time. Insofar as the processes in the measuring instrument need to be considered in evaluating the observation result, they too must be understood classically-intuitively as modifications of an object given in space. Here the limits that are imposed on such a representation by the uncertainty relations truly naturally occur as indeterminacies. They must themselves be interpreted within the intuitive picture, and within it they are only intelligible in that the observer lacks the knowledge of the precise value of certain physical quantities forming a necessary part of the picture, such as the position or the momentum of a particle.[13] From which it follows that *for this mode of observation* the interaction between measuring instrument and object remains uncontrollable within the scope of the uncertainty relations.

Therefore, the single realisation that the classical-intuitive conceptions of natural events proceeding in space and time are drawn upon in quantum mechanics only as analogies for the description of nature, which depending on the observational context of the observer are curtailed in one way or another, but that qua such analogies they

[13]The German here reads: 'daß für gewisse dem Bild notwendig zukommende physikalische Größen, wie etwa dem [sic] Ort oder dem [sic] Impuls einer Partikel, dem Betrachter die Kenntnis ihres genauen Werts mangelt' (*eds.*).

are still indispensable for the processing of observations—this realisation indeed suffices to make intelligible the peculiarities of the quantum mechanical approach: from it follows the necessity of employing the rules of the quantum mechanical formalism and those of the classical theories side-by-side, [the necessity] of making the 'cut' before the measurement that discontinues the quantum mechanical tracing of a process in favour of the classical interpretation of the processes in the measuring apparatus. It shows to what extent the unintuitiveness of the quantum mechanical formalism is consistent with the adherence to the intuitive classical pictures, and how the uncontrollability of the disturbance of the measuring apparatus is consistent with the seamless causality of natural events, which assigns to each process, in an in-principle verifiable way, a cause from which it necessarily follows.

Chapter III. Transcendental Idealism

§16. The Antinomies and Their Consequences

The question remains whether and how this result can be included in a philosophical view of nature—in particular, how one can assess in its light the philosophical claim of a priori principles for the knowledge of nature. I consider this question on the basis of the critical philosophy.

The crucial natural-philosophical principles that Kant has proposed as a priori judgements are persistence of substance, causality, and interaction. He has intimately bound up the introduction of these principles with the doctrine of transcendental idealism, according to which they cannot provide adequate knowledge of reality 'in itself' but only a limited knowledge of nature that stops at the conceiving of 'phenomena'.

Through a lack of clarity in the reasons for this doctrine, Kant himself has placed difficulties in the way of their understanding. Alongside the proof from the doctrine of the antinomies—that in the intuitive forms of space and time only a limited applicability of the categories and their underlying general philosophical notion of oneness and lawfulness is possible—enters the erroneous assumption that already the a priori character of these notions robs them of the objective significance of determining reality in itself. This second reason for transcendental idealism—the doctrine of formal idealism—in truth amounts to denying the a priori principles the character of knowledge in the strict sense of the word. It has often diverted attention from the import of the antinomies. Only Fries, through the refutation of formal idealism,[14] has cleared the view to the consequences of the antinomies, and [has] thereby especially clearly let emerge in what sense the natural-philosophical principles uncovered by Kant can only find application as *analogies*.

[14] J. Fr. Fries: "Neue oder anthropologische Kritik der Vernunft", second edition, page XXIV ff. Heidelberg 1828. New edition Verlag "Öffentliches Leben". Berlin 1935 (Fries 1828/1968, pp. XXIV ff.).

Consider, for example, the principle of the causal connection of natural processes,[15] that is, the idea that for every event a cause can be found from which it follows by necessity. On closer inspection, applying this concept of causal connections to natural processes leaves the physicist with only a strangely empty formal schema in his hands. To begin with, in order to obtain a complete overview of causal relationships, he must find the relations of cause and effect in the connection between what comes earlier and what comes later, even within any ever-so-short process. In order to really *understand* natural circumstances, the physical description of natural-law relationships must therefore advance so far as to causally link the state of a system at one point in time with the previous and subsequent evolution of the process. In this connection, however, it becomes apparent that a strict application of the concept pair 'cause–effect' is impossible, for there are no temporally contiguous states and hence for no state of a system can one specify another that has directly brought it about or has been caused by it.

The formalism of physics does justice to these circumstances by expressing natural-law relationships in the form of differential equations. In these equations one does not describe the state variables at various times as dependent upon one other, rather the equations put the time derivatives of these variables into functional relationships with the variables themselves. Take for example the attractive force that is exerted by a larger mass on smaller bodies, and the law of falling bodies that accordingly determines the motion of these bodies. This law associates the distance of the falling body from the attracting mass at any instant in time to the acceleration it has at that point. Thus it causally links the presence of the falling body at this point with the acceleration imparted to it there. Now what is this acceleration, which—if anything at all—is produced as the immediate effect of the falling body's instantaneous state and of the attracting mass? It is defined by a differential quotient, the derivative of the fall velocity with respect to time; so it describes the change this velocity experiences at the point in question. However, this change of velocity is obviously not an independent physical process, which as an effect could be contraposed to another one, its cause. Rather, it only specifies a *relation* that obtains between the instantaneous velocity and the subsequent course of nature. And moreover, it is a relation whose foundations cannot even be uniquely determined. For where and when does this change in velocity actually take place? At the considered instant itself—when the body is at the specified location—and at this location alone nothing can be ascertained of either its velocity or, a fortiori, of the change thereof—not even through the most precise observation. For this we require a wider space-time region, in order to apprehend both at all. But any region that one could choose for this purpose, any spatio-temporal neighbourhood of the given state of the falling body, already goes beyond the region necessary for determining the acceleration, and reveals additional changes in velocity arising according to the differential equation from the attractive force that acts upon the falling body during the further path of the falling motion, depending on the positions it has reached then. It is therefore not at all possible to

[15]For the following considerations compare Nelson (1917, pp. 324 ff.).

isolate the effect that is exerted on the falling body by the attractive force at a specific point so that one could unambiguously demarcate from each other the processes within the course of events that stand directly in the relation of cause and effect to each other.

And one comes up against precisely the same difficulty when one sets out to determine exactly the physical state of a system itself. Necessary variables that come under consideration in this, like those for the mass density of a body or for the velocity of a process, must on the one hand be specified for every single point in space, resp. time, but on the other hand can only be specified by taking limits that introduce into the discussion a finite spatial or temporal neighbourhood of these points. The question, to which part of the physical system now mass density or velocity pertain, here again remains unanswered, and so disappears from physical consideration: the point in space and time to which these data are formally assigned does not harbour the bearer of these properties; otherwise one would be able to determine them from it alone. However, no spatially and temporally extended part of the system under investigation comes into question either—at any rate not as a whole—since at any finite distance from the point under consideration it features mass densities and velocities that are independent of the desired quantities and unnecessary for their determination. The limiting process for determining mass density and velocity could also have started from an arbitrarily much smaller part and still led to the correct result. One can thus say nothing as to the specification of the properties of physical objects or events that determine these as they are constituted in themselves; rather, the alleged properties of physical systems in truth only specify certain relations between the parts of the system, without these parts being themselves unambiguously specifiable.

This peculiar situation pervading every exact description of nature in manifold modifications, reveals that the conception of things in space with time-varying states that stand in seamless causal relationships with each other is of but limited applicability. Knowledge of nature shows us not a reality that is fully determined according to its own inner properties, but only relational networks that are unanalysable [unauflösbar] in the sense that, for these relations, one can give no foundations that are unambiguous and determined in themselves.

The concepts of substance and causality, and with them those of things in space and of their states evolving in accordance with natural laws, are not thereby eliminated from physical consideration. It has only become apparent that they do not describe adequately the natural events, but are rather used as mere analogies. Qua such analogies, however, they have their use—indeed they are indispensable for the development of physical knowledge. It is enough to make a ‘cut’ somewhere in the investigation of the relational network across which the relational connections cannot be traced, and in this limited perspective one conceives of things in space and of causal links for processes. As long as one is dealing only with finite, spatially and temporally extended physical systems whose own inner structure need not be considered beyond a certain limit, one escapes the difficulties of the limiting process that make impossible the full application of the concepts of cause and of substance. In fact, however, in every observation we actually start with finite, extended systems that can be conceived as things in space and causal processes in time. On the other hand, this very interpretation of sensation further leads to the challenge of better determining these things and processes beyond the arbitrary boundary of the 'cut', and hence to the insight into the limits that are drawn in this process for the appropriate use of these natural concepts.

§17. Critical Philosophy and Quantum Mechanics

The relation of these considerations to the characteristic natural-philosophical features of quantum mechanics is obvious. When one compares the requirements of critical philosophy developed here with the principles of quantum mechanics, one obtains the following result:

The critical philosophy claims:

a) That in the formation of natural knowledge, order comes to the manifold of sensation in this and only this way: that the categories, applied to intuitively determined processes in space and time, provide the theoretical schema for interpreting sensation.
b) That this order is complete insofar as any empirical datum can be completely incorporated into the law-like relationships between natural processes and explained by them. Any observation for which this explanation has not yet succeeded, especially with respect to the seamless causal linking of processes, remains an unsolved, meaningful problem for physical research.
c) That nevertheless the categories themselves are not completely applicable, insofar as the ordering of the contents of sensation—especially when it aims toward the goal of a thorough mastery of nature beyond any arbitrarily set boundary—remains necessarily within the investigation of certain relational networks, and does not reach the oneness of a reality determined in itself. Accordingly, the categories that are based on this notion of the oneness of reality provide the guide to the interpretation of sensation merely as *analogies*.

All three claims resonate in a surprising way with the discoveries of quantum mechanics:

Regarding a): Bohr's persistent reference to the indispensability of classical concepts even in quantum mechanical investigations shows that the deep break with classical theories has not affected the concepts through whose application the manifold of sensations is ordered into physical experience. True, the Bohrian 'classical concepts' include a series of empirical-physical features that are foreign to the Kantian categories; moreover, it is not yet at all proven to what degree the classical pictures do justice to the natural-philosophical consequences of the categories. But these differences pertain to problems that belong to a natural-philosophical interpretation of the classical theories, and in this sense go beyond the present investigations. What is essential here is that quantum mechanics behaves conservatively exactly at the point where it has frequently been lauded for triumphing over traditional notions: the fundamental concepts that mediate the transition from sensation to natural knowledge are not touched by quantum mechanics, despite its revolutionary upheavals. While this represents no justification for the philosophical approach that sees in these concepts the expression of a priori rational knowledge, it yet represents an empirical confirmation, which is all the more significant given that the creators of quantum mechanics themselves thought, in the first stages of its development, that they had lost entirely the connection with the classical concepts and the *understanding* of natural processes bound up with it.

Regarding b): In spite of the provably insurmountable limits that quantum mechanics has derived for the prediction of future observations, also this second postulate of critical philosophy—in particular, the demand for the universal applicability of the law of causality—does not stand in conflict with quantum mechanics. Rather, it has turned out that the inevitability of these limits could itself be ensured only by proving that the quantum mechanical formalism has already reached the causal closure demanded here—in any case with respect to the status of the uncertainty relations relevant here—and that *for this reason* it is no longer susceptible of extension.

Regarding c): Compared to classical physics, quantum mechanics represents a tightening of the limitations mentioned under c). While the antinomies demonstrate that the classical concepts of intuitive systems construed in space and time lose their apparent autonomy on closer inspection and dissolve into a network of relations whose foundations ultimately remain indefinite, quantum mechanics carries this relative character of natural description still one remarkable step further. It does away with the notion that these relational networks should be determined at any rate through objective circumstances of things in space and time, and shows them in turn to depend on the manner in which the observer obtains knowledge of the system. But even if this proof goes significantly beyond the outcome of the doctrine of the antinomies, it can still be understood from the perspective of the latter—and conversely, allows this less far-reaching claim to emerge more clearly. For the basis of all the difficulties facing the understanding of quantum mechanics lies—as has been shown—in the notion that every description of nature captures things objectively (in the strong sense of the word) existing in space and evolving in time, and that it has to describe them in terms of the properties pertaining to them, and thereby unambiguously. This notion disappears by itself if one takes as a starting point that the physical description at best advances up to the formal characterisation of a relational network with indeterminate foundations, but not to the representation of substances determinate in themselves. Even in the physical description of such a relational network it could indeed be imaginable and possible that empirical research *unambiguously* determines this system of relations. But this is in no way necessary. For in no case is the description of this relational network an adequate rendering of an in-itself determinate, unified reality; at best it shows—inasmuch as it proves itself to be knowledge—only one side of reality, which is insufficient to capture it fully. And then it is entirely possible that this incomplete description—even if it is knowledge—is not uniquely determined, but brings to light these or other relationships according to the observational procedures, to the questions from which research proceeds. Whether the description of nature is equivocal in this way or whether it possesses the uniqueness that was assumed in classical physics can only be decided through physical research.

§18. The Splitting of Truth

These considerations lead to an even deeper connection between the results of quantum mechanics and the reflections of the critical philosophy. The proof in transcendental idealism that natural knowledge is inadequate for capturing reality but rather only picks out, in an incomplete way, relational networks whose foundations remain indeterminate within the scope of this knowledge, opens the way for the possibility of different mutually independent yet mutually compatible modes of confronting reality through perception. Only with insight into this possibility is the understanding of the actual structure of human perception disclosed, which—no matter how one might force it—is irreconcilable with the postulate of a universal science comprising all areas of perception. Alongside physical knowledge comes that of the psychological nature as an autonomous and equally legitimate science; alongside these two come evaluative, ethical and aesthetic perspectives whose claims to objectivity find no place in the natural sciences without the latter thereby excluding them from the realm of knowledge.

This splitting of truth into different worldviews—as Apelt calls the separation of different realms of knowledge established by Fries[16]—has been incorporated and extended by the advancing physical research in a peculiar way. The relative character of the quantum mechanical description of nature leads to this, that already in the purely physical treatment of natural systems various representations appear side by

---

[16] Apelt (1904). The similarity of Apelt's considerations to quantum mechanical arguments emerges clearly from the images with which the distinctness of possible realms of knowledge are described. As Apelt writes: 'Human knowledge does not resemble a level surface that one can completely survey with a single glance from any high vantage point; rather it is more like a hilly country, a complete image of which one must assemble only little by little from partial views. There are multiple heights, multiple vantage points one upon the other, each of which presents a different view and where something now shows, now hides itself'. And de Broglie represents the complementarity of position and momentum measurements thus: 'There are, so to speak, two planes that we cannot see sharply at the same time. We might make a comparison: let there be a figure whose various parts are drawn on two close parallel planes Π and Π'. If we observe the figure through a none-too-precise optical instrument, we can focus it on a plane between Π and Π' and obtain an image that still reasonably resembles the figure. We then have the impression that the figure is drawn in one plane. But if we use a very good instrument, then it cannot sharply depict Π and Π' at the same time. The more we focus it on Π, the worse we see the parts drawn on Π' and conversely; we are thus forced to recognise that the figure does not lie in one plane. Classical mechanics corresponds to the imprecise instrument; with it we have the impression that we can determine simultaneously position and velocity of the particle exactly. But with the new mechanics, which corresponds to the precise instrument, we come to realise that the spatio-temporal localisation and the energetic description are two different planes of reality that one cannot simultaneously see precisely' (de Broglie 1929, p. 7 ff.).

side, none of which claims absolute validity—rather, which are all valid only relative to the respective context of observation, and precisely because of that can exist in harmony with one another despite their differences. From this point of view, the natural-philosophical novelty of quantum mechanics is describable thus: the splitting of truth goes deeper than philosophy and natural science had previously assumed. It penetrates into the physical knowledge of nature itself; instead of merely delimiting its scope against other possibilities for grasping reality, it separates various equally legitimate representations within the physical description that cannot be unified into a single picture of nature.

In fundamental discussions of quantum mechanics given by Heisenberg,[17] the trains of thought of this doctrine of the splitting of truth also come out explicitly. Under the impact of the advancing physical research and its achievements—which have brought century-old questions to a close and have satisfied claims about the perception of nature—he has also considered and emphasised the other, frequently neglected aspect of the perception of nature: that in spite of such successes, any progress in the explanation of nature is accompanied by a renunciation—the renunciation 'of an understanding of the world in the original sense'. Instead of this understanding, in physics we have only a formal grasp of connections in nature, which along with physical knowledge creates room for other types and means of perception as well.

In this connection, the link between the trains of thought of transcendental idealism and the natural-philosophical foundations of quantum mechanics in no way speaks against the physical progress from classical theory to quantum mechanics being explicitly grounded in guidance through experience and being independent of philosophical speculations. *Experience* has decided against the widespread and previously completely unshaken approach, that an objective method of describing nature independent of the context of observation can be approximated with arbitrary precision. And the very fact that we are dealing here with a result of advancing *empirical* research very crucially differentiates the discoveries of quantum mechanics from the antinomies that ground transcendental idealism, and which thoroughly depend on mathematical and philosophical considerations.

Even the founders of the critical philosophy, who discovered the doctrine of transcendental idealism and explored its implications, were far away from predicting the development that led from classical physics to quantum mechanics. It is understandable that they did not do so, but, despite all insight into the diversity of possible contexts of knowledge, stopped before the physical view of nature and saw in it a unified, self-sufficient knowledge that is independent of relations with the observer and of his manner of observing. For in light of the fact that belief in the world's oneness undergirds all knowledge, it is not the unity and self-sufficiency of a worldview, but precisely the splitting of reality into different worldviews that is strange and in need of explanation. And a reason to take this splitting to go even deeper within the domain of natural sciences than the antinomies lead one to recognise, was not available until the discovery of quantum mechanics. Until that point physics

[17] Heisenberg (1933).

appeared to be on the best route to carry out the programme—if not completely then at least to an arbitrary approximation—of an objective view of nature independent of the respective context of observation.

Therefore, even if it remains the undeniable merit of physical research to have advanced the understanding of the natural-philosophical foundations of our knowledge of nature by a decisive step, this advance means just as little a break with the prior philosophical development as quantum mechanics represents a break with classical physics. Rather, closer examination reveals that despite all prima facie discrepancies with the apparent conclusions also of the critical philosophy, the crucial discoveries of quantum mechanics fit consistently together with the principles of that philosophy, and through these their significance for the knowledge of nature becomes intelligible.

References

Apelt EF (1904) Über Begriff und Aufgabe der Naturphilosophie. Abhandlungen der Fries'schen Schule (Neue Folge) 1(1):89–134

Bohr N (1931) Atomtheorie und Naturbeschreibung. Springer, Berlin

de Broglie L (1928) La mécanique ondulatoire. Gauthier-Villars, Paris

de Broglie L (1929) Einführung in die Wellenmechanik. Akademische Verlagsgesellschaft, Leipzig. Translation of de Broglie (1928) by Peierls R

Dirac P (1930) Die Prinzipien der Quantenmechanik. S. Hirzel, Leipzig

Eisler R (1930) Kant-Lexikon: Nachschlagewerk zu Kants sämtlichen Schriften, Briefen und handschriftlichem Nachlass. Georg Olms, Berlin

Fries JF (1828/1968) Neue oder anthropologische Kritik der Vernunft. Sämtliche Schriften, vol 4, pp 31–478

Fries JF (1967–2011) Sämtliche Schriften. 30 vols. Scientia, Aalen. Edited by König G and Geldsetzer L

Heisenberg W (1930) Die physikalischen Prinzipien der Quantentheorie. S. Hirzel, Leipzig

Heisenberg W (1933) Zur Geschichte der physikalischen Naturerklärung. Berichte der Verhandlungen der Sächsischen Akademie der Wissenschaften zu Leipzig, Mathematisch-Physikalische Klasse 85:29–40

Hermann G (1935a) Die naturphilosophischen Grundlagen der Quantenmechanik. Abhandlugen der Fries'schen Schule 6(2):75–152

Hermann G (1935b) Die naturphilosophischen Grundlagen der Quantenmechanik. Die Naturwissenschaften 23(42):718–721

Herrmann K (ed) (2017) Grete Henry-Hermann: Philosophie–Mathematik–Quantenmechanik. Springer, Berlin

Von Laue M (1934) Über Heisenbergs Ungenauigkeitsbeziehungen und ihre erkenntnistheoretische Bedeutung. Die Naturwissenschaften 22(26):439–441

Nelson L (1917) Kritik der praktischen Vernunft. Veit, Leipzig

von Neumann J (1932) Mathematische Grundlagen der Quantenmechanik. Springer, Berlin

Popper KR (1934) Logik der Forschung. Springer, Vienna

Popper KR, von Weizsäcker CF (1934) Zur Kritik der Ungenauigkeitsrelationen. Die Naturwissenschaften 22(48):807–808

Schrödinger E (1934) Über die Unanwendbarkeit der Geometrie im Kleinen. Die Naturwissenschaften 22(31):518–521

Von Weizsäcker CF (1931) Ortsbestimmung eines Elektrons durch ein Mikroskop. Zeitschrift für Physik 70(1–2):114–130

Index

E. Crull and G. Bacciagaluppi (eds.), *Grete Hermann - Between Physics and Philosophy*, Studies in History and Philosophy of Science 42,
DOI 10.1007/978-94-024-0970-3

CPSIA information can be obtained
at www.ICGtesting.com
Printed in the USA
BVOW06*0214131017
497578BV00006B/22/P

9 789402 409680